Textbook of
Microbiology

1nd Edition

Dr. A .K. Kushwaha

M.Sc., B.Ed., M.A., Ph.D.

Department of Botany,

Sarla Dwivedi P.G College Akbarpur Kanpur (D) U.P India

209101

© *Reserved*

All disputes subject Kanpur Dehat (UP) Jurisdiction only.

PREFACE

Microbiology is the study of microscopic organisms, such as bacteria, viruses, archaea, fungi and protozoa. This discipline includes fundamental research on the biochemistry, physiology, cell biology, ecology, evolution and clinical aspects of microorganisms, including the host response to these agents.

This book will serve as an introduction to Microbiology to the beginners in the field. Actually the book is in intended to fulfil the long felt need of student of graduate and postgraduate level of all universities. The syllabi of all the universities have been kept in view during the preparation of the manuscript of this text. This work may also serve as laboratory manual. The present text provides a background of facts, terminology, general principle and specific disease of common crop plants and animals of world.

The text has been written in simple language and profusely illustrated with self-explanatory diagrams and coloured photographs. The figures are provided with detailed Legends. Some of the diagrams, of the diseased specimens have been drawn author himself, and most of the figure have been quoted from the authentic works of various authors and online websites.

I have already mentioned the book is primary and elementary text for degree and postgraduate students and not a text for researchers, and therefore, The Bibliography has been kept at minimum.

In conclusion, It is a pleasant aspect that I have now the opportunity to express my sincere appreciation and gratitude for all of them. In the first place, I would like to record my gratitude to my Father Shri C.L Kushwaha and my mother Smt. Shivkanti kushwaha, for being the outstanding advisor.

I gratefully acknowledge Soni Kushwaha, Dr. S.P Singh, Dr. A.P Saxena, and Dr. J.P Shukla, Dr. Amit pandey, Dr. Jitendra shukla, Dr. Nitin Sachan and Dr. Umakant prajapati for his contributions and valuable feedback which made him a backbone of this book.

The suggestion and healthy criticism of the book will be of much value to me for the improvement of the next edition.

A.K. Kushwaha
anilrania@gmail.com

CONTENTS

Chapter.....................Page no.

1. **MICROBIOLOGY AND THEIR HISTORY** ....1-8.

2. **MICROSCOPY**.........9-21
   - Staining Techniques
   - Introduction to Microscopes
   - Types of Microscopes
   - **Limitations**

3. **DISTRIBUTION OF MICROORGANISMS** .22-29
   - Microorganisms in soil
   - Microorganisms in water
   - Microbes of the air
   - Associated with man
   - In association with insects

4. **CLASSIFICATION AND IDENTIFICATION METHODS OF MICROORGANISMS**.30-41
   - Classification of Prokaryotes
   - Evolution of Prokaryotes
   - Categories of microorganisms in ecology

5. **THE METHODS IN MICROBIOLOGY** ...42-46
   - Pure cultures
   - Sterilization
   - Staining techniques

6. **PROKARYOTIC CELLS AND EUKARYOTIC CELLS**...................47-53.

7. **NUCLEIC ACIDS** ....54-90.
   - The Chemical Nature of DNA
   - RNA, a Different Nucleic Acid
   - The Secondary Structure of DNA
   - Basics of DNA Replication
   - The Leading and Lagging Strands
   - Telomere Replication
   - Central Dogma and Transcription

8. **THE BACTERIA**....91-115.
   - General Characteristics
   - Bacteria Morphology:
   - Reproduction in Bacteria

9. **BACTERIAL GENETICS** 116-127
   - Genetic organization
   - Mutations
   - Plasmids:
   - Types of Transposable Genetic Elements

10. **NUTRITION AND GROWTH OF BACTERIA** ....128-153.
    - Nutritional Requirements of Cells
    - Growth Factors
    - The Effect of Oxygen

- The Effect of pH on Growth
- The Effect of Temperature on Growth
- Water Availability
- Methods in bacteriology
- Culture Medium:
- Sterilisation vs. disinfection
- Staining of bacteria

11. CULTIVATION OF BACTERIA IN CULTURE MEDIA..................154-173.

12. ACTINOMYCETES.................................174-181

- Classification
- Importance of actinomycetes
- Actinomycosis

13. *PSEUDOMONAS, AND VIBRIO, XANTHOMONAS* ....182-196.

- Classification history
- Diseases
- Treatment

14. ENTEROBACTERIACEAE..........................197-209

- *Salmonella,*
- *Escherichia,*
- *Shigella*
- *Klebsiella*

15. *RICKETTSIA* ....210-215.

- Cell Structure and Metabolism
- Genome Structure
- Pathology
- Treatment

16. ARCHAEBACTERIA.................................216-225

- Origin and evolution
- Types of Archaebacteria
- *Lokiarcheota*
- *Methanobrevibacter smithii*

17. MYCOPLASMAS...................................226-233

- Structure of Mycoplasmas:
- Reproduction in Mycoplasma:
- Transmission of Mycoplasma:
- Diseases Caused by Mycoplasma:

18. THE CHLAMYDIA 234-241

- **Chlamydial Infection**
- Treatment

19. VIRUSES 242-264
- Virus history
- Viral Morphology
- Replication of viruses

20. BACTERIOPHAGES.................................265-270

21. TOBACCO MOSAIC VIRUS (TMV)......... 271-277

22. POTATO VIRUS.........278-284

- Potato virus Y,
- Potato virus X (PVX)
- Wild potato mosaic virus (WPMV

23. MYCOVIRUSES ……..285-291

- Kuru virus,
- Measles (rubeola) virus,
- Oncogenic or cancercausing viruses
- Viroids

24. CYANOPHAGES………292-294

25. TYPES OF VIRAL INFECTION………………..295-301

- Respiratory Viral Infections
- Viral Skin Infections
- Foodborne Viral Infections
- Sexually Transmitted Viral Infections
- Other Viral Infections
- Antiviral Medication and Other Treatment
- Viruses and Cancer
- Viral Illness Prevention

26. REOVIRUSES…………302-305

- **Rotavirus**
- **African horse sickness**
- **Bluetongue virus**
- **Colorado tick fever**

27. RETROVIRUS……….306-315

- Human immunodeficiency viruses (HIV)
- AIDS

28. ISOLATION AND PURIFICATION OF VIRUSES AND COMPONENTS……….316-325

29. THE MYCOSES……….326-327

30. SUPERFICIAL MYCOSES OR DERMATOPHYTOSIS….328-336

31. CANDIDIASIS ………337-343

32. MUCORMYCOSIS….344-348

33. ASPERGILLOSIS……349-353

34. PREDACEOUS FUNGI.354-357

- Nematode trapping fungi
- Endoparasitic Fungi

35. BIOFERTILIZER ……358-364

36. MYCORRHIZA ……....365-372

- Types of Mycorrhizal Fungi

37. IMMUNOLOGY AND VACCINE………………..373-391

38. MICROBIOLOGY OF AIR……………………….392-401

39. WATER MICROBIOLOGY………..402-404

40. SOIL MICROORGANISMS…….405-408

41. ENVIRONMENTAL MICROBIOLOGY……….409-411

42. FOOD MICROBIOLOGY…………..412-425

43. INDUSTRIAL MICROBIOLOGY………...426-431

44. PETROLEUM MICROBIOLOGY………...432-439

45. SCOPE AND APPLICATIONS OF MICROBIOLOGY …...440-445

46. MICROBIOLOGY MCQ & ANSWERS…………………....446-471

47. TERMINOLOGY………472-475

REFERENCES

1. MICROBIOLOGY AND THEIR HISTORY

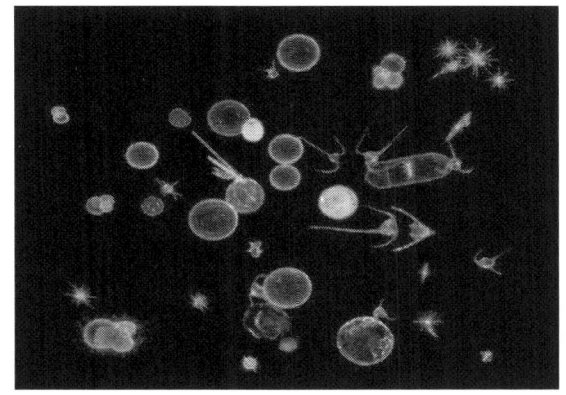

Microbiology is the study of the biology of microscopic organisms - viruses, bacteria, algae, fungi, slime molds, and protozoa. The methods used to study and manipulate these minute and mostly unicellular organisms differ from those used in most other biological investigations. Recombinant DNA technology uses microorganisms, particularly bacteria and viruses, to amplify DNA sequences and generate the encoded products. Moving genes from one microorganism to another, or amplifying them within microorganisms, permits application of microbial skills to solve medical and environmental problems. Many microorganisms are unique among living things in their ability to use gaseous nitrogen from the air for their nutritional requirements, or to degrade complex macromolecules in such materials as wood. By rearranging the genes that control these and other processes, scientists seek to engineer microorganisms that will process wastes, fertilize agricultural land, produce desirable biomolecules, and solve other problems inexpensively and safely. The 17th-century discovery of living forms existing invisible to the naked eye was a significant milestone in the history of science, for from the 13th century onward it had been postulated that "invisible" entities were responsible for decay and disease. The word *microbe* was coined in the last quarter of the 19th century to describe these organisms, all of which were thought to be related. As microbiology eventually developed into a specialized science, it was found that microbes are a very large group of extremely diverse organisms.

Daily life is interwoven inextricably with microorganisms. In addition to populating both the inner and outer surfaces of the human body, microbes abound in the soil, in the seas, and in the air. Abundant, although usually unnoticed, microorganisms provide ample evidence of their presence—sometimes unfavourably, as when they cause decay of materials or spread diseases, and sometimes favourably, as when they ferment sugar to wine and beer, cause bread to rise, flavour cheeses, and produce valued products such as antibiotics and insulin. Microorganisms are of incalculable value to Earth's ecology, disintegrating animal and plant remains and converting them to simpler substances that can be recycled in other organisms.

A Brief History of Microbiology

Microbiology has had a long, rich history, initially cantered in the causes of infectious diseases but now including practical applications of the science. Many individuals have made significant contributions to the development of microbiology.

- 1590: Mounted two lenses in a tube to produce the first compound microscope. (Hans and Zacharias Janssen, Dutch lens grinders) 1660: Published "Micrographia", containing drawings and detailed observations of biological materials made with the best compound microscope. (Robert Hooke)
- 1677: Observed "little animals" (Antony Leeuwenhoek)
- 1796: First scientific Small pox vaccination (Edward Jenner)
- 1850: Advocated washing hands to stop the spread of disease (Ignaz Semmelweis)
- 1861: Disproved spontaneous generation (Louis Pasteur)
- 1862: Supported Germ Theory of Disease (Louis Pasteur)
- 1867: Practiced antiseptic surgery (Joseph Lister)
- 1876: First proof of Germ Theory of Disease with Bacillus anthracis discovery (Robert Koch)
- 1877: Published his method for fractional sterilization (John Tyndall)
- 1879: Neisseria gonorrhoeoe, the first human pathogen identified. (Albert Neisser)
- 1880: Finds malarial parasites in erythrocytes of infected individuals. (C. L.Alphonse Laveran)
- 1881: Growth of Bacteria on solid media (Robert Koch)
- 1882: Outlined Kochs postulates (Robert Koch)
- 1882: Developed acid-fast Stain (Paul Ehrlich)
- 1882: Mycobacterium tuberculosis isolated. (Robert Koch)
- 1883: Independently discovered Corynebacterium diphtheriae. (Edward Theodore Klebs and Fredrich Loeffler)
- 1883: Pioneered developments in microscopy such as immersion lenses and apochromatic lenses which reduce chromatic aberration. (Carl Zeiss and Ernst Abbe)
- 1884: Developed Gram Stain (Christian Gram)
- 1884: Process of phagocytosis described. (Ilya Ilich Metchnikoff)
- 1885: First Rabies vaccination (Louis Pasteur)
- 1885: Discovered cure for syphilis (Paul Ehrlich)
- 1885: E.coli identified. (Theodor Escherich) 1887: Invented Petri Dish (Julius Richard Petri)
- 1888: Toxin of Cornyebacterium diphtheriae discovered. (Emile Roux and Alexandre Yersin) 1889: Discovered that bacteria can be agglutinated by serum. (A. Charrin and J. Roger) 1889: First pure culture of the strict anaerobic pathogen, the Clostridium tetani. (Kitasato) 1890: Discovery of diphtheria antitoxin serum. (Emil von Behring and Shibasaburo Kitasato)
- 1891: Proposed that antibodies are responsible for immunity. (Paul Ehrlich)
- 1892: Discovered virus of tobacco mosaic disease (Dmitri Iosifovich Ivanovski)
- 1892: Clostridium perfringens identified. (William Welch and George Nuttall)
- 1893: First account of a zoonotic disease, established that ticks carry Babesia microti. (Theobald Smith and F.L. Kilbourne)
- 1894: Endotoxin identified in Vibrio cholerae. (Richard Pfeiffer)

- 1894: Yersinia (Pasteurella) pestis isolated. (Alexandre Yersin)
- 1897: Killed vaccine against plague. (Waldemar Haffkine)
- 1897: Killed vaccine against typhoid fever (Almwroth Wright and David Sample)
- 1899: Recognized viral dependence on cells for reproduction (Martinus Beijerinck)
- 1899: Showed that the malarial parasite undergoes a cycle of development in mosquitoes and that the disease is transmitted by the bite of female mosquitoes. (Ronald Ross)
- 1900: Proved that mosquitoes carries the yellow fever agent (Walter Reed)
- 1901: Complement fixation test developed. (Jules Bordet and Octave Gengou)
- 1903: Leishmania donovani observed. (William Leishman)
- 1905: Treponema pallidum identified. (Fritz R. Schaudinn and Erich Hoffman)
- 1909: Causative agent of Rocky Mountain spotted fever, Rickettsia identified. (Howard Ricketts)
- 1909: Trypanosoma cruzi identified. (Carlos Chagas) 1910: Systematic and scientific studies of dermatophytes, medium for the growth of pathogenic fungi. (Raymond Sabouraud) 1911: Experimental proof of an infectious etiologic agent of cancer. (Francis Peyton Rous)
- 1915: First discovery of bacteriophage. (Frederick Twort)
- 1917: Coined the name "bacteriophage." (Felix d Herrelle)
- 1919: Blood agar used as a medium to study the hemolytic reactions for the genus Streptococcus. (James Brown)
- 1924: BCG to immunize against the tuberculosis. (Albert Calmette and Camille Guerin)
- 1926: Distinguishes between bacteria and viruses, establishing virology as a separate area of study. (Thomas Rivers)
- 1928: Discovered transformation in bacteria and establishes the foundation of molecular genetics. (Frederick Griffith)
- 1928: Discovered Penicillin (Alexander Fleming)
- 1931: Constructed the first electron microscope. (Ernst Ruska)
- 1931: Devise a technique of cultivating viruses in eggs. (Alice Woodruff and Ernest Goodpasture)
- 1933: Described a method of producing streptococcal antigens and sera for use in precipitin tests. (Rebecca Lancefield)
- 1934: First typing of a strain of bacteria with bacteriophage. (Alice Evans) 1938: Vaccine against yellow fever. (Max Theiler)
- 1940: Isolate the antibiotic from Fleming's mold cultures and demonstrate that it can cure infections. (Howard Florey and Ernest Chain)
- 1940: Bacterial product recognized to mediate resistance to an antibacterial agent (Penicillin) in E.coli. (Ernest Chain and E.P. Abraham)
- 1940: Discovered actinomycin, the first antibiotic obtained pure from an actinomycete. (Selman Waksman and H. Boyd Woodruff)
- 1941: Demonstrated that penicillin is non-toxic to human. (Charles Fletcher) 1941: Viral hemagglutination described. (George Hirst)
- 1942: Birth of immunofluorescence. (Albert H. Coons, H.J. Creech, R.N. Jones, and E. Berliner)
- 1942: Identify adjuvants that can significantly boost antibody production. (Jules Freund and Katherine McDermott)

- 1944: First to demonstrate successful treatment of tuberculosis with streptomycin. (W. H. Feldman and H. C. Hinshaw)
- 1949: Technique to grow polio virus in test tube cultures of human tissues. (John Franklin Enders, Thomas H. Weller and Frederick Chapman Robbins) 1952: Used the term plasmid to describe extranuclear genetic elements that replicate autonomously. (Joshua Lederberg)
- 1952: Transduction discovered in Salmonella typhimurium. (Joshua Lederberg and Norton Zinder) 1953: Killed polio vaccine. (Jonas Salk)
- 1953: First useful fungal antibiotic, NYSTATIN developed. (Elizabeth Lee Hazen and Rachel Fuller Brown)
- 1957: Interferon discovered. (Alick Isaacs and Jean Lindemann)
- 1957: Proposed that a slow virus is responsible for the wasting disease kuru. (D. Carleton Gajdusek)
- 1958: Antibody labeling agent, flourescein isothiocyanate (FITC) developed. Begining of RIA and ELISA. (Joseph H. Burkhalter and Robert Seiwald
-) 1959: Transferable drug resistance discovered in Shigella. (O. Sawada)
- 1963: Described the "Australia Antigen" (hepatitis B antigen). (Baruch Blumberg)
- 1963: Vaccine against Hepatitis B. (Baruch Blumberg and Irving Millman) 1966: Established standards for antibiotic susceptibility testing based on disc diffusion procedure. (William Kirby and Alfred Bauer)
- 1967: Viroids discovered. (Theodor O. Diener)
- 1968: Limulus lysate assay for endotoxin detection. (Levin and Bang)
- 1969: DNA hybridization used to classify members of family Enterobacteriaceae. (Don Brenner)
- 1970: Restriction endonucleases, important tool in genetic engineering discovered. (Hamilton Smith and Kent W. Wilcox) 1970: Independently discovered reverse transcriptase in RNA viruses. (Howard Temin and David Baltimore)
- 1972: Recombinant DNA molecule from viral and bacterial DNA constructed. (Paul Berg)
- 1975: Sexual reproduction in the fungus described. (Kyung and Kwon-Chung) 1975: Monoclonal antibodies by Hybridoma technique. (Georg Kohler and Cesar Milstein)
- 1976: Proto-oncogenes identified. (J. Michael Bishop and Harold Varmus) 1976: Plasmodium falciparum cultivated in vitro. (William Trager and Jim Jensen) 1977: Developed a method to sequence DNA (W. Gilbert & F. Sanger)
- 1979: Smallpox (variola) is declared officially eliminated
- 1982: Prions discovered. (Stanley Prusiner)
- 1983: Polymerase Chain Reaction invented (Kary Mullis)
- 1983: Discovery of the immunodeficiency virus (HIV). (Luc Montaigner and Robert Gallo)
- 1984: Helicobacter pylori identified. (Barry Marshall)
- 1985: First anti-retorviral AZT discovered. (Robert Gallo, Dani Bolognesi, Sam Broder)
- 1995: First microbial genomic sequence published (Haemophilus influenzae) (TIGR)

Early history of microbiology. Historians are unsure who made the first observations of microorganisms, but the microscope was available during the mid-1600s, and an English scientist named **Robert**

Fig.1. Robert Hooke Microscope

1665 - Robert Hooke, through observation of thin slices of cork, Hooke reported that life's smallest structures were like "little boxes." Hooke discovered the beginning of cell theory - the theory that all living things are composed of cells.

1668 - Francesco Redi, proved that maggots did not generate spontaneously. He filled two jars with meat. One was left unsealed, and flies were able to fly in a lay eggs on the meat, developing into larvae. The other was sealed where flies were not allowed to enter at the meat was left without maggots or larvae. Still, people were not convinced so Redi again filled another jar with meat but instead covered it with net. No larvae appeared in the jar even though air was present.

Fig. 1.2. - *Francesco Redi,* **expperiment**

1673-1723 - Anton Van Leeuwenhoek, "the Father of Microbiology." Leeuwenhoek, in 1673 observed live microorganisms through magnifying lenses that he created. Because of Leeuwenhoek's discovery, many scientist believed there were some forms of life that arose spontaneously. Leeuwenhoek wrote about "animalcules," the structures he saw underneath his magnifying lens. Leeuwenhoek, also made drawings of what he found, as those drawings were identified as representation of bacteria and protozoa.

1745 - Lazzaro Spallanzani, was convinced that air entered Needham's solution after the broth was boiled, creating microorganisms in the broth. Spallanzani filled a flask with broth, boiled it, and then covered it to prove that microorganisms did not enter once the flask was covered. Needham claimed that spontaneous generation had been destroyed by the heat and was kept out of the flasks by the seals.

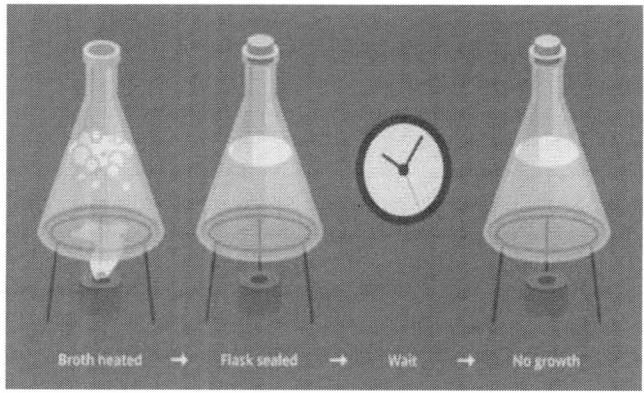

Fig1.4. Spallanzani's observations were still criticized by there not being enough oxygen in the sealed flask for microorganisms to live.

1858 - Rudolf Virchow, questioned spontaneous generation with biogenesis. He hypothesized that living cells come from preexisting living cells. However, Virchow could not offer any evidence of his hypothesis so the debate continued until French scientist Louis Pasteur in 1861.

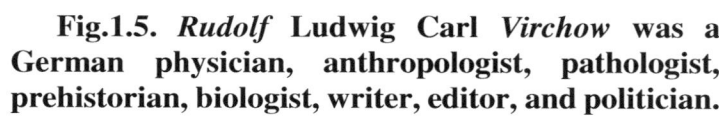

Fig.1.5. *Rudolf* **Ludwig Carl** *Virchow* **was a German physician, anthropologist, pathologist, prehistorian, biologist, writer, editor, and politician.**

Louis Pasteur (1822–1895) expanded upon Spallanzani's findings by exposing boiled broths to the air in vessels that contained a filter to prevent all particles from passing through to the growth medium. He also did this in vessels with no filter at all, with air being admitted via a curved tube that prevented dust particles from coming in contact with the broth. By boiling the broth beforehand, Pasteur ensured that no microorganisms survived within the broths at the beginning of his experiment. Nothing grew in the broths in the course of Pasteur's experiment. This meant that the living organisms that grew in such broths came from outside, as spores on dust, rather than spontaneously generated within the broth. Thus, Pasteur dealt the death blow to the theory of spontaneous generation and supported germ theory instead.

German scientist **Robert Koch** was born in Clausthal in the Harz Mountains, then part of the Kingdom of Hanover, as the son of a mining official. He studied medicine at the University of Göttingen and graduated in 1866. He then served in the Franco-Prussian War and later became district medical officer in Wollstein (Wolsztyn), Prussian Poland. Working with very limited resources, he became one of the founders of bacteriology, the other major figure being Louis Pasteur.

After Casimir Davaine demonstrated the direct transmission of the anthrax bacillus between cows, Koch studied anthrax more closely. He invented methods to purify the bacillus from blood samples and grow pure cultures. He found that, while it could not survive outside a host for long, anthrax built persisting endospores that could last a long time. These endospores, embedded in soil, were the cause of unexplained "spontaneous" outbreaks of anthrax. Koch published his findings in 1876 and was rewarded with a job at the Imperial Health Office in Berlin in 1880. In 1881, he urged for the sterilization of surgical instruments using heat.

Fig.1.7. *Robert Koch: An image of Robert Koch, a pioneering microbiologist. Koch's research and methods helped link the causal nature of microbes to certain diseases, including anthrax.*

Probably as important as his work on tuberculosis, for which he was awarded a Nobel Prize in 1905, are Koch's postulates. These postulates stated that to establish that an organism is the cause of a disease, it must be found in all cases of the disease examined. Additionally, it must be absent in healthy organisms prepared and maintained in a pure culture capable of producing the original infection, even after several generations in culture retrievable from an inoculated animal and cultured again. By using his methods, Koch's pupils found the organisms responsible for diphtheria, typhoid, pneumonia, gonorrhoea, cerebrospinal meningitis, leprosy, bubonic plague, tetanus, and syphilis.

Perhaps the key method Koch developed was the ability to isolate pure cultures, explained in brief here. Pure cultures of multicellular organisms are often more easily isolated by simply picking out a single individual to initiate a culture. This is a useful technique for pure culture of fungi, multicellular algae, and small metazoa. Developing pure culture techniques is crucial to the observation of the specimen in question. The most common method to isolate individual microbes and produce a pure culture is to prepare a streak plate. The streak plate method is a way to physically separate the microbial population and is done by spreading the inoculate back and forth with an inoculating loop over the solid agar plate. Upon incubation, colonies will arise and single cells will have been isolated from the biomass.

The development of microbiology. In the late 1800s and for the first decade of the 1900s, scientists seized the opportunity to further develop the germ theory of disease as enunciated by Pasteur and proved by Koch. There emerged a **Golden Age of Microbiology** during which many agents of different infectious diseases were identified. Many of the etiologic agents of microbial disease were discovered during that period, leading to the ability to halt epidemics by interrupting the spread of microorganisms.

Despite the advances in microbiology, it was rarely possible to render life-saving therapy to an infected patient. Then, after World War II, the **antibiotics** were introduced to medicine. The incidence of pneumonia, tuberculosis, meningitis, syphilis, and many other diseases declined with the use of antibiotics.

Work with viruses could not be effectively performed until instruments were developed to help scientists see these disease agents. In the 1940s, the **electron microscope** was developed and perfected. In that decade, cultivation methods for viruses were also introduced, and the knowledge of viruses developed rapidly. With the development of vaccines in the 1950s and 1960s, such viral diseases as polio, measles, mumps, and rubella came under control.

Modern microbiology. Modern microbiology reaches into many fields of human endeavor, including the development of pharmaceutical products, the use of quality-control methods in food and dairy product production, the control of disease-causing microorganisms in consumable waters, and the industrial applications of microorganisms. Microorganisms are used to produce vitamins, amino acids, enzymes, and growth supplements. They manufacture many foods, including fermented dairy products (sour cream, yogurt, and buttermilk), as well as other fermented foods such as pickles, sauerkraut, breads, and alcoholic beverages.

One of the major areas of applied microbiology is **biotechnology.** In this discipline, microorganisms are used as living factories to produce pharmaceuticals that otherwise could not be manufactured. These substances include the human hormone insulin, the antiviral substance interferon, numerous blood-clotting factors and clotdissolving enzymes, and a number of vaccines. Bacteria can be reengineered to increase plant resistance to insects and frost, and biotechnology will represent a major application of microorganisms in the next century.

****

2.

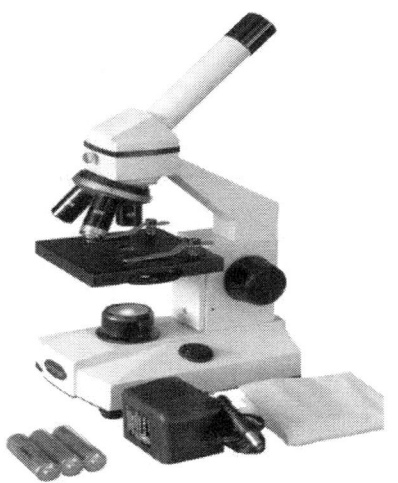

MICROSCOPY

Staining Techniques

Because microbial cytoplasm is usually transparent, it is necessary to stain microorganisms before they can be viewed with the light microscope. In some cases, staining is unnecessary, for example when microorganisms are very large or when motility is to be studied, and a drop of the microorganisms can be placed directly on the slide and observed. A preparation such as this is called a **wet mount.** A wet mount can also be prepared by placing a drop of culture on a cover-slip (a glass cover for a slide) and then inverting it over a hollowed-out slide. This procedure is called the **hanging drop.**

In preparation for staining, a small sample of microorganisms is placed on a slide and permitted to air dry. The smear is heat fixed by quickly passing it over a flame. **Heat fixing** kills the organisms, makes them adhere to the slide, and permits them to accept the stain.

Simple stain techniques. Staining can be performed with basic dyes such as crystal violet or methylene blue, positively charged dyes that are attracted to the negatively charged materials of the microbial cytoplasm. Such a procedure is the **simple stain procedure.** An alternative is to use a dye such as nigrosin or Congo red, acidic, negatively charged dyes. They are repelled by the negatively charged cytoplasm and gather around the cells, leaving the cells clear and unstained. This technique is called the **negative stain technique.**

Differential stain techniques. The **differential stain technique** distinguishes two kinds of organisms. An example is the **Gram stain technique.** This differential technique separates bacteria into two groups, Gram-positive bacteria and Gram-negative bacteria. Crystal violet is first applied, followed by the mordant iodine, which fixes the stain (Figure.2). Then the slide is washed with alcohol, and the Gram-positive bacteria retain the crystal-violet iodine stain; however, the Gram-negative bacteria lose the stain. The Gram-negative bacteria subsequently stain with the safranin dye, the counterstain, used next. These bacteria appear red under the oil-immersion lens, while Gram-positive bacteria appear blue or purple, reflecting the crystal violet retained during the washing step.

Another differential stain technique is the **acid-fast technique.** This technique differentiates species of *Mycobacterium* from other bacteria. Heat or a lipid solvent is used to carry the first stain, carbolfuchsin, into the cells. Then the cells are washed with a dilute acid-alcohol solution. *Mycobacterium* species resist the effect of the acid-alcohol and retain the carbolfuchsin stain (bright red). Other bacteria lose the stain and take on the subsequent methylene blue stain (blue). Thus, the acid-fast bacteria appear bright red, while the nonacid-fast bacteria appear blue when observed under oil-immersion microscopy.

Other stain techniques seek to identify various bacterial structures of importance. For instance, a special stain technique highlights the **flagella** of bacteria by coating the flagella with dyes or metals to increase their width. Flagella so stained can then be observed.

A special stain technique is used to examine bacterial **spores.** Malachite green is used with heat to force the stain into the cells and give them color. A counterstain, safranin, is then used to give color to the nonsporeforming bacteria. At the end of the procedure, spores stain green and other cells stain red.

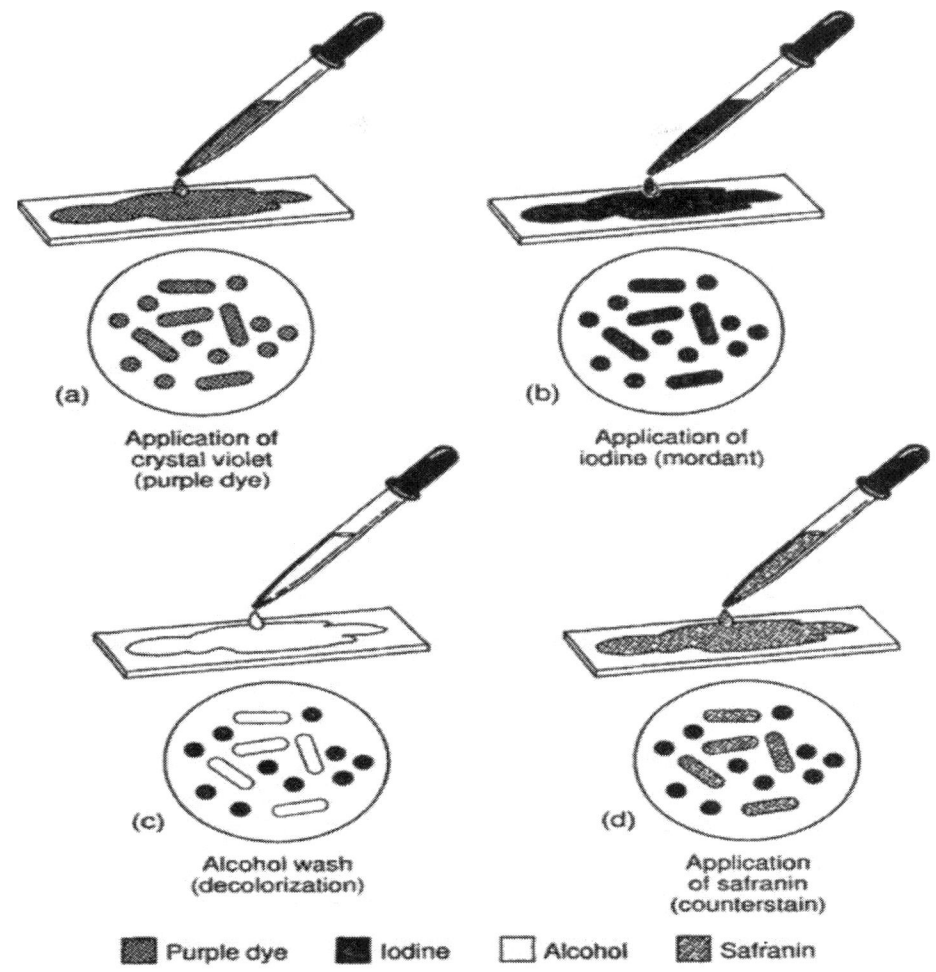

Fig.2.The Gram stain procedure used for differentiating bacteria into two groups.

Another differential stain technique is the **acid-fast technique.** This technique differentiates species of *Mycobacterium* from other bacteria. Heat or a lipid solvent is used to carry the first stain, carbolfuchsin, into the cells. Then the cells are washed with a dilute acid-alcohol solution. *Mycobacterium* species resist the effect of the acid-alcohol and retain the carbolfuchsin stain (bright red). Other bacteria lose the stain and take on the subsequent methylene blue stain (blue). Thus, the acid-fast bacteria appear bright red, while the nonacid-fast bacteria appear blue when observed under oil-immersion microscopy.

Other stain techniques seek to identify various bacterial structures of importance. For instance, a special stain technique highlights the **flagella** of bacteria by coating the flagella with dyes or metals to increase their width. Flagella so stained can then be observed.

A special stain technique is used to examine bacterial **spores.** Malachite green is used with heat to force the stain into the cells and give them color. A counterstain, safranin, is then used to give color to the nonsporeforming bacteria. At the end of the procedure, spores stain green and other cells stain red.

Introduction to Microscopes

Since microorganisms are invisible to the unaided eye, the essential tool in microbiology is the microscope. One of the first to use a microscope to observe microorganisms was **Robert Hooke**, the English biologist who observed algae and fungi in the 1660s. In the 1670s, **Anton van Leeuwenhoek**, a Dutch merchant, constructed a number of simple microscopes and observed details of numerous forms of protozoa, fungi, and bacteria. During the 1700s, microscopes were used to further elaborate on the microbial world, and by the late 1800s, the sophisticated light microscopes had been developed. The electron microscope was developed in the 1940s, thus making the viruses and the smallest bacteria (for example, rickettsiae and chlamydiae) visible.

Microscopes permit extremely small objects to be seen, objects measured in the metric system in micrometers and nanometers. A **micrometer** (μm) is equivalent to a millionth of a meter, while a **nanometer** (nm) is a billionth of a meter. Bacteria, fungi, protozoa, and unicellular algae are normally measured in micrometers, while viruses are commonly measured in nanometers. A typical bacterium such as *Escherichia coli* measures about two micrometers in length and about one micrometer in width.

Types of Microscopes

Various types of microscopes are available for use in the microbiology laboratory. The microscopes have varied applications and modifications that contribute to their usefulness.

The light microscope.

The common light microscope used in the laboratory is called a **compound microscope** because it contains two types of lenses that function to magnify an object. The lens closest to the eye is called the **ocular**, while the lens closest to the object is called the **objective.** Most microscopes have on their base an apparatus called a **condenser**, which condenses light rays to a strong beam. A **diaphragm** located on the condenser controls the amount of light coming through it. Both coarse and fine adjustments are found on the light microscope (Figure).

To magnify an object, light is projected through an opening in the stage, where it hits the object and then enters the objective. An image is created, and this image becomes an object for the ocular lens, which remagnifies the image. Thus, the **total magnification** possible with the microscope is the magnification achieved by the objective multiplied by the magnification achieved by the ocular lens.

A compound light microscope often contains four **objective lenses:** the scanning lens (4X), the low-power lens (10X), the high-power lens (40 X), and the oil-immersion lens (100 X). With an ocular lens that magnifies 10 times, the total magnifications possible will be 40 X with the scanning lens, 100 X with the low-power lens, 400 X with the high-power lens, and 1000 X with the oil-immersion lens. Most microscopes are **parfocal.** This term means that the microscope remains in focus when one switches from one objective to the next objective.

The ability to see clearly two items as separate objects under the microscope is called the **resolution** of the microscope. The resolution is determined in part by the wavelength of the light used for observing. Visible light has a wavelength of about 550 nm, while ultraviolet light has a wavelength of about 400 nm or less. The resolution of a microscope increases as the wavelength decreases, so ultraviolet light allows one to detect objects not seen with visible light. The **resolving power** of a lens refers to the size of the smallest object that can be seen with that lens. The resolving power is based on the wavelength of the light used and the numerical aperture of the lens. The **numerical aperture (NA)** refers to the widest cone of light that can enter the lens; the NA is engraved on the side of the objective lens.

*If the user is to see objects clearly, sufficient light must enter the objective lens. With modern microscopes, entry to the objective is not a problem for scanning, low-power, and high-power lenses. However, the oil-immersion lens is exceedingly narrow, and most light misses it. Therefore, the object is seen poorly and without resolution. To increase the resolution with the oil-immersion lens, a drop of **immersion oil** is placed between the lens and the glass slide (Figure 2.1).* Immersion oil has the same light-bending ability (index of refraction) as the glass slide, so it keeps light in a straight line as it passes through the glass slide to the oil and on to the glass of the objective, the oil-immersion lens. With the increased amount of light entering the objective, the resolution of the object increases, and one can observe objects as small as bacteria. Resolution is important in other types of microscopy as well.

Other light microscopes. In addition to the familiar compound microscope, microbiologists use other types of microscopes for specific purposes. These microscopes permit viewing of objects not otherwise seen with the light microscope.

An alternative microscope is the **dark-field microscope**, which is used to observe live spirochetes, such as those that cause syphilis. This microscope contains a special condenser that scatters light and causes it to reflect off the specimen at an angle. A light object is seen on a dark background.

A second alternative microscope is the **phase-contrast microscope.** This microscope also contains special condensers that throw light "out of phase" and cause it to pass through the object at different speeds. Live, unstained organisms are seen clearly with this microscope, and internal cell parts such as mitochondria, lysosomes, and the Golgi body can be seen with this instrument.

The **fluorescent microscope** uses ultraviolet light as its light source. When ultraviolet light hits an object, it excites the electrons of the object, and they give off light in various shades of color. Since ultraviolet light is used, the resolution of the object increases. A laboratory technique called the fluorescent-antibody technique employs fluorescent dyes and antibodies to help identify unknown bacteria.

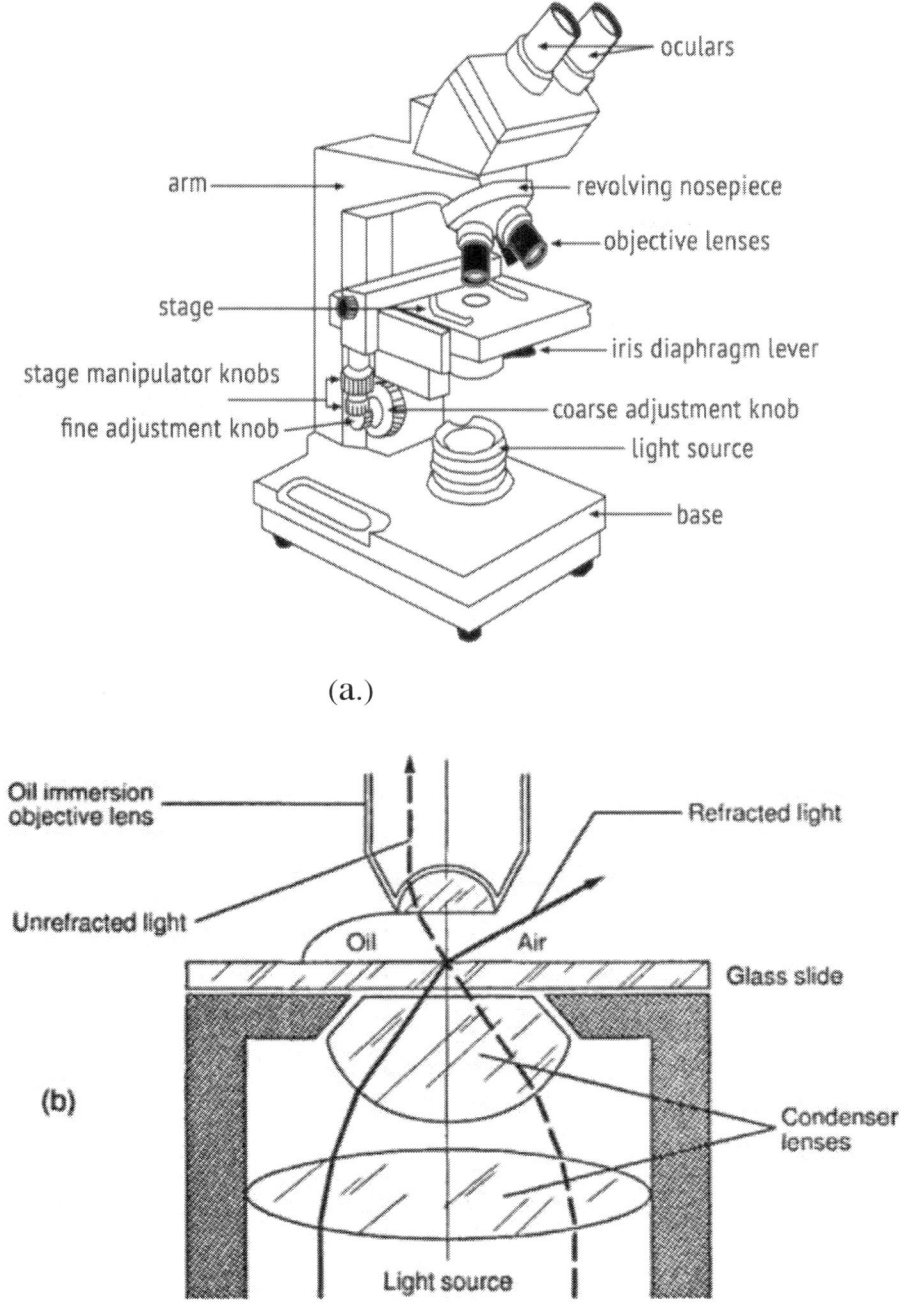

Fig.2.1.

Light microscopy. (a) The important parts of a common light microscope. (b) How immersion oil gathers more light for use in the microscope.

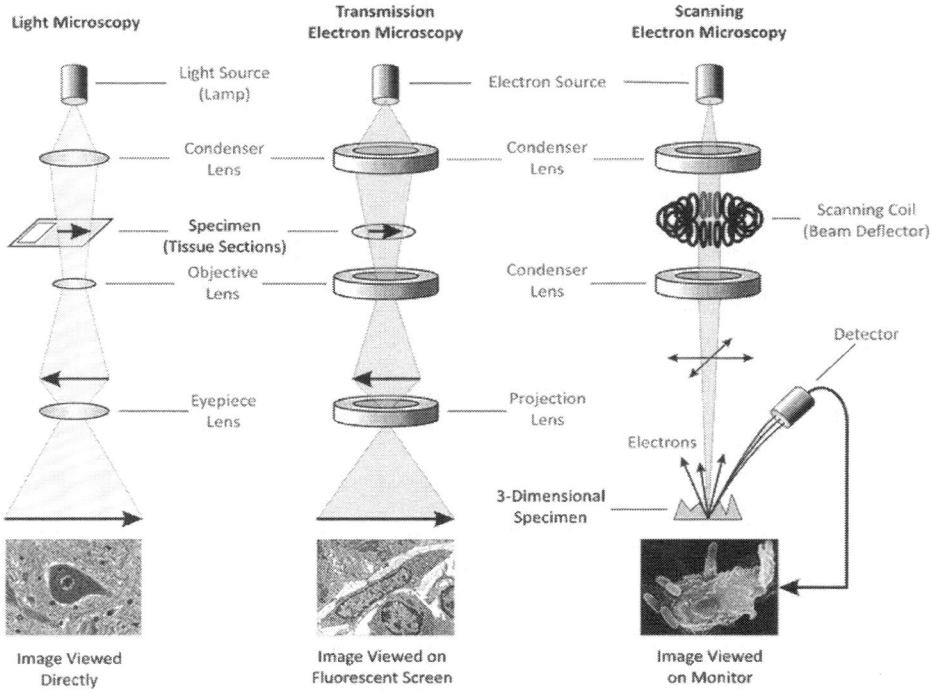

Fig.2.2 Differences between Light Microscope and Electron Microscope

Electron microscope

- An electron microscope is a microscope that uses a beam of accelerated electrons as a source of illumination.
- It is a special type of microscope having a high resolution of images, able to magnify objects in nanometres, which are formed by controlled use of electrons in vacuum captured on a phosphorescent screen.
- Ernst Ruska (1906-1988), a German engineer and academic professor, built the first Electron Microscope in 1931, and the same principles behind his prototype still govern modern EMs.

Electron microscopes use signals arising from the interaction of an electron beam with the sample to obtain information about structure, morphology, and composition.

1. The electron gun generates electrons.

2. Two sets of condenser lenses focus the electron beam on the specimen and then into a thin tight beam.

3. To move electrons down the column, an accelerating voltage (mostly between 100 kV-1000 kV) is applied between tungsten filament and anode.

4. The specimen to be examined is made extremely thin, at least 200 times thinner than those used in the optical microscope. Ultra-thin sections of 20-100 nm are cut which is already placed on the specimen holder.

5. The electronic beam passes through the specimen and electrons are scattered depending upon the thickness or refractive index of different parts of the specimen.

6. The denser regions in the specimen scatter more electrons and therefore appear darker in the image since fewer electrons strike that area of the screen. In contrast, transparent regions are brighter.

7. The electron beam coming out of the specimen passes to the objective lens, which has high power and forms the intermediate magnified image.

8. The ocular lenses then produce the final further magnified image.

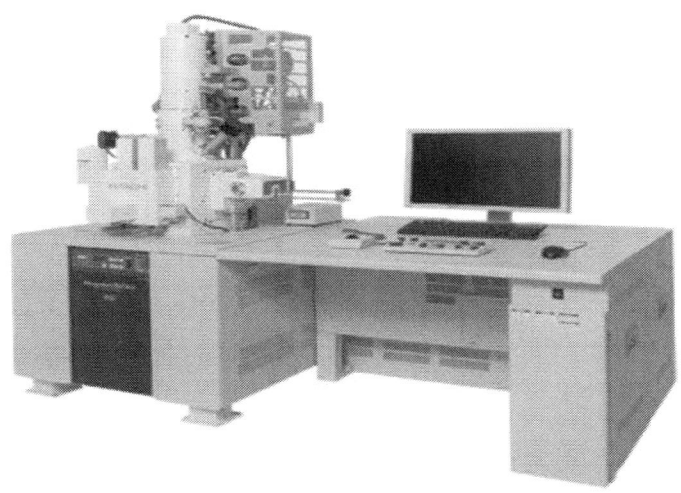

Fig.2.3 Electron Microscopes

There are two types of electron microscopes, with different operating styles:

1. **The transmission electron microscope (TEM)**

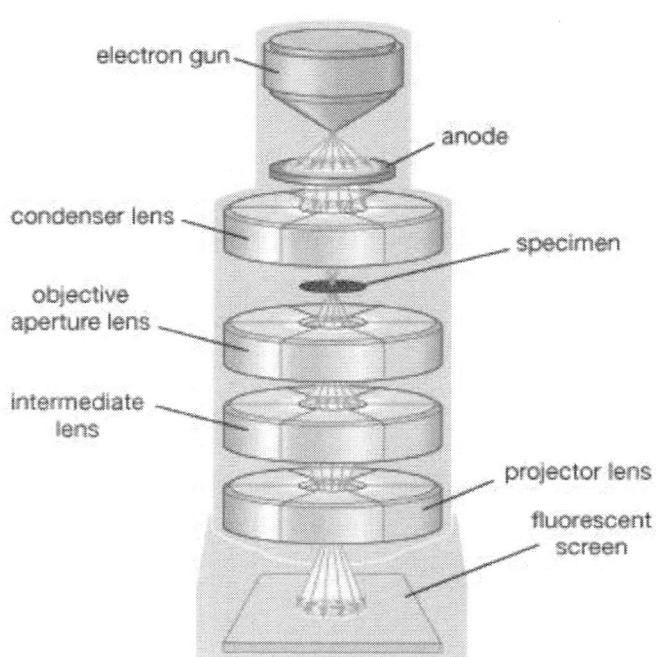

Fig.2.4. Transmission electron microscope

- The transmission electron microscope is used to view thin specimens through which electrons can pass generating a projection image.
- The TEM is analogous in many ways to the conventional (compound) light microscope.
- TEM is used, among other things, to image the interior of cells (in thin sections), the structure of protein molecules (contrasted by metal shadowing), the organization of molecules in viruses and cytoskeletal filaments (prepared by the negative staining technique), and the arrangement of protein molecules in cell membranes (by freeze-fracture).

2. The scanning electron microscope (SEM)

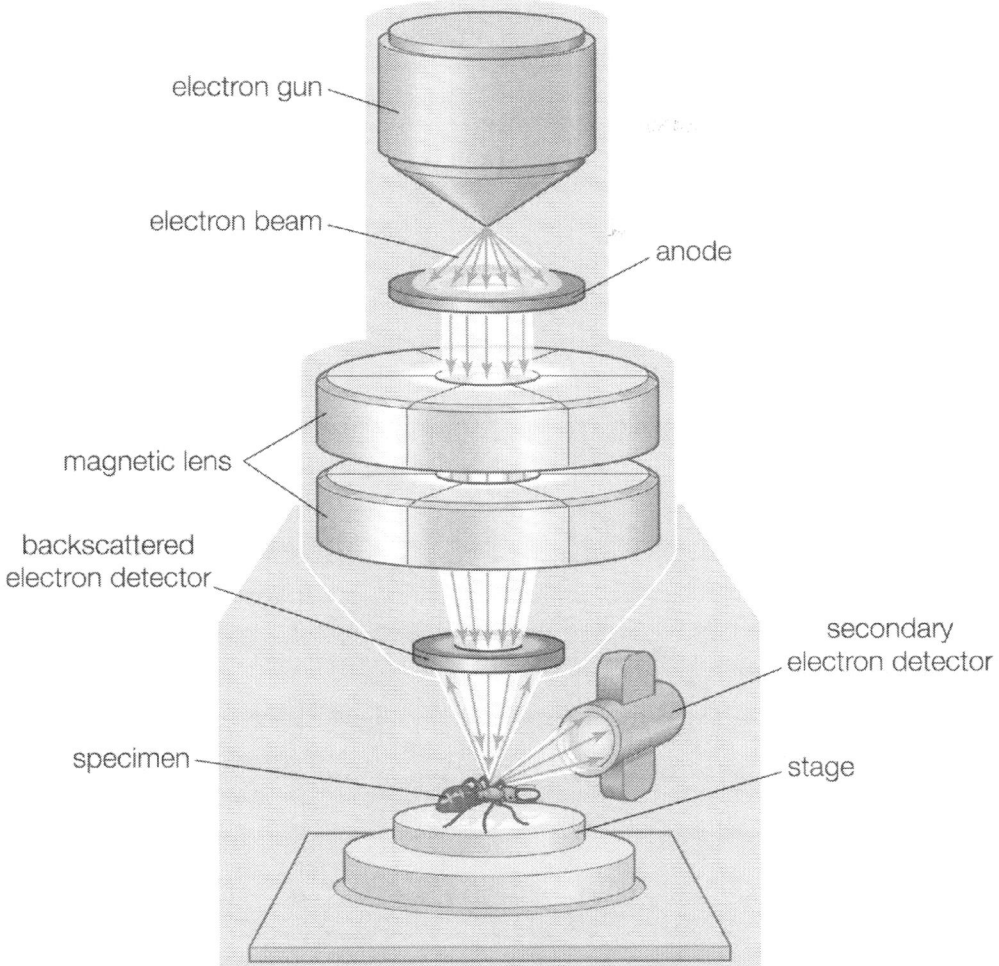

Fig.2.5. Scanning electron microscope

- Conventional scanning electron microscopy depends on the emission of secondary electrons from the surface of a specimen.
- Because of its great depth of focus, a scanning electron microscope is the EM analog of a stereo light microscope.
- It provides detailed images of the surfaces of cells and whole organisms that are not possible by TEM. It can also be used for particle counting and size determination, and for process control.
- It is termed a scanning electron microscope because the image is formed by scanning a focused electron beam onto the surface of the specimen in a raster pattern.

Parts of Electron microscope

EM is in the form of a tall vacuum column which is vertically mounted. It has the following components:

1. **Electron gun**

- The electron gun is a heated tungsten filament, which generates electrons.

2. **Electromagnetic lenses**

- **Condenser lens** focuses the electron beam on the specimen. A second condenser lens forms the electrons into a thin tight beam.
- The electron beam coming out of the specimen passes down the second of magnetic coils called the **objective lens**, which has high power and forms the intermediate magnified image.
- The third set of magnetic lenses called **projector (ocular) lenses** produce the final further magnified image.
- Each of these lenses acts as an image magnifier all the while maintaining an incredible level of detail and resolution.

3. **Specimen Holder**

- The specimen holder is an extremely thin film of carbon or collodion held by a metal grid.

4. **Image viewing and Recording System.**

- The final image is projected on a fluorescent screen.
- Below the fluorescent screen is a camera for recording the image.

Applications

- Electron microscopes are used to investigate the ultrastructure of a wide range of biological and inorganic specimens including microorganisms, cells, large molecules, biopsy samples, metals, and crystals.

- Industrially, electron microscopes are often used for quality control and failure analysis.
- Modern electron microscopes produce electron micrographs using specialized digital cameras and frame grabbers to capture the images.
- Science of microbiology owes its development to the electron microscope. Study of microorganisms like bacteria, virus and other pathogens have made the treatment of diseases very effective.

Advantages

- Very high magnification
- Incredibly high resolution
- Material rarely distorted by preparation
- It is possible to investigate a greater depth of field
- Diverse applications

Limitations

- The live specimen cannot be observed.
- As the penetration power of the electron beam is very low, the object should be ultra-thin. For this, the specimen is dried and cut into ultra-thin sections before observation.
- As the EM works in a vacuum, the specimen should be completely dry.
- Expensive to build and maintain
- Requiring researcher training
- Image artifacts resulting from specimen preparation.
- This type of microscope is a large, cumbersome extremely sensitive to vibration and external magnetic fields.

****

3.

DISTRIBUTION OF MICROORGANISMS

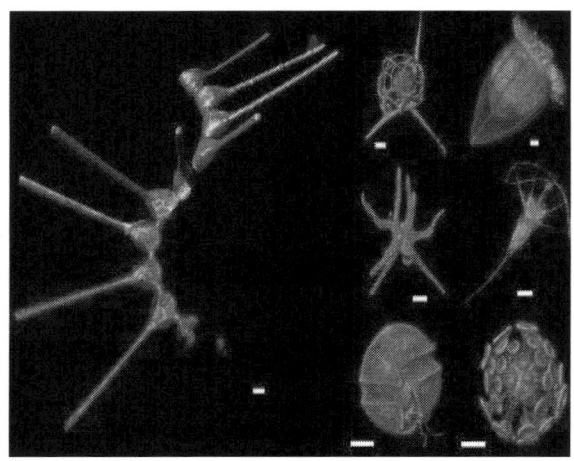

Investigators have proved that microorganisms may be found practically everywhere except fire. Leeuwenhoek found microorganisms in water, pepper infusion, and scrapings from his teeth as well as from several other sources.

MICROORGANISMS IN SOIL

It is well known that cultivated soils possess enormous numbers of microscopie organisms. Bacteria, actinomycetes, yeasts, and moulds make up the greatest population of microscopic life. The common bacterial genera found in the soil are *Achromobacter, Actinomyces, Arthrobacter, Bacillus, Clostridium, Flavobacterium, Micrococcus* and *Pseudomonas*. The richer the soil in organic matter, the greater will be the number of micro-organisms present. Most of the microscopic agents found in soil are beneficial as they increase the fertility of the soil. Occasionally disease producing microorganisms are found in the soil. In soil contaminated with human sewage the likelyhood of harmful microorganisms increases. In moist soil the micrubial population may reach 100 million or more cells per gram of soil, in dry desert soil there may be a few thousand cells per gram of soil. However, if this desert soil is moistened with water it will soon be teaming with bacteria and fungi. The soil fungiare present both as spores and as actively growing mycelium. Ithas been estimated that one gram of soil may contain 100 metres of mycelium. Some of the microbes may be free living in the soil while others may be leading a symbiotic life in association with the roots of the plants growing therein. The Rhizobium, a bacteriumis found in the root nodules whereas there are several mycorrhizal fungi.

MICROORGANISMS IN WATER

Water occurring in nature contains dissolved salts and gases, especially sea and mineral waters. Water covers 70% of the earth's surface, and thus, it is the most essential habitat of life. The overall volume of inland waters is estimated at 7.5×10^5 km^3, of seas and oceans at 1.4×10^9 km^3, and of glaciers and continental glaciers at 1.8×10^7 km. Water makes up the most crucial component of living organisms (70-90% of cell mass) and fulfils a purpose in taking part in various biological reactions and processes.

Types of Waters Inhabited by Microorganisms

The biotopes of water microorganisms may be underground and/or surface waters as well as bottom sediments.

- The underground waters (mineral and thermal springs, ground waters) - due to their oligotrophic character (nutrient - deficient) are usually inhabited by a sparse microflora that is represented by a low number of species with almost a complete lack of higher plants or animals.
- The surface waters such as streams, rivers, lakes and sea waters are inhabited by a diverse flora and fauna. Microorganisms in those waters are a largely varied group. Next to the typical water species, other microorganisms from soil habitats and sewage derived from living and industrial pollution occur.
- Bottom sediments are a transient type of habitat i.e. the soil-water habitat that is almost always typically oxygen-free in which the processes of

anaerobic decomposition by microorganisms cause the release of hydrogen sulphide and methane into water. In the bottom sediment, anaerobic putrefying microflora, cellulolytic bacteria and the anaerobic chemoautotrophs develop.

Groups of Water Organisms

Microorganisms occupy surface waters in all of the zones; they may be suspended in water (plankton), cover stationary underwater objects, plants etc. (periphyton), or live in bottom sediments (benthos).

a. Plankton

The group of organisms that passively float in water not being able to resist the movement and the flow of water mass is called plankton or bioseston. These are of following types:

1 Phytoplankton

Phytoplankton are mainly microscopic algae and blue-green algae. It is a varied community in terms of the systematics and mainly composed of forms smaller than 50 μm. Sea phytoplankton are dominated by diatoms and dinophyta, whereas fresh water phytoplankton are dominated one by cryptophytes, diatoms, green algae, and blue-green algae.

2 Zooplankton

Zooplankton are small water animals that occur in plankton. There are three systematic groups that occur in fresh waters: rotifers, branchiopods and copepods. The sea water plankton is composed of copepods, ctenophores, urochordata, arrow-worms as well as some species of snails. Most of them are filtrators (condense suspended particles) or predators.

3 Protozoa plankton

Protozoa plankton consists of protozoa which occupy the open water zones like flagellates and ciliates. They are the main consumers of bacteria. Moreover, most ciliates feed upon flagellates, algae and smaller ciliates. The protozoa itself feeds the zooplankton.

4 The heterotrophic bacteria plankton

The heterotrophic bacteria plankton occupy waters which are abundant in organic compounds. The amount of bacteria in open waters varies between 10^5-10^7 cells/ml.

5 Virus plankton

Virus plankton is composed of viruses which are the smallest element of plankton. Their numbers may be very high (from 108 in 1ml) in various fresh and sea water habitats. Viruses are, next to the protozoa, a crucial factor in bacteria mortality.

b. Periphyton

Periphyton occupy the shore line zones. They are a group of organisms that create outgrowths upon various objects and underwater plants. Most of the time, they usually consist of small algae - diatoms, green algae and bacteria. Moreover, various settled or semi-settled protozoa, eelwarms, oligochaetes, insect larva, and even crustaceans make up the periphyton biocenosis. Periphyton has a characteristic complex biocenosis and many ecological relationships can be observed between its components.

c. Benthos

The bottom habitat is occupied by a group of organisms called the benthos. The muddy bottom contains an abundance of organic compounds that are created as a result of dead matter decomposition (fallen parts of plants and animals). At great depths the bottom is free from any plants which, due to a lack of light cannot grow. However, the absence of oxygen supports the development of, among others, an oxygen-free putrid microflora. Among the benthos microflora the most numerous are bacteria and fungi (decomposers) as well as some animals (detritophages). Both of the above groups are responsible for decomposition of the organic matter. Benthos of shallow reservoirs may also contain some algae.

IN SPACE

Interest in outer space has resulted in conjectures as to the possibility of the survival of bacteria and other microorganisms in inter planetary regions. Samplings of air taken at fairly high altitudes have shown the presence of viable microorganisms. In an experiment using desert soils as the source of the bacteria, certain of these forms grew under conditions thought to exist on Mars. These included an almost complete lack of oxygen and temperatures approaching freezing point.

MICROBES OF THE AIR

Several types of microbes and their spores are common in the outdoor air as well as indoor air. The degree of bacterial, viral and fungal contaminants in the outdoor air depends upon the density of human and animal population, nature of the ground soil, temperature and humidity. Aeroplane survey show that the bacteria in the upper air are mainly of aerobic spore bearing bacilli and less proportion microbes like *Achromobacter*, *Sarcina* and *Micrococcus*, etc.

The indoor air-borne microbes may be in three forms

(1) Attached to dust particles.

(2) In the gross droplets expelled from the nose and mouth.

(3) In the 'droplet nuclei' formed from evaporation of smaller droplets expelled from nose and mouth.

Degree of air-contamination may be measured by:

1. **Sedimentation** ie, exposing a series of agar plates in different parts of a room for given time.

2. **Impaction.** It means collection of particles on a solid or semisolid surface. There are several devices, however, the simplest one is by drawing measured volumes of air drawn through a funnel and allowed to play on agar plates.

3. **Impingement.** This is the term used for collection of particles in a liquid medium. The serial dilution of liquid allows a wide range of concentration of airborne microbes.

4. **Filteration.** Several members of microbial world can be collected on membrane filters and can be seen and identified.

5. **Precipitation.** By this method particles are precipitated from the air by thermal or electrostatic means, permitting the organisms to deposit on agar coated cylinders.

6. **Sampling by size.** This is done by using multistage stacked plate impactor. Six petridishes containing nutrient agar are stacked in a single pile. Over each of them is a sieve perforated by 400 holes diminishing in diameter from top to bottom plate. Air is drawn through at a pressure at the rate of 1 cubic foot per minute passing from one plate to the next and increasing in velocity from above downwards. The larger particles settle on the first two plates and smallest those of about 1 um on the bottom two plates.

7. **Animal method of sampling.** This can be done using animals as guinea-pig for tubercle bacillus, which form nodules in the lung. The method is also employed for trapping of viruses in the respiratory tract of some animals.

8. Culturing the sample. The suitable type of medium should be used for the particular type of species being looked for. The viruses are best collected into a fluid medium and inoculated into tissue cultures of fertile eggs. The number of plaques in monolayer tissue culture enables the original number of particles to be determined.

ASSOCIATED WITH MAN

The surface of one's body harbours many varieties of bacteria, some of which may be harmful if one sustains a wound thus introducing these microorganisms into the tissue and may result into severe infection. The nose, throat, mucous membranes of the mouth and digestive tract of human beings and animals are always inhabited by many different species of microorganisms. Most of these are classified as the normal flora of the individual but at times pathogenic microorganisms establish themselves in one's nose, throat, or digestive tract, where they multiply and produce disease.

IN FOOD

Since most foods that we eat are heated only few microorganisms are likely to be present. However, if cooked foods are allowed to remain in the open air, dust particles containing microorganisms may carry them into the food. Most of these organisms are harmless. Harmful bacteria may occasionally get into foods and cause illness in persons who eat the contaminated foods.

IN RUMEN OF CATTLE

Studies on the microflora and microfauna of the rumen (first stomach) of cattle have shown the presence of various microorganisms in large number that apparently play an important role in the digestion of the great quantities of cellulose ingested by the cow and other animals.

IN ASSOCIATION WITH INSECTS

Various insects harbour many kinds of microorganisms. Some microorganisms are found on the feet and bodies of flies and other insects, and some are present within, their body cavities and organ structures. Flies, mosquitoes, ticks and other insects may, therefore, be responsible at times for carrying a disease agent. Fortunately, most of the time the microorganisms carried by insects are the same as those harmless ones found in water, soil, and the air. Insects may suffer from fatal bacterial infections,

LITHOTROPHIC AND MINE DWELLERS

The yeasts, flagellates, moulds and chemoorganotrophic bacteria are the most common which occur in lithotrophic and mine environments. The various acidophilic iron and/or sulphur oxidizing bacteria are responsible for the oxidation of mineral sulphides in leach biotypes. At ambient temperatures the most common as well as the most important bacteria are *Thiobacillus ferroaxidans, Thiobacillus thiooxidans* and *Leptospirillum ferrooxidans.*

****

4.

CLASSIFICATION AND IDENTIFICATION METHODS OF MICROORGANISMS

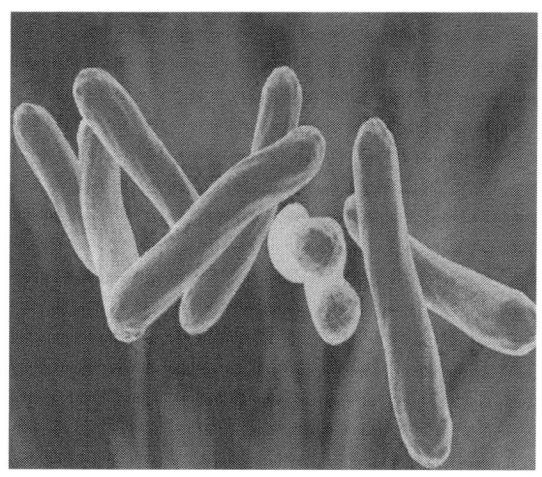

Microorganisms are very diverse. They include bacteria, fungi, algae, and protozoa; microscopic plants (green algae); and animals such as rotifers and planarians. Most microorganisms are unicellular (single-celled), but this is not universal.

Single-celled microorganisms were the first forms of life to develop on earth, approximately 3 billion–4 billion years ago. Further evolution was slow, and for about 3 billion years in the Precambrian eon, all organisms were microscopic. So, for most of the history of life on earth the only forms of life were microorganisms. Bacteria, algae, and fungi have been identified in amber that is 220 million years old, which shows that the morphology of microorganisms has changed little since the Triassic period. When at the end of the 19th century information began to accumulate about the diversity within the bacterial world, scientists started to include the bacteria in phylogenetic schemes to explain how life on earth may have developed. Some of the early phylogenetic trees of the prokaryote world were morphology-based. Others were based on the then-current ideas on the presumed conditions on our planet at the time that life first developed.

Microorganisms tend to have a relatively rapid evolution. Most microorganisms can reproduce rapidly, and microbes such as bacteria can also freely exchange genes through conjugation, transformation, and transduction, even between widely-divergent species. This horizontal gene transfer, coupled with a high mutation rate and many other means of genetic variation, allows microorganisms to swiftly evolve (via natural selection) to survive in new environments and respond to environmental stresses.

The relationship between the three domains (Bacteria, Archaea, and Eukaryota) is of central importance for understanding the origin of life. Most of the metabolic pathways, which comprise the majority of an organism's genes, are common between Archaea and Bacteria, while most genes involved in genome expression are common between Archaea and Eukarya. Within prokaryotes, archaeal cell structure is most similar to that of Gram-positive bacteria.

Phenotypic Methods of Classifying and Identifying Microorganisms

Classification seeks to describe the diversity of bacterial species by naming and grouping organisms based on similarities. Microorganisms can be classified on the basis of cell structure, cellular metabolism, or on differences in cell components such as DNA, fatty acids, pigments, antigens, and quinones.

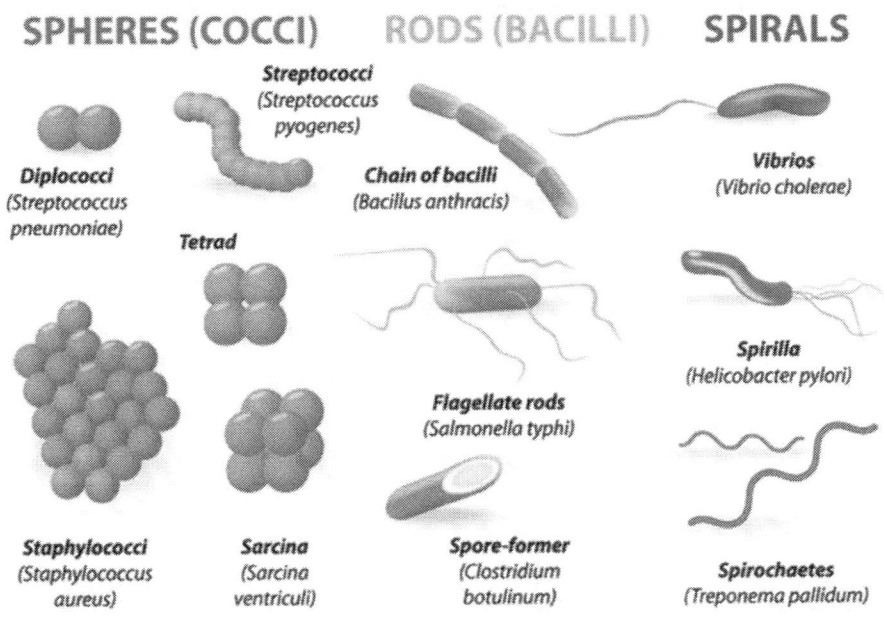

Fig.4. Bacterial Morphology

Bacterial Morphology: Basic morphological differences between bacteria. The most often found forms and their associations.

There are some basic differences between Bacteria, Archaea, and Eukaryotes in cell morphology and structure which aid in phenotypic classification and identification:

Bacteria: lack membrane-bound organelles and can function and reproduce as individual cells, but often aggregate in multicellular colonies. Their genome is usually a single loop of DNA, although they can also harbor small pieces of DNA called plasmids. These plasmids can be transferred between cells through bacterial conjugation. Bacteria are surrounded by a cell wall, which provides strength and rigidity to their cells.

Archaea: In the past, the differences between bacteria and archaea were not recognized and archaea were classified with bacteria as part of the kingdom Monera. Archaea are also single-celled organisms that lack nuclei. Archaea in fact differ from bacteria in both their genetics and biochemistry. While bacterial cell membranes are made from phosphoglycerides with ester bonds, archaean membranes are made of ether lipids.

Eukaryotes: Unlike bacteria and archaea, eukaryotes contain organelles such as the cell nucleus, the Golgi apparatus, and mitochondria in their cells. Like bacteria, plant cells have cell walls and contain organelles such as chloroplasts in addition to the organelles in other eukaryotes.

The Gram stain, developed in 1884 by Hans Christian Gram, characterizes bacteria based on the structural characteristics of their cell walls. The thick layers of peptidoglycan in the "Gram-positive" cell wall stain purple, while the thin "Gram-negative" cell wall appears pink. By combining morphology and Gram-staining, most bacteria can be classified as belonging to one of four groups (Gram-positive cocci, Gram-positive bacilli, Gram-negative cocci, and Gram-negative bacilli). Some organisms are best identified by stains other than the Gram stain, particularly mycobacteria or Nocardia, which show acid-fastness on Ziehl–Neelsen or similar stains. Other organisms may need to be identified by their growth in special media, or by other techniques, such as serology.

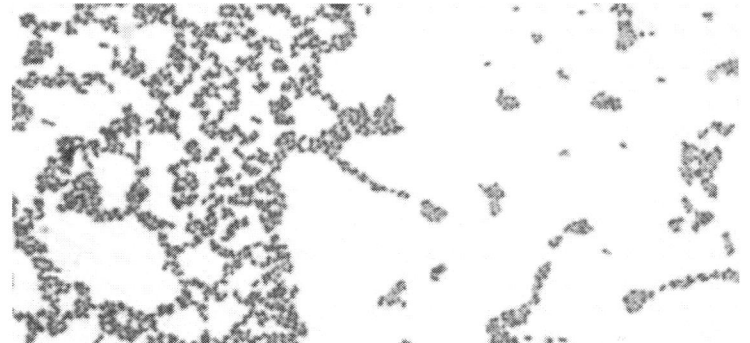

Fig. 4.1. Gram-positive bacteria: Streptococcus mutans visualized with a Gram stain.

While these schemes allowed the identification and classification of bacterial strains, it was unclear whether these differences represented variation between distinct species or between strains of the same species. This uncertainty was due to the lack of distinct structures in most bacteria, as well as lateral gene transfer between unrelated species. Due to lateral gene transfer, some closely related bacteria can have very different morphologies and metabolisms. To overcome this uncertainty, modern bacterial classification emphasizes molecular systematics, using genetic techniques such as guanine cytosine ratio determination, genome-genome hybridization, as well as sequencing genes that have not undergone extensive lateral gene transfer, such as the rRNA gene.

Classification of Prokaryotes

Prokaryotic organisms were the first living things on earth and still inhabit every environment, no matter how extreme.

Learning Objectives

Discuss the origins of prokaryotic organisms in terms of the geologic timeline

Evolution of Prokaryotes

In the recent past, scientists grouped living things into five kingdoms (animals, plants, fungi, protists, and prokaryotes) based on several criteria such as: the absence or presence of a nucleus and other membrane-bound organelles, the absence or presence of cell walls, multicellularity, etc. In the late 20th century, the pioneering work of Carl Woese and others compared sequences of small-subunit ribosomal RNA (SSU rRNA) which resulted in a more fundamental way to group organisms on earth. Based on differences in the structure of cell membranes and in rRNA, Woese and his colleagues proposed that all life on earth evolved along three lineages, called domains. The domain Bacteria comprises all organisms in the kingdom Bacteria, the domain

Archaea comprises the rest of the prokaryotes, and the domain Eukarya comprises all eukaryotes, including organisms in the kingdoms Animalia, Plantae, Fungi, and Protista.

Fig 4.2. Prokaryotes in extreme environments: Certain prokaryotes can live in extreme environments such as the Morning Glory pool, a hot spring in Yellowstone National Park. The spring's vivid blue colour is from the prokaryotes that thrive in its very hot waters.

The current model of the evolution of the first, living organisms is that these were some form of prokaryotes, which may have evolved out of protobionts. In general, the eukaryotes are thought to have evolved later in the history of life. However, some authors have questioned this conclusion, arguing that the current set of prokaryotic species may have evolved from more complex eukaryotic ancestors through a process of simplification. Others have argued that the three domains of life arose simultaneously, from a set of varied cells that formed a single gene pool.

Two of the three domains, Bacteria and Archaea, are prokaryotic. Based on fossil evidence, prokaryotes were the first inhabitants on Earth, appearing 3.5 to 3.8 billion years ago during the Precambrian Period. These organisms are abundant and ubiquitous; that is, they are present everywhere. In addition to inhabiting moderate environments, they are found in extreme conditions: from boiling springs to permanently frozen environments in Antarctica; from salty environments like the Dead Sea to environments under tremendous pressure, such as the depths of the ocean; and from areas without oxygen, such as a waste management plant, to radioactively-contaminated regions, such as Chernobyl. Prokaryotes reside in the human digestive system and on the skin, are responsible for certain illnesses, and serve an important role in the preparation of many foods.

Phylogenetic Analysis

The molecular approach to microbial phylogenetic analysis revolutionized our thinking about evolution in the microbial world.

Microbial phylogenetics: The study of the evolutionary relatedness among various groups of microorganisms.

Microbial phylogenetics is the study of the evolutionary relatedness among various groups of microorganisms. The molecular approach to microbial phylogenetic analysis revolutionized our thinking about evolution in the microbial world. The purpose of phylogenetic analysis is to understand the past evolutionary path of organisms. Even though we will never know for certain the true phylogeny of any organism, phylogenetic analysis provides best assumptions, thereby providing a framework for various disciplines in microbiology. Due to the technological innovation of modern molecular biology and the rapid advancement in computational science, accurate inference of the phylogeny of a gene or organism seems possible in the near future.

Gene sequences can be used to reconstruct the bacterial phylogeny. These studies indicate that bacteria diverged first from the archaeal/eukaryotic lineage. The term "bacteria" was traditionally applied to all microscopic, single-cell prokaryotes. However, molecular systematics showed prokaryotic life to consist of two separate domains, originally called Eubacteria and Archaebacteria, but now called Bacteria and Archaea that evolved independently from an ancient common ancestor. The archaea and eukaryotes are more closely related to each other than to the bacteria. Due to the relatively recent introduction of molecular systematics and a rapid increase in the number of genome sequences that are available, bacterial classification remains a changing and expanding field. For example, a few biologists argue that the Archaea and Eukaryotes evolved from Gram-positive bacteria.

While morphological or metabolic differences allowed the identification and classification of bacterial strains, it was unclear whether these differences represented variation between distinct species or between strains of the same species. This uncertainty was due to the lack of distinctive structures in most bacteria, as well as lateral gene transfer between unrelated species. The developing technology of nucleic acid sequencing, together with the recognition that sequences of building blocks in informational macromolecules can be used as 'molecular clocks' that contain historical information, led to the development of the three-domain model (Archaea – Bacteria – Eucaryota) in the late 1970's, primarily based on small subunit ribosomal RNA sequence comparisons pioneered by Carl Woese and George Fox.

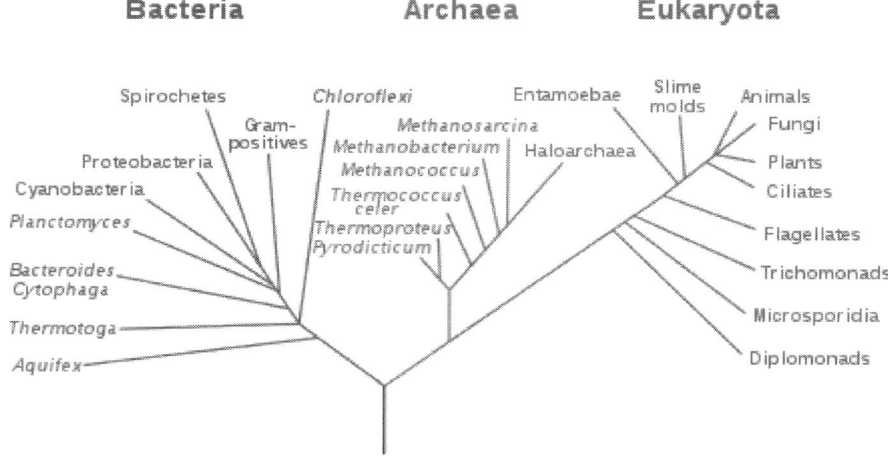

Fig.4.3. A speculatively rooted tree for rRNA genes, showing the three life domains: bacteria, archaea, and eukaryota. The black branch at the bottom of the phylogenetic tree connects the three branches of living organisms to the last universal common ancestor.

As more genome sequences become available, scientists have found that determining these relationships is complicated by the prevalence of lateral gene transfer (LGT) among archaea and bacteria. Due to lateral gene transfer, some closely related bacteria can have very different morphologies and metabolisms. To overcome this uncertainty, modern bacterial classification emphasizes molecular systematics, using genetic techniques such as guanine cytosine ratio determination, genome-genome hybridization, as well as sequencing genes that have not undergone extensive lateral gene transfer, such as the rRNA gene.

As with bacterial classification, identification of microorganisms is increasingly using molecular methods. Diagnostics using such DNA-based tools, such as polymerase chain reaction, are increasingly popular due to their specificity and speed, compared to culture-based methods. However, even using these improved methods, the total number of bacterial species is not known and cannot even be estimated with any certainty. Following present classification, there are a little less than 9,300 known species of prokaryotes, which includes bacteria and archaea. but attempts to estimate the true level of bacterial diversity have ranged from 10^7 to 10^9 total species – and even these diverse estimates may be off by many orders of magnitude.

There are four steps in general phylogenetic analysis of molecular sequences: (i) selection of a suitable molecule or molecules (phylogenetic marker), (ii) acquisition of molecular sequences, (iii) multiple sequence alignment (MSA), and (iv) phylogenetic treeing and evaluation.

Multilocus sequence analysis (MLSA) represents the novel standard in microbial molecular systematics. In this context, MLSA is implemented in a relatively straightforward way, consisting essentially in the concatenation of several sequence partitions for the same set of organisms, resulting in a "supermatrix" which is used to infer a phylogeny by means of distance-matrix or optimality criterion-based methods. This approach is expected to have an increased resolving power due to the large number of characters analyzed and a lower sensitivity to the impact of conflicting signals (i.e. phylogenetic incongruence) that result from eventual horizontal gene transfer events. The strategies used to deal with multiple partitions can be grouped in three broad categories: the total evidence, separate analysis, and combination approaches. The concatenation approach that dominates MLSAs in the microbial molecular systematics literature is known to systematists working with plants and animals as the "total molecular evidence" approach. It has been used to solve difficult phylogenetic questions such as the relationships among the major groups of cetaceans, that of microsporidia and fungi, or the phylogeny of major plant lineages. The total molecular evidence approach has been criticized because by directly concatenating all available sequence alignments. The evidence of conflicting phylogenetic signals in the different data partitions is lost along with the possibility to uncover the evolutionary processes that gave rise to such contradictory signals.

Nongenetic Categories for Medicine and Ecology

In medicine, microorganisms are identified by morphology, physiology, and other attributes; in ecology by habitat, energy, and carbon source.

BACTERIA

Although most bacteria are harmless, even beneficial, quite a few are pathogenic. Each pathogenic species has a characteristic spectrum of interactions with its human hosts.

Conditionally, pathogenic bacteria are only pathogenic under certain conditions; such as a wound that allows for entry into the blood, or a decrease in immune function. Bacterial infections can also be classified by location in the body, for example, the vagina, lungs, skin, spinal cord and brain, and urinary tract.

When identifying bacteria in the laboratory, the following chatacteristics are used: Gram staining, shape, presence of a capsule, bonding tendency (singly or in pairs), motility, respiration, growth medium, and whether it is intra- or extracellular.

Culture techniques are designed to grow and identify particular bacteria, while restricting the growth of the others in the sample. Often these techniques are designed for specific specimens: for example, a sputum sample will be treated to identify organisms that cause pneumonia. Once a pathogenic organism has been isolated, it can be further characterised by its morphology, growth patterns (aerobic or anaerobic), patterns of hemolysis, and staining.

VIRUSES

Similar to the classification systems used for cellular organisms, virus classification is the subject of ongoing debate due to their pseudo-living nature. Essentially, they are non-living particles with some chemical characteristics similar to those of life; thus, they do not fit neatly into an established biological classification system.

Viruses are mainly classified by phenotypic characteristics,such as:

- Morphology
- Nucleic acid type
- Mode of replication
- Host organisms

Type of disease they cause

Currently there are two main schemes used for the classification of viruses: (1) the International Committee on Taxonomy of Viruses (ICTV) system; and (2) the Baltimore classification system, which places viruses into one of seven groups. To date, six orders have been established by the ICTV:

Caudovirales

Herpesvirales

Mononegavirales

Nidovirales

Picornavirales

Tymovirales

These orders span viruses with varying host ranges, only some of which infect human hosts.

Baltimore classification is a system that places viruses into one of seven groups depending on a combination of:

- Their nucleic acid (DNA or RNA)
- Strandedness (single or double)
- Sense method of replication

Other classifications are determined by the disease caused by the virus or its morphology, neither of which is satisfactory as different viruses can either cause the same disease or look very similar. In addition, viral structures are often difficult to determine under the microscope. Classifying viruses according to their genome means that those in a given category will all behave in a similar fashion, offering some indication of how to proceed with further research.

Other organisms invariably cause disease in humans, such as obligate intracellular parasites that are able to grow and reproduce only within the cells of other organisms.

CATEGORIES OF MICROORGANISMS IN ECOLOGY

In ecology, microorganisms are classified by the type of habitat they require, or trophic level, energy source and carbon source.

Habitat Type

Biologists have found that microbial life has an amazing flexibility for surviving in extreme environments that would be completely inhospitable to complex organisms. Some even concluded that life may have begun on Earth in hydrothermal vents far under the ocean's surface.

An *extremophile* is an organism that thrives in physically or geochemically extreme conditions, detrimental to most life on Earth. Most known extremophiles are microbes. The domain *Archaea* contains renowned examples, but extremophiles are present in numerous and diverse genetic lineages of both bacteria and archaeans. In contrast, organisms that live in more moderate environments may be termed *mesophiles* or *neutrophiles*.

There are many different classes of extremophiles, each corresponding to the way its environmental niche differs from mesophilic conditions. Many extremophiles fall under multiple categories and are termed *polyextremophiles*. Some examples of types of extremophiles:

- *Acidophile*: an organism with optimal growth at levels of pH 3 or below
- *Xerophile*: an organism that can grow in extremely dry, desiccating conditions; exemplified by the soil microbes of the Atacama Desert

- *Halophile*: an organism requiring at least 0.2M concentrations of salt (NaCl) for growth
- *Thermophile*: an organism that can thrive at temperatures between 45–122 °C

Trophic level, energy source and carbon source

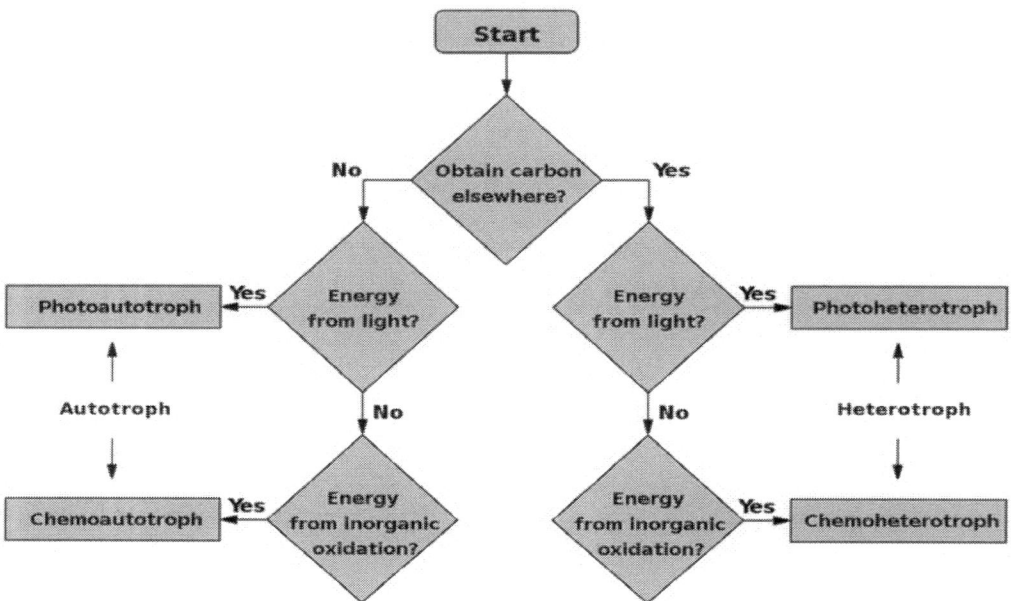

Fig.4.4. The nutritional modes of an organism: A flowchart to determine if a species is autotroph, heterotroph, or a subtype.

Phototrophs: carry out photon capture to acquire energy. They use the energy from light to carry out various cellular metabolic processes. They are not obligatorily photosynthetic. Most of the well-recognized phototrophs are *autotrophs*, also known as *photoautotrophs*, and can fix carbon.

Photoheterotrophs: produce ATP through photophosphorylation but use environmentally-obtained organic compounds to build structures and other bio-molecules.

Photolithoautotroph: an autotrophic organism that uses light energy, and an inorganic electron donor (e.g., H_2O, H_2, H_2S), and CO_2 as its carbon source.

Chemotrophs: obtain their energy by the oxidation of electron donors in their environments.

Chemoorganotrophs: organisms which oxidize the chemical bonds in organic compounds as their energy source and attain the carbon molecules they need for cellular function. These oxidized organic compounds include sugars, fats and proteins.

Chemoorganoheterotrophs (or organotrophs) exploit reduced-carbon compounds as energy sources, such as carbohydrates, fats, and proteins from plants and animals. *Chemolithoheterotrophs* (or *lithotrophic heterotrophs*) utilize inorganic substances to produce ATP, including hydrogen sulfide and elemental sulfur.

Lithoautotroph: derives energy from reduced compounds of mineral origin. May also be referred to as *chemolithoautotrophs*, reflecting their autotrophic metabolic pathways. Lithoautotrophs are exclusively microbes and most are bacteria. For lithoautotrophic bacteria, only inorganic molecules can be used as energy sources.

Mixotroph: Can use a mix of different sources of energy and carbon. These may be alternations between photo- and chemotrophy, between litho- and organotrophy, between auto- and heterotrophy or a combination of them. Can be either eukaryotic or prokaryotic. (https://courses.lumenlearning.com)

****

5. THE METHODS IN MICROBIOLOGY

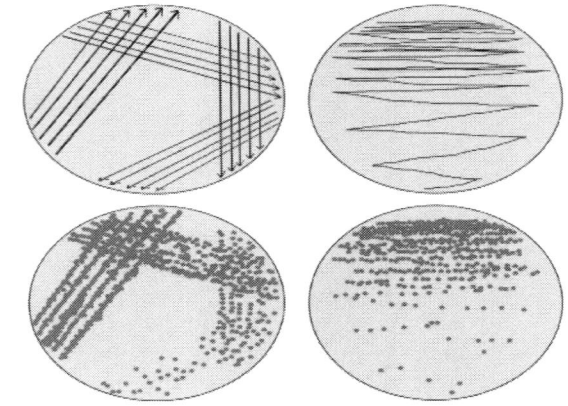

The subject of microbiology is so vast that a large number of techniques and methods are used to study the various aspects of microbiology. The techniques to observe them have been described in the First Chapter under the sub-heading invention of microscope. Some techniques have been described under the chapters Identification and differentiation of bacteria', Bacterial cultivation' and Reproduction and Growth of Bacteria'. Some of the techniques related to viruses have been described separately. However, some basic techniques applicable to study all types of microbes belonging to bacteria, protozoa, algae, fungi and viruses are given below. While studying bacteria under microscope a very thin film is prepared on a glass slide, by taking a drop of bacterial suspension in water (suspension with turbid appearance). This thin film is referred to as smear.

Pure cultures

The growth of a single type of microorganism is regarded as pure culture or axenic culture in contrast to mixed cultures containing more than one kind of microorganisms. **The streak plate and pour plate methods** are the most common methods to separate the mixed cultures. The special wires called **inoculation needles or inoculation loops** are used to transfer the inoculum (specimen) to the medium. The **inoculation needle** is made free from microbes by heating in the flame of Bunsen burner or spirit lamp and the technique is referred to as incineration, The inoculation is done under aseptic conditions i.e. in the environment free from undesirable microorganisms, using an inoculation hood with **U.V. light** arrangement or a laminar flow. The serial dilution has been described separately in the chapter referred above. The characteristics of various isolated bacterial colonies are also taken into consideration on the basis of form (puncti form, circular, filaments, irregular, rhizoid and spindle), elevation (flat, raised, convex, pulvinate, umbonate and undulate), and margin (entire, undulate, lobate, irregular, filamentous and curled). The techniques required on obtaining a pure culture are refrigeration by decreasing the environmental temperature below 5°C. The freezing may be done to preserve by placing the culture tubesin acetone or dry-ice baths at-70°C or liquid nitrogen at -200°C. The freeze-drying or lyophilization may be done at-70 ^{0}C by drawing off the water from the cell by high vacuum pump. The American Type Culture Collection (ATCC) is an international source of maintaining the preserved pure cultures. There is a good collection of fungal microbes at CMI (Common Wealth Mycological Institute of Kew, in England) at present known as International Mycologieal Institute, Bakeham Lane, Egham, Surrey, TW209TY, United Kingdom. This is an institute of CAB International, which is an inter-governmental organization providing services world-wide to agriculture, forestry, human health and the management of natural resources. Institute of Microbial Technology (IMTECH), Chandigarh, at present holds a collection of about 3000 cultures in its stock, according to the institutes annual report of 1993- 94.

Sterilization

Sterilization refers to any process that eliminates, removes, kills, or deactivates all forms of life (in particular referring to microorganisms such as fungi, bacteria, viruses, spores, unicellular eukaryotic organisms such as Plasmodium, etc.) and other biological agents like prions present in a specific surface, object or fluid, for example food or biological culture media.Sterilization can be achieved through various means, including heat, chemicals, irradiation, high pressure, and filtration. Sterilization is distinct from disinfection, sanitization, and pasteurization, in that those methods reduce rather than eliminate all forms of life and biological agents present. After sterilization, an object is referred to as being sterile or aseptic.

Types of sterilization

Although there are many techniques of sterilization available out there, heat and chemical method is widely employed to sterile metals and instument. Apart from that, there are numerous techniques out there. In this article, you can go through the most commonly used sterilization techniques.

Heat Method: This is the most common method of sterilization. The heat is used to kill the microbes in the substance. The extent of sterilization is affected by the temperature of the heat and duration of heating. On the basis of type of heat used, heat methods are categorized into-

(i) Wet Heat/Steam Sterilization- In most labs, this is a widely used method which is done in autoclaves.. Autoclaves use steam heated to 121–134 °C under pressure. This is a very effective method that kills/deactivates all microbes, bacterial spores and viruses. Autoclaving kills microbes by hydrolysis and coagulation of cellular proteins, which is efficiently achieved by intense heat in the presence of water. The intense heat comes from the steam. Pressurized steam has a high latent heat and at 100°C it holds 7 times more heat than water at the same temperature. In general, Autoclaves can be compared with a typical pressure cooker used for cooking except in the trait that almost all the air is removed from the autoclave before the heating process starts. Wet heat sterilization techniques also include boiling and pasteurization.

(ii) Dry heat sterilization- In this method, specimens containing bacteria are exposed to high temperatures either by flaming, incineration or a hot air oven. Flaming is used for metallic devices like needles, scalpels, scissors, etc. Incineration is used especially for inoculating loops used in microbe cultures. The metallic end of the loop is heated to red hot on the flame. The hot air oven is suitable for dry material like powders, some metal devices, glassware, etc.

Filtration is the quickest way to sterilize solutions without heating. This method involves filtering with a pore size that is too small for microbes to pass through. Generally filters with a pore diameter of 0.2 um are used for the removal of bacteria. Membrane filters are more commonly used filters over sintered or seitz or candle filters. It may be noted that viruses and phage are much smaller than bacteria, so the filtration method is not applicable if these are the prime concern.

Radiation sterilization: This method involves exposing the packed materials to radiation (UV, X-rays, gamma rays) for sterilization. The main difference between different radiation types is their penetration and hence their effectiveness. UV rays have low penetration and thus are less effective, but it is relatively safe and can be used for small area sterilization. X-rays and gamma rays have far more penetrating power and thus are more effective for sterilization on a large scale. It is, however, more dangerous and thus needs special attention. UV irradiation is routinely used to sterilize the interiors of biological safety cabinets between uses. X-rays are used for sterilizing large packages and pallet loads of medical devices. Gamma radiation is commonly used for sterilization of disposable medical equipment, such as syringes, needles, cannulas and IV sets, and food.

Chemical method of sterilization: Heating provides a reliable way to get rid of all microbes, but it is not always appropriate as it can damage the material to be sterilized. In that case, chemical methods for sterilization is used which involves the use of harmful liquids and toxic gases without affecting the material. Sterilization is effective using gases because they penetrate quickly into the material like steam. There are a few risks, and the chances of explosion and cost factors are to be considered.

The commonly used gases for sterilization are a combination of ethylene oxide and carbon-dioxide. Here Carbon dioxide is added to minimize the chances of an explosion. Ozone gas is another option which oxidize most organic matter. Hydrogen peroxide, Nitrogen dioxide, Glutaraldehyde and formaldehyde solutions, Phthalaldehyde, and Peracetic acid are other examples of chemicals used for sterilization. Ethanol and IPA are good at killing microbial cells, but they have no effect on spores.

Staining techniques

Some of the staining techniques have been described under Identification and differentiation of bacteria', the **standard wet mount** is prepared by placing or suspending the microbes in a drop of water. But there are problems to see the small microbes like bacteria, to eliminate any problem to bring the small microbes under a sharp focus an alternative device called **hanging drop slide preparation** is generally employed. The deep-well projection slides can be used for demonstrating protozoa, algae, fungal spores (both non-motile and motile) and agglutination including blood typing and crystal formation. The negative smear preparation is often employed to study the microbes which are difficult to stain. Therefore a little of nigrosin or India ink is applied on a clean slide. The microbes to be studied are mixed with the inoculation needle with this chemical. The nigrosin or India ink does not stick to the microbes rather just surround the microbes which shine as rods, cocci or spiral against a dark gray-purple back ground when the microscopic examination of the smear is made. Two common stains employed in the laboratory are simple stain and differential stain. The most commonly employed dyes are the cationic dyes. There is positive charge on the coloured portion of the cationic dyes.

One of the often used stain called methylene blue is methylene chloride (CH CL). The colour or pigment containing part of the molecule of CH_2CL_2 is called chromatophore whereas the non-pigment containing ion is referred to as auxochrome. On coming in contact with the dye methylene blue the negative electrical charge present on the surface of the cell will make it to react with positively charged methylene ion called chromatophore. Therefore, such dyes are known as basic dycs. The endospores of bacteria like *Bacillus* and *Clostridium* are stained with malchite green to stain the endospores, because the endospores are chemically resistant to most of the dyes. The **Gram stain** and **Acid-fast** stain are two similar methods. The acid fast stain is used to identify *Mycobacterium tuberculosis* and *M. leprae*. The acid-fast cells are **carbol fuchsin** red whereas the methylene blue is used as the counter stain. In a mixture of acid fast (***Mycobacterium***) and non-acid fast (***Streptococcus***) on the same slide will give the appearance of red bacilli to mycobacteria and blue cocci to streptococcus. This type of stain differentiates the different types of cells whereas the endospore and capsular stains can differentiate the different parts of the same cell. The dyes can be broadly categorised under 3 types as: **(1) acid or anionic dye,** in which the charge on the ions of dye is negative; **(2) basic or cationic dye** in which charge on the dye is positive as referred above; and **(3) the neutral dye** which is a complex salt of dye acid with dye base as **eosinate of methylene blue**.

****

6. PROKARYOTIC CELLS AND EUKARYOTIC CELLS

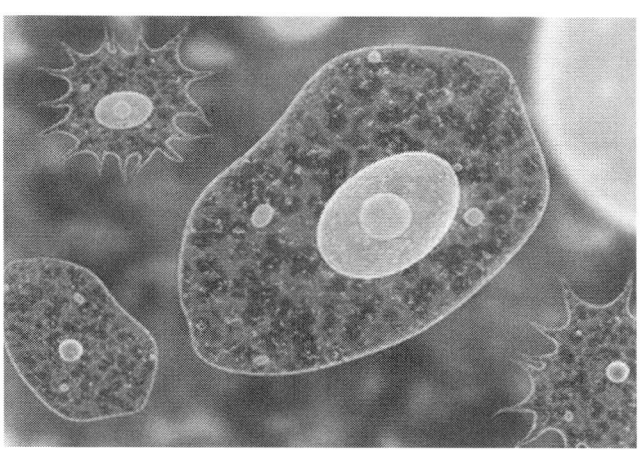

Prokaryotes are simple, small cells, whereas **eukaryotic cells** are complex, large structured and are present in trillions which can be single celled or multicellular. Prokaryotic cells do not have a **well-defined nucleus** but DNA molecule is located in the cell, termed as **nucleoid**, whereas eukaryotic cells have a **well-defined nucleus**, where genetic material is stored. Based on the structure and functions, cells are broadly classified as Prokaryotic cell and Eukaryotic cell

Prokaryotic Cells are the most primitive kind of cells and lack few features as compared to the eukaryotic cell. **Eukaryotic cells** have evolved from prokaryotic cells only but contain different types of organelles like Endoplasmic reticulum, Golgi body, Mitochondria etc, which are specific in their functions. But features like growth, response, and most importantly giving birth to the young ones are the commonly shared by all living organisms.

In the following content, we will discuss the general difference between the two types of cells. As these 'cells' are considered as the structural and functional unit of life, whether it's a single cell organism like bacteria, protozoa, or multicellular organisms like plants and animals.

Comparison Chart

Basis For Comparison	Prokaryotic Cells	Eukaryotic Cells
Size	0.5-3um	2-100um
Kind of Cell	Single-cell	Multicellular
Cell Wall	Cell wall present, comprise of peptidoglycan or mucopeptide (polysaccharide).	Usually cell wall absent, if present (plant cells and fungus), comprises of cellulose (polysaccharide).
Presence of Nucleus	Well-defined nucleus is absent, rather 'nucleoid' is present which is an open region containing DNA.	A well-defined nucleus is present enclosed within nuclear memebrane.
Shape of DNA	Circular, double-stranded DNA.	Linear, double-stranded DNA.
Mitochondria	Absent	Present
Ribosome	70S	80S
Golgi Apparatus	Absent	Present
Endoplasmic Reticulum	Absent	Present
Mode of Reproduction	Asexual	Most commonly sexual
Cell Divison	Binary Fission, (conjugation, transformation, transduction)	Mitosis
Lysosomes and Peroxisomes	Absent	Present
Chloroplast	(Absent) scattered in the cytoplasm.	Present in plants, algae.
Transcription and Translation	Occurs together.	Transcription occurs in nucleus and translation in cytosol.
Organelles	Organelles are not membrane bound, if present any.	Organelles are membrane bound and are specific in function.

Replication	Single origin of replication.	Multiple origins of replication.
Number of Chromosomes	Only one (not true called as a plasmid).	More than one.
Examples	Archaea, Bacteria.	Plants and Animals.

Definition of Prokaryotic Cells

Pro means 'old,' and *karyon* means 'nucleus,' So as the name suggest the history of the evolution of prokaryotic cells is at least **3.5 billion years** old, but they are still important to us in many aspects like they are **used** in industries for fermentation (Lactobacillus, Streptococcus), for research work, etc. In comparison to eukaryotic cells, they lack few organelles and are not advanced as eukaryotes.

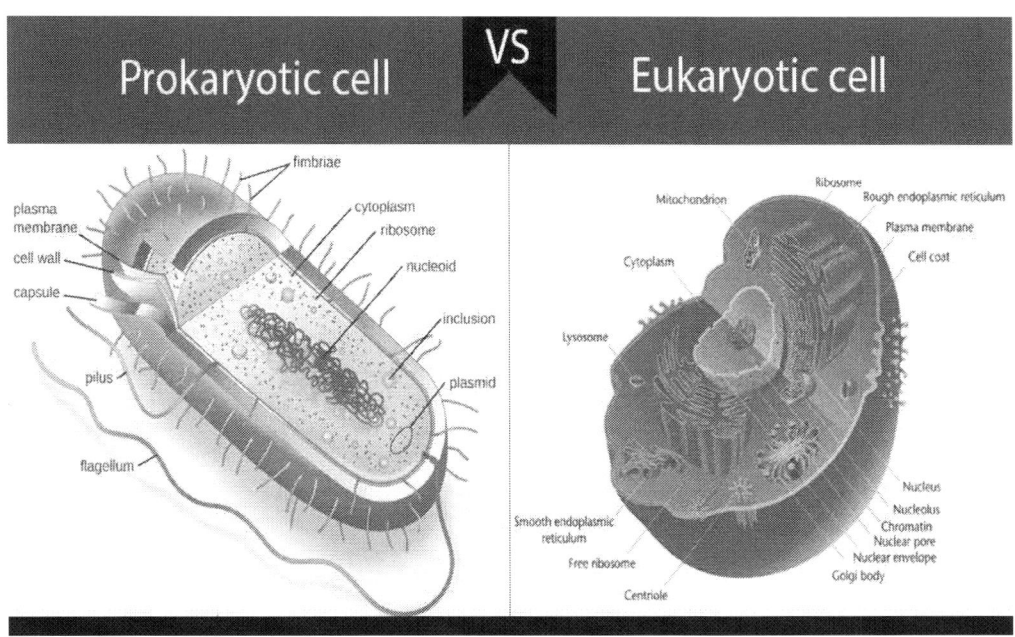

Fig.6. Differences between Prokaryotic and Eukaryotic Cells

Generalized structure of Prokaryotic cell consists of the following:

1. **Glycocalyx:** This layer function as a receptor, the adhesive also provide protection to the cell wall.
2. **Nucleoid:** It is the location of the genetic material (DNA), large DNA molecule is condensed into the small packet.

3. **Pilus:** Hair like hollow attachment present on the surface of bacteria, and is used to transfers of DNA to other cells during cell-cell adhesion.

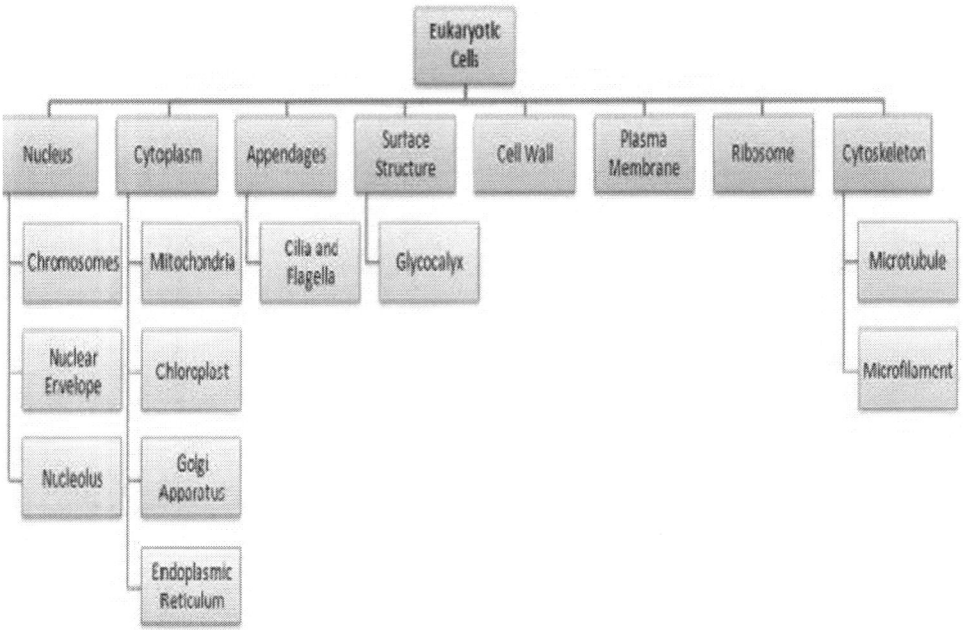

4. **Mesosomes:** It is the extension of the cell membrane, unfolded into the cytoplasm their role is during the cellular respiration.
5. **Flagellum:** Helps in movement, attached to the basal body of the cell.
6. **Cell Wall:** It provides rigidity and support for the cell.
7. **Fimbriae:** Helps in attachment to the surface and other bacteria while mating. These are small hair-like structure.
8. **Inclusion/Granule**s: It helps in storage of carbohydrates, glycogen, phosphate, fats in the form of particles which can be used when needed.
9. **Ribosomes:** Tiny particles which help in protein synthesis.
10. **Cell membrane:** Thin layer of protein and lipids, surrounds cytoplasm and regulate the flow of materials inside and outside the cells.
11. **Endospore:** It helps cell in surviving during harsh conditions.

In terms of peptidoglycan present in the cell wall, prokaryotes can be divided into Gram-positive and Gram -negative bacteria. The former contains a large amount of peptidoglycan in their cell wall while the latter have the thin layer.

Definition of Eukaryotic Cells

Eu means 'new,' and *karyon* means 'nucleus,' so these are the advanced type of cells found in plant, animals, and fungi. Eukaryotic cells have a well-defined nucleus and different organelles to perform different functions within the cell, though working is complex to understand. This kind of cells are found in algae, fungi, protozoa, plants, and animals and can be single-celled, colonial or multicellular. Among them, fungi and protists (algae and protozoa) are the major kingdoms.

The general structure of Eukaryotic cells contain:

- **Nucleus**: Eukaryotic cells have a well-defined nucleus where DNA (genetic material) is stored, it helps in the production of protein synthesis and ribosomes also. The chromosome is present inside the nucleus, which is surrounded by the **nuclear envelope**. It is a bi-lipid layer and controls the passage of ions and molecules.
- **Cytoplasm**: It is the location where other organelles are located, and other metabolic activities of the cell also take place here. It consists of –
    - **Mitochondria**: It is called 'the powerhouse of the cell,' and is responsible for making ATP. Mitochondria has its own DNA and ribosomes.
    - **Chloroplast**: These are found in algae and plants, it is one of the most important organelles in the plant which helps in converting energy sunlight into chemical energy through photosynthesis. They resemble mitochondria.
    - **Golgi Apparatus**: It consists of a stack of many flattened, disc-shaped sacs known as cisternae. The exact nature of Golgi varies, but it helps in the packaging of materials and in secreting them.
        - Lysosomes and Vacuoles – The most important function of Endoplasmic reticulum and Golgi apparatus is the synthesis of Lysosomes, which helps in digestion of intracellular molecules with the help of the enzyme called hydrolase.
        - Vacuoles are the membrane-bound cavities containing fluid as well as solid materials, and they engulf materials through endocytosis.

- **Endoplasmic Reticulum**: It transport lipids, proteins, and other materials through the cell. They are of two types of smooth endoplasmic reticulum and rough endoplasmic reticulum.

- **Appendages**: Cilia and Flagella are locomotory attachments, helps in the movement of a cell towards positive stimuli. Cilia are shorter than flagella and numerous.
- **Surface structure**: Glycocalyx is a kind of polysaccharide, and it is the outermost layer of the cell which helps in cell adherence, protection and in receiving signals from other cells.
- **Cell Wall**: Cell Wall provides shapes, rigidity, and support to the cell. Compositions of the cell wall may vary of different organisms but which can be of either cellulose, pectin, chitin or peptidoglycan.
- **Cytoplasmic Membrane/Plasma Membrane**: It is a thin semipermeable, surrounding the cytoplasm, it acts as the barrier of the cell which regulates entry and exit of the substances inside and outside the cell. This layer is made up of two layers of phospholipids embedded with proteins. In Plant cell, this layer is present below the cell wall whereas in the Animal cell it is the outermost layer.
- **Ribosomes**: Though small in size but are present in numbers, they help in protein synthesis. Eukaryotes have 80S ribosomes which are further divided into two subunits which are 40S and 60S (S stands for Sedverg unit).
- **Cytoskeleton**: It is supporting framework of the cells, which is of two types Microtubules and Microfilaments. Microtubules have a diameter of about 24 nanometers (nm), made up of a protein called tubulin, while Microfilaments has a diameter of 6nm, made of the protein called actin. Microtubules are the largest filament and Microfilament the smallest one.

Key Difference between Prokaryotic Cells and Eukaryotic Cells

Following are the substantial difference between Prokaryotic Cells and Eukaryotic Cell:

1. **Prokaryotic cells** are the primitive kind of cell, whose size varies from **0.5-3μm**, they are generally found in single-cell organisms, while **Eukaryotic cells** are the modified cell structure containing different components in it, their size varies from **2-100μm**, they are found in multicellular organisms.
2. **Organelles** like mitochondria, ribosomes, Golgi body, endoplasmic reticulum, cell wall, chloroplast, etc. are **absent in prokaryotic cells**, while

these organelles are found in eukaryotic organisms. Though cell wall and chloroplast are not found in the animal cell, it is present in the green plant cell, few bacteria, and algae.

3. The main difference between Prokaryotic cells and the Eukaryotic cell is the **nucleus**, which is not well defined in prokaryotes whereas it is well structured, compartmentalized and functional in eukaryotes.

4. Cell organelles are present which are **membrane-bound** and have individual functions in eukaryotic cells; many organelles are absent in prokaryotic cells.

5. In prokaryotes, the **cell division** takes place through conjugation, transformation, and transduction but in eukaryotes, it is through the process of cell division.

6. The process of **transcription and translation** occurs together, and there is a single origin of replication in the prokaryotic cell. On the other hand, there are multiple origins of replication and transcription occurs in nucleus and translation in the cytosol.

7. Genetic Material (DNA) is **circular** and double-stranded in Prokaryotes, but in Eukaryotes, it is **linear** and double-stranded.

8. Prokaryotes reproduce **asexually**; commonly Prokaryotes have a **sexual** mode of reproduction.

9. Prokaryotes are the simplest, smallest and most abundantly found cells on earth; Eukaryotes are larger and complex cells.

****

7. NUCLEIC ACIDS

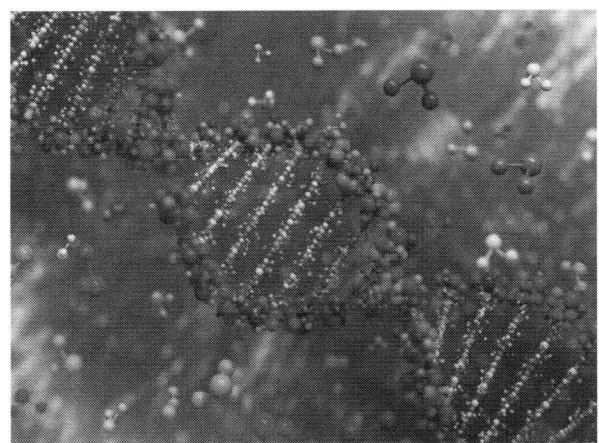

1. Introduction

DNA was first observed by a German biochemist named Frederich Miescher in 1869. But for many years, researchers did not realize the importance of this molecule. It was not until 1953 that James Watson, Francis Crick, Maurice Wilkins and Rosalind Franklin figured out the structure of DNA — a double helix — which they realized could carry biological information.

Watson, Crick and Wilkins were awarded the Nobel Prize in Medicine in 1962 "for their discoveries concerning the molecular structure of nucleic acids and its significance for information transfer in living material." Franklin was not included in the award, although her work was integral to the research.

Elemental analysis of nucleic acids showed the presence of phosphorus, in addition to the usual C, H, N & O. Unlike proteins, nucleic acids contained no sulfur. Complete hydrolysis of chromosomal nucleic acids gave inorganic phosphate, 2-deoxyribose (a previously unknown sugar) and four different heterocyclic bases (shown in the following diagram). To reflect the unusual sugar component, chromosomal nucleic acids are called deoxyribonucleic acids, abbreviated DNA. Analogous nucleic acids in which the sugar component is ribose are termed ribonucleic acids, abbreviated RNA. The acidic character of the nucleic acids was attributed to the phosphoric acid moiety.

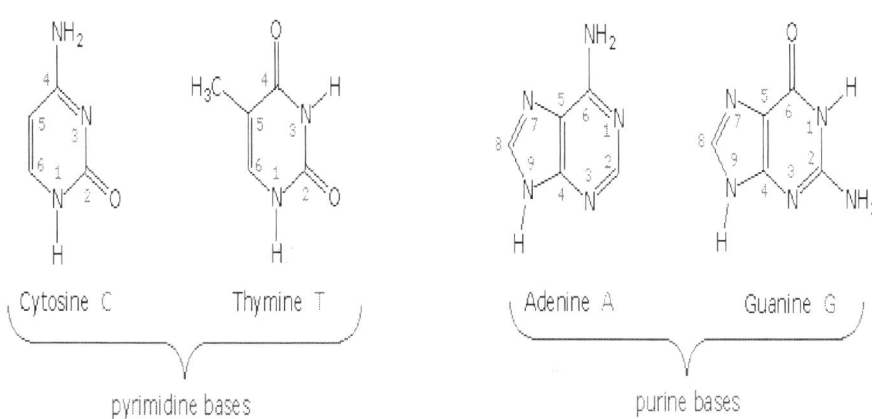

The two monocyclic bases shown here are classified as **pyrimidines**, and the two bicyclic bases are **purines**. Each has at least one N-H site at which an organic substituent may be attached. They are all polyfunctional bases, and may exist in tautomeric forms.
Base-catalyzed hydrolysis of DNA gave four **nucleoside** products, which proved to be N-glycosides of 2'-deoxyribose combined with the heterocyclic amines. Structures and names for these nucleosides will be displayed above by clicking on the heterocyclic base diagram. The base components are colored green, and the sugar is black. As noted in the 2'-deoxycytidine structure on the left, the numbering of the sugar carbons makes use of primed numbers to distinguish them from the heterocyclic base sites. The corresponding N-glycosides of the common sugar ribose are the building blocks of RNA, and are named adenosine, cytidine, guanosine and uridine (a thymidine analog missing the methyl group).From this evidence, nucleic acids may be formulated as alternating copolymers of phosphoric acid (**P**) and nucleosides (**N**), as shown:

~ P – N – P – N'– P – N''– P – N'''– P – N ~

At first the four nucleosides, distinguished by prime marks in this crude formula, were assumed to be present in equal amounts, resulting in a uniform structure, such as that of starch. However, a compound of this kind, presumably common to all organisms, was considered too simple to hold the hereditary information known to reside in the chromosomes. This view was challenged in 1944, when Oswald Avery and colleagues demonstrated that bacterial DNA was likely the genetic agent that carried information from one organism to another in a process called "transformation". He concluded that *"nucleic acids must be regarded as possessing biological specificity, the chemical basis of which is as yet undetermined."* Despite this finding, many scientists continued to believe that chromosomal proteins, which differ across species, between individuals, and even within a given organism, were the locus of an organism's genetic information. It should be noted that single celled organisms like bacteria do not have a well-defined nucleus. Instead, their single chromosome is associated with specific proteins in a region called a "nucleoid". Nevertheless, the DNA from bacteria has the same composition and general structure as that from multicellular organisms, including human beings.

Views about the role of DNA in inheritance changed in the late 1940's and early 1950's. By conducting a careful analysis of DNA from many sources, Erwin Chargaff found its composition to be species specific. In addition, he found that the amount of adenine (A) always equaled the amount of thymine (T), and the amount of guanine (G) always equaled the amount of cytosine (C), regardless of the DNA source. As set forth in the following table, the ratio of (A+T) to (C+G) varied from 2.70 to 0.35. The last two organisms are bacteria.

Nucleoside Base Distribution in DNA

Organism	Base Composition (mole %)				Base Ratios		Ratio (A+T)/(G+C)
	A	G	T	C	A/T	G/C	
Human	30.9	19.9	29.4	19.8	1.05	1.00	1.52
Chicken	28.8	20.5	29.2	21.5	1.02	0.95	1.38
Yeast	31.3	18.7	32.9	17.1	0.95	1.09	1.79
Clostridium perfringens	36.9	14.0	36.3	12.8	1.01	1.09	2.70
Sarcina lutea	13.4	37.1	12.4	37.1	1.08	1.00	0.35

In a second critical study, Alfred Hershey and Martha Chase showed that when a bacterium is infected and genetically transformed by a virus, at least 80% of the viral DNA enters the bacterial cell and at least 80% of the viral protein remains outside. Together with the Chargaff findings this work established DNA as the repository of the unique genetic characteristics of an organism.

2. The Chemical Nature of DNA

The polymeric structure of DNA may be described in terms of monomeric units of increasing complexity. In the top shaded box of the following illustration, the three relatively simple components mentioned earlier are shown. Below that on the left, formulas for phosphoric acid and a nucleoside are drawn. Condensation polymerization of these leads to the DNA formulation outlined above. Finally, a 5'-monophosphate ester, called a **nucleotide** may be drawn as a single monomer unit, shown in the shaded box to the right. Since a monophosphate ester of this kind is a strong acid (pK_a of 1.0), it will be fully ionized at the usual physiological pH (ca.7.4). Names for these DNA components are given in the table to the right of the diagram.

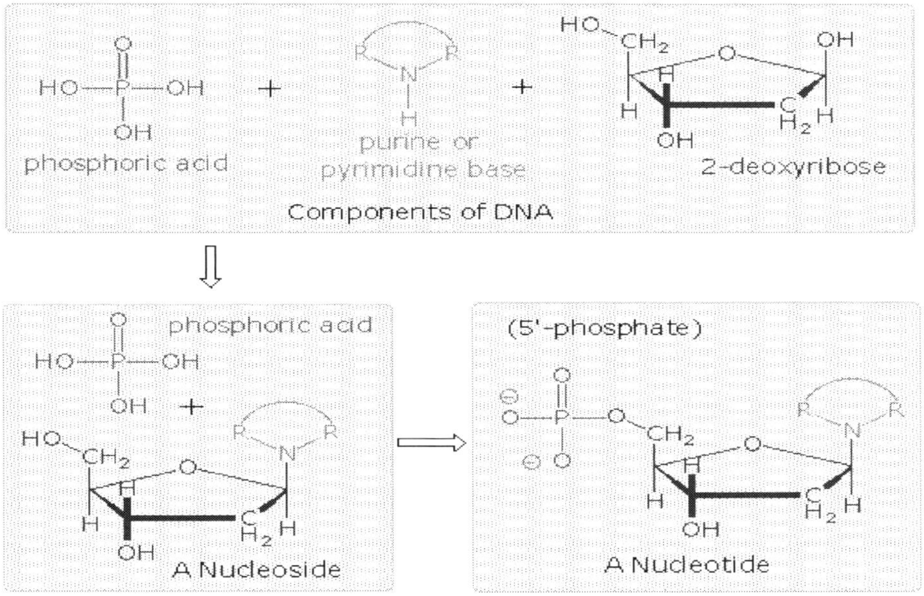

Isomeric 3'-monophospate nucleotides are also known, and both isomers are found in cells. They may be obtained by selective hydrolysis of DNA through the action of nuclease enzymes. Anhydride-like di- and tri-phosphate nucleotides have been identified as important energy carriers in biochemical reactions, the most common being ATP (adenosine 5'-triphosphate).

Names of DNA Base Derivatives		
Base	**Nucleoside**	**5'-Nucleotide**
Adenine	2'-Deoxyadenosine	2'-Deoxyadenosine-5'-monophosphate
Cytosine	2'-Deoxycytidine	2'-Deoxycytidine-5'-monophosphate
Guanine	2'-Deoxyguanosine	2'-Deoxyguanosine-5'-monophosphate
Thymine	2'-Deoxythymidine	2'-Deoxythymidine-5'-monophosphate

A complete structural representation of a segment of the DNA polymer formed from 5'-nucleotides.

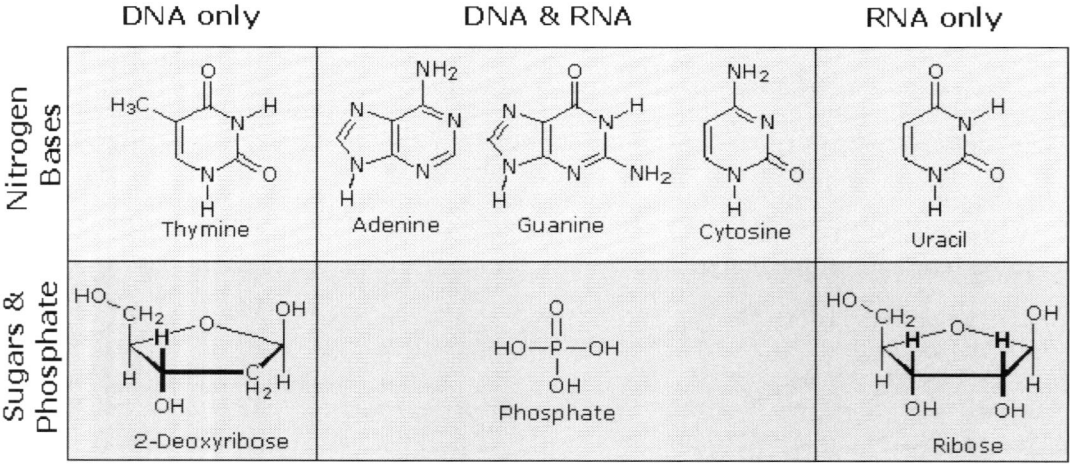

- First, the remaining P-OH function is quite acidic and is completely ionized in biological systems.

- Second, the polymer chain is structurally directed. One end (5') is different from the other (3').

- Third, although this appears to be a relatively simple polymer, the possible permutations of the four nucleosides in the chain become very large as the chain lengthens.
- Fourth, the DNA polymer is much larger than originally believed. Molecular weights for the DNA from multicellular organisms are commonly 10^9 or greater.

Information is stored or encoded in the DNA polymer by the pattern in which the four nucleotides are arranged. To access this information the pattern must be "read" in a linear fashion, just as a bar code is read at a supermarket checkout. Because living organisms are extremely complex, a correspondingly large amount of information related to this complexity must be stored in the DNA. Consequently, the DNA itself must be very large, as noted above. Even the single DNA molecule from an *E. coli* bacterium is found to have roughly a million nucleotide units in a polymer strand, and would reach a millimeter in length if stretched out. The nuclei of multicellular organisms incorporate chromosomes, which are composed of DNA combined with nuclear proteins called histones. The fruit fly has 8 chromosomes, humans have 46 and dogs 78 (note that the amount of DNA in a cell's nucleus does not correlate with the number of chromosomes). The DNA from the smallest human chromosome is over ten times larger than *E. coli* DNA, and it has been estimated that the total DNA in a human cell would extend to 2 meters in length if unraveled. Since the nucleus is only about 5μm in diameter, the chromosomal DNA must be packed tightly to fit in that small volume. In addition to its role as a stable informational library, chromosomal DNA must be structured or organized in such a way that the chemical machinery of the cell will have easy access to that information, in order to make important molecules such as polypeptides. Furthermore, accurate copies of the DNA code must be created as cells divide, with the replicated DNA molecules passed on to subsequent cell generations, as well as to progeny of the organism. The nature of this DNA organization, or secondary structure, will be discussed in a later section.

3. RNA, a Different Nucleic Acid

The high molecular weight nucleic acid, DNA, is found chiefly in the nuclei of complex cells, known as **eucaryotic cells**, or in the nucleoid regions of **procaryotic cells**, such as bacteria. It is often associated with proteins that help to pack it in a usable fashion. In contrast, a lower molecular weight, but much more abundant nucleic acid, **RNA**, is distributed throughout the cell, most commonly in small numerous organelles called **ribosomes**. Three kinds of RNA are identified, the largest subgroup (85 to 90%) being ribosomal RNA, **rRNA**, the major component of ribosomes, together with proteins. The size of rRNA molecules varies, but is generally less than a thousandth the size of DNA. The other forms of RNA are messenger RNA , **mRNA**, and transfer RNA , **tRNA**. Both have a more transient existence and are smaller than rRNA. All these RNA's have similar constitutions, and differ from DNA in two important respects. As shown in the following diagram, the sugar component of RNA is ribose, and the pyrimidine base uracil replaces the thymine base of DNA. The RNA's play a vital role in the transfer of information (transcription) from the DNA library to the protein factories called ribosomes, and in the interpretation of that information (translation) for the synthesis of specific polypeptides.

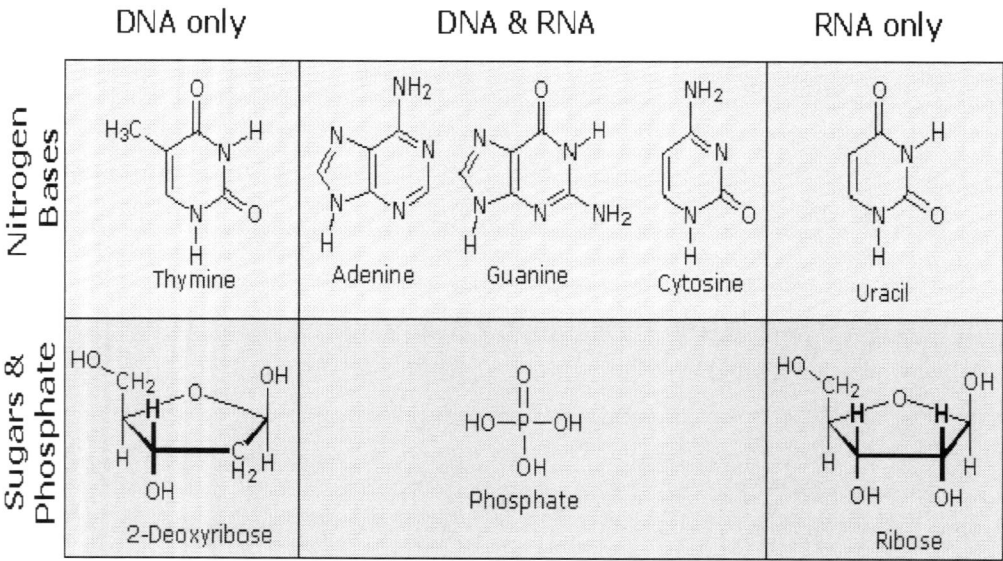

4. The Secondary Structure of DNA

In the early 1950's the primary structure of DNA was well established, but a firm understanding of its secondary structure was lacking. Indeed, the situation was similar to that occupied by the proteins a decade earlier, before the alpha helix and pleated sheet structures were proposed by **Linus Pauling.** Many researchers grappled with this problem, and it was generally conceded that the molar equivalences of base pairs (A & T and C & G) discovered by Chargaff would be an important factor. **Rosalind Franklin**, working at King's College, London, obtained X-ray diffraction evidence that suggested a long helical structure of uniform thickness. Francis Crick and James Watson, at Cambridge University, considered hydrogen bonded base pairing interactions, and arrived at a double stranded helical model that satisfied most of the known facts, and has been confirmed by subsequent findings.**Base Pairing** Careful examination of the purine and pyrimidine base components of the nucleotides reveals that three of them could exist as hydroxy pyrimidine or purine tautomers, having an aromatic heterocyclic ring. Despite the added stabilization of an aromatic ring, these compounds prefer to adopt amide-like structures. These options are shown in the following diagram, with the more stable tautomer drawn in blue.

4-amino-2-hydroxypyrimidine
cytosine

2,4-dihydroxypyrimidines
R=H uracil ; R=CH$_3$ thymine

2-amino-6-hydroxypurine
guanidine

A simple model for this tautomerism is provided by 2-hydroxypyridine. As shown on the left below, a compound having this structure might be expected to have phenol-like characteristics, such as an acidic hydroxyl group. However, the boiling point of the actual substance is 100° C greater than phenol and its acidity is 100 times less than expected (pKa = 11.7). These differences agree with the 2-pyridone tautomer, the stable form of the zwitterionic internal salt. Further evidence supporting this assignment will be displayed by clicking on the diagram. Note that this tautomerism reverses the hydrogen bonding behavior of the nitrogen and oxygen functions (the N-H group of the pyridone becomes a hydrogen bond donor and the carbonyl oxygen an acceptor).

2-hydroxypyridine ⇌ [...] ↔ **2-pyridone**

H-bond acceptor / H-bond donor .... H-bond donor / H-bond acceptor

Infrared: 1650, 1610, 1581 cm^{-1} .... 1658, 1584, 1539 cm^{-1}

^{13}C nmr: 165, 142, 135, 120, 107 ppm. .... 163, 140, 139, 120, 106, 37 ppm.

The additional evidence for the pyridone tautomer, that appears diagram, consists of **infrared and carbon nmr** absorptions associated with and characteristic of the amide group. The data for 2-pyridone is given on the left. Similar data for the N-methyl derivative, which cannot tautomerize to a pyridine derivative, is presented on the right.

Once they had identified the favored base tautomers in the nucleosides, Watson and Crick were able to propose a complementary pairing, via hydrogen bonding, of guanosine (G) with cytidine (C) and adenosine (A) with thymidine (T). This pairing, which is shown in the following diagram, explained Chargaff's findings beautifully, and led them to suggest a double helix structure for DNA. Before viewing this double helix structure itself, it is instructive to examine the base pairing interactions in greater detail. The G#C association involves three hydrogen bonds (colored pink), and is therefore stronger than the two-hydrogen bond association of A#T. These base pairings might appear to be arbitrary, but other possibilities suffer destabilizing steric or electronic interactions. By clicking on the diagram two such alternative couplings will be shown. The C#T pairing on the left suffers from carbonyl dipole repulsion, as well as steric crowding of the oxygens. The G#A pairing on the right is also destabilized by steric crowding (circled hydrogens).

Hydrogen Bonded Base Pairs

G≡C   A=T

A simple mnemonic device for remembering which bases are paired comes from the line construction of the capital letters used to identify the bases. A and T are made up of intersecting straight lines. In contrast, C and G are largely composed of curved lines. The RNA base uracil corresponds to thymine, since U follows T in the alphabet.

The Double Helix

After many trials and modifications, Watson and Crick conceived an ingenious double helix model for the secondary structure of DNA. Two strands of DNA were aligned anti-parallel to each other, i.e. with opposite 3' and 5' ends, as shown in part **a** of the following diagram. Complementary primary nucleotide structures for each strand allowed intra-strand hydrogen bonding between each pair of bases. These complementary strands are colored red and green in the diagram. Coiling these coupled strands then leads to a double helix structure, shown as cross-linked ribbons in part **b** of the diagram. The double helix is further stabilized by hydrophobic attractions and pi-stacking of the bases. A space-filling molecular model of a short segment is displayed in part **c** on the right. The helix shown here has ten base pairs per turn, and rises 3.4 Å in each turn. This right-handed helix is the favored conformation in aqueous systems, and has been termed the **B-helix**. As the DNA strands wind around each other, they leave gaps between each set of phosphate backbones. Two alternating grooves result, a wide and deep **major groove** (*ca.* 22Å wide), and a shallow and narrow **minor groove** (*ca.* 12Å wide). Other molecules, including polypeptides, may insert into these grooves, and in so doing perturb the chemistry of DNA. Other helical structures of DNA have also been observed, and are designated by letters (e.g. **A** and **Z**).

The Double Helix Structure for DNA

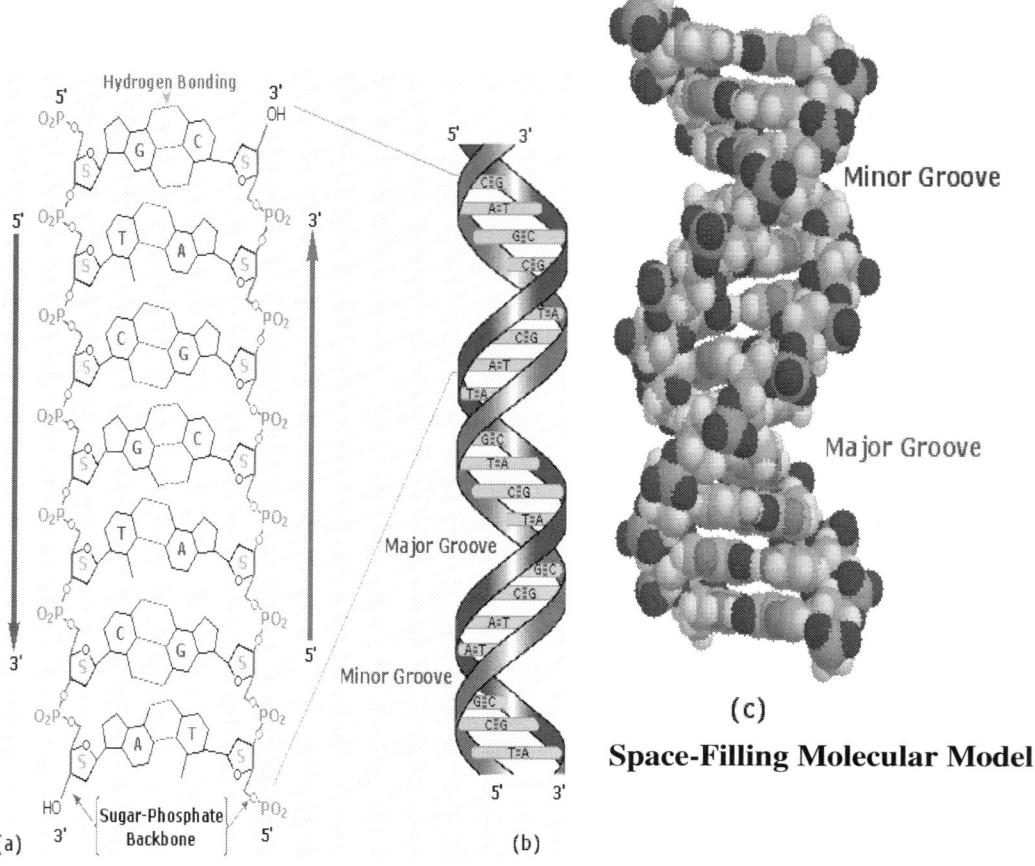

A model of a short DNA segment may be examined by

> Excellent sites, incorporating Chime and Jmol models for visualizing DNA, has been created by: Eric Martz, Univ. Mass. Amherst. Frieda Reichsman, Univ. Mass. Amherst.

Fig.7.1

1. **Basics of DNA Replication**

Watson and Crick's discovery that DNA was a two-stranded double helix provided a hint as to how DNA is replicated. During cell division, each DNA molecule has to be perfectly copied to ensure identical DNA molecules to move to each of the two daughter cells. The double-stranded structure of DNA suggested that the two strands might separate during replication with each strand serving as a template from which the new complementary strand for each is copied, generating two double-stranded molecules from one.

Models of Replication

There were three models of replication possible from such a scheme: conservative, semi-conservative, and dispersive. In conservative replication, the two original DNA strands, known as the parental strands, would re-basepair with each other after being used as templates to synthesize new strands; and the two newly-synthesized strands, known as the daughter strands, would also basepair with each other; one of the two DNA molecules after replication would be "all-old" and the other would be "all-new". In semi-conservative replication, each of the two parental DNA strands would act as a template for new DNA strands to be synthesized, but after replication, each parental DNA strand would basepair with the complementary newly-synthesized strand just synthesized, and both double-stranded DNAs would include one parental or "old" strand and one daughter or "new" strand. In dispersive replication, after replication both copies of the new DNAs would somehow have alternating segments of parental DNA and newly-synthesized DNA on each of their two strands.

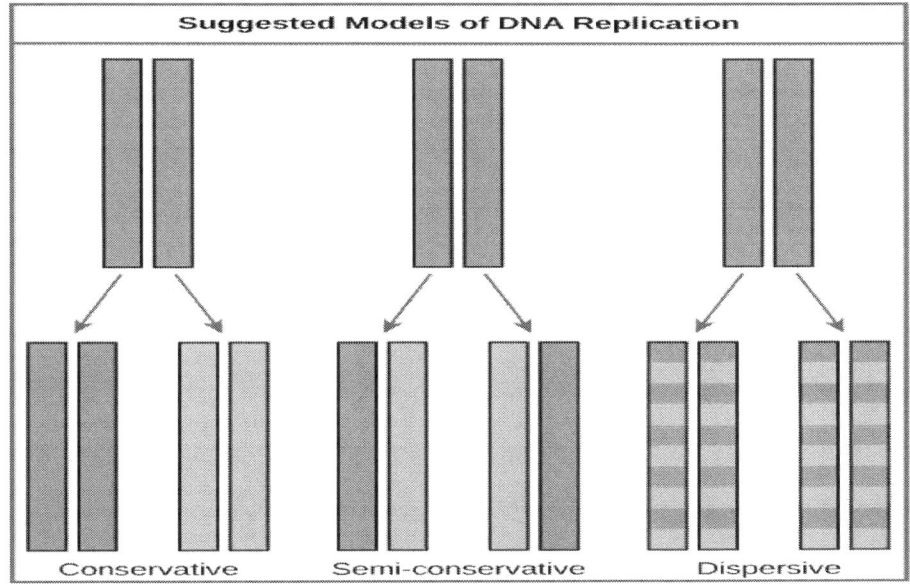

Fig.7.1. Models of DNA Replication

Suggested Models of DNA Replication: The three suggested models of DNA replication. Grey indicates the original parental DNA strands or segments and blue indicates newly-synthesized daughter DNA strands or segments.

To determine which model of replication was accurate, a seminal experiment was performed in 1958 by two researchers: Matthew Meselson and Franklin Stahl.

Meselson and Stahl

Meselson and Stahl were interested in understanding how DNA replicates. They grew *E. coli* for several generations in a medium containing a "heavy" isotope of nitrogen (^{15}N) that is incorporated into nitrogenous bases and, eventually, into the DNA. The *E. coli* culture was then shifted into medium containing the common "light" isotope of nitrogen (^{14}N) and allowed to grow for one generation. The cells were harvested and the DNA was isolated. The DNA was centrifuged at high speeds in an ultracentrifuge in a tube in which a cesium chloride density gradient had been established. Some cells were allowed to grow for one more life cycle in ^{14}N and spun again.

Meselson and Stahl: Meselson and Stahl experimented with E. coli grown first in heavy nitrogen (^{15}N) then in ligher nitrogen (^{14}N.) DNA grown in ^{15}N (red band) is heavier than DNA grown in ^{14}N (orange band) and sediments to a lower level in the cesium chloride density gradient in an ultracentrifuge. When DNA grown in ^{15}N is switched to media containing ^{14}N, after one round of cell division the DNA sediments halfway between the ^{15}N and ^{14}N levels, indicating that it now contains fifty percent ^{14}N and fifty percent ^{15}N.. In subsequent cell divisions, an increasing amount of DNA contains ^{14}N only. These data support the semi-conservative replication model.

During the density gradient ultracentrifugation, the DNA was loaded into a gradient (Meselson and Stahl used a gradient of cesium chloride salt, although other materials such as sucrose can also be used to create a gradient) and spun at high speeds of 50,000 to 60,000 rpm. In the ultracentrifuge tube, the cesium chloride salt created a density gradient, with the cesium chloride solution being denser the farther down the tube you went. Under these circumstances, during the spin the DNA was pulled down the ultracentrifuge tube by centrifugal force until it arrived at the spot in the saltsolution. At the point, the molecules stopped sedimenting and formed a stable band. By looking at the relative positions of bands of molecules run in the same gradients, you can determine the relative densities of different molecules. The molecules that form the lowest bands have the highest densities.

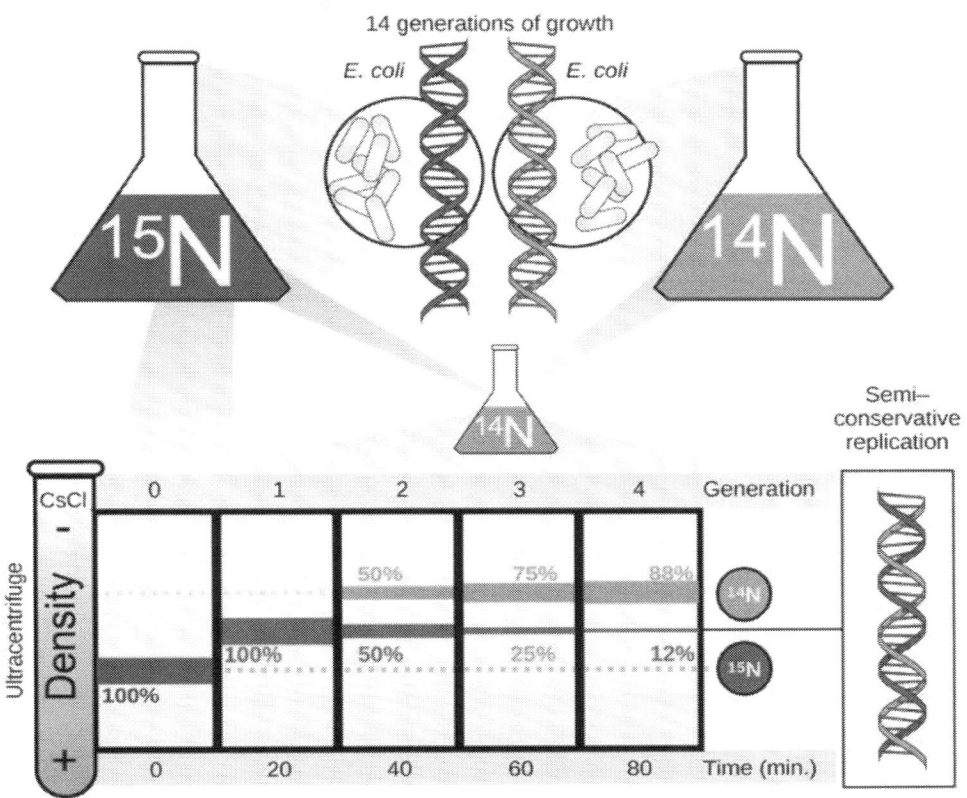

Fig.7.2. Meselson and Stahl experimented

DNA from cells grown exclusively in ^{15}N produced a lower band than DNA from cells grown exclusively in ^{14}N. So DNA grown in ^{15}N had a higher density, as would be expected of a molecule with a heavier isotope of nitrogen incorporated into its nitrogenous bases. Meselson and Stahl noted that after one generation of growth in ^{14}N (after cells had been shifted from ^{15}N), the DNA molecules produced only single band intermediate in position in between DNA of cells grown exclusively in ^{15}N and DNA of cells grown exclusively in ^{14}N. This suggested either a semi-conservative or dispersive mode of replication. Conservative replication would have resulted in two bands; one representing the parental DNA still with exclusively ^{15}N in its nitrogenous bases and the other representing the daughter DNA with exclusively ^{14}N in its nitrogenous bases. The single band actually seen indicated that all the DNA molecules contained equal amounts of both ^{15}N and ^{14}N.

The DNA harvested from cells grown for two generations in ^{14}N formed two bands: one DNA band was at the intermediate position between ^{15}N and ^{14}N and the other corresponded to the band of exclusively ^{14}N DNA. These results could only be explained if DNA replicates in a semi-conservative manner. Dispersive replication would have resulted in exclusively a single band in each new generation, with the band slowly moving up closer to the height of the ^{14}N DNA band. Therefore, dispersive replication could also be ruled out.

Meselson and Stahl's results established that during DNA replication, each of the two strands that make up the double helix serves as a template from which new strands are synthesized. The new strand will be complementary to the parental or "old" strand and the new strand will remain basepaired to the old strand. So each "daughter" DNA actually consists of one "old" DNA strand and one newly-synthesized strand. When two daughter DNA copies are formed, they have the identical sequences to one another and identical sequences to the original parental DNA, and the two daughter DNAs are divided equally into the two daughter cells, producing daughter cells that are genetically identical to one another and genetically identical to the parent cell.

DNA Replication in Prokaryotes

Prokaryotic DNA is replicated by DNA polymerase III in the 5' to 3' direction at a rate of 1000 nucleotides per second.

DNA Replication in Prokaryotes

DNA replication employs a large number of proteins and enzymes, each of which plays a critical role during the process. One of the key players is the enzyme DNA polymerase, which adds nucleotides one by one to the growing DNA chain that are complementary to the template strand. The addition of nucleotides requires energy; this energy is obtained from the nucleotides that have three phosphates attached to them, similar to ATP which has three phosphate groups attached. When the bond between the phosphates is broken, the energy released is used to form the phosphodiester bond between the incoming nucleotide and the growing chain. In prokaryotes, three main types of polymerases are known: DNA pol I, DNA pol II, and DNA pol III. DNA pol III is the enzyme required for DNA synthesis; DNA pol I and DNA pol II are primarily required for repair.

There are specific nucleotide sequences called origins of replication where replication begins. In *E. coli*, which has a single origin of replication on its one chromosome (as do most prokaryotes), it is approximately 245 base pairs long and is rich in AT sequences. The origin of replication is recognized by certain proteins that bind to this site. An enzyme called helicase unwinds the DNA by breaking the hydrogen bonds between the nitrogenous base pairs. ATP hydrolysis is required for this process. As the DNA opens up, Y-shaped structures called replication forks are formed. Two replication forks at the origin of replication are extended bi-directionally as replication proceeds. Single-strand binding proteins coat the strands of DNA near the replication fork to prevent the single-stranded DNA from winding back into a double helix. DNA polymerase is able to add nucleotides only in the 5' to 3' direction (a new DNA strand can be extended only in this direction). It also requires a free 3'-OH group to which it can add nucleotides by forming a phosphodiester bond between the 3'-OH end and the 5' phosphate of the next nucleotide. This means that it cannot add nucleotides if a free 3'-OH group is not available. Another enzyme, RNA primase, synthesizes an RNA primer that is about five to ten nucleotides long and complementary to the DNA, priming DNA synthesis. A primer provides the free 3'-OH end to start replication. DNA polymerase then extends this RNA primer, adding nucleotides one by one that are complementary to the template strand.

DNA Replication in Prokaryotes: A replication fork is formed when helicase separates the DNA strands at the origin of replication. The DNA tends to become more highly coiled ahead of the replication fork. Topoisomerase breaks and reforms DNA's phosphate backbone ahead of the replication fork, thereby relieving the pressure that results from this supercoiling. Single-strand binding proteins bind to the single-stranded DNA to prevent the helix from re-forming. Primase synthesizes an RNA primer. DNA polymerase III uses this primer to synthesize the daughter DNA strand. On the leading strand, DNA is synthesized continuously, whereas on the lagging strand, DNA is synthesized in short stretches called Okazaki fragments. DNA polymerase I replaces the RNA primer with DNA. DNA ligase seals the gaps between the Okazaki fragments, joining the fragments into a single DNA molecule.

The replication fork moves at the rate of 1000 nucleotides per second. DNA polymerase can only extend in the 5' to 3' direction, which poses a slight problem at the replication fork. As we know, the DNA double helix is anti-parallel; that is, one strand is in the 5' to 3' direction and the other is oriented in the 3' to 5' direction. One strand (the leading strand), complementary to the 3' to 5' parental DNA strand, is synthesized continuously towards the replication fork because the polymerase can add nucleotides in this direction. The other strand (the lagging strand), complementary to the 5' to 3' parental DNA, is extended away from the replication fork in small fragments known as Okazaki fragments, each requiring a primer to start the synthesis. Okazaki fragments are named after the Japanese scientist who first discovered them.

The leading strand can be extended by one primer alone, whereas the lagging strand needs a new primer for each of the short Okazaki fragments. The overall direction of the lagging strand will be 3' to 5', while that of the leading strand will be 5' to 3'. The sliding clamp (a ring-shaped protein that binds to the DNA) holds the DNA polymerase in place as it continues to add nucleotides. Topoisomerase prevents the over-winding of the DNA double helix ahead of the replication fork as the DNA is opening up; it does so by causing temporary nicks in the DNA helix and then resealing it. As synthesis proceeds, the RNA primers are replaced by DNA. The primers are removed by the exonuclease activity of DNA pol I, while the gaps are filled in by deoxyribonucleotides. The nicks that remain between the newly-synthesized DNA (that replaced the RNA primer) and the previously-synthesized DNA are sealed by the enzyme DNA ligase that catalyzes the formation of phosphodiester linkage between the 3'-OH end of one nucleotide and the 5' phosphate end of the other fragment.

The table summarizes the enzymes involved in prokaryotic DNA replication and the functions of each.

Prokaryotic DNA Replication: Enzymes and Their Function	
Enzyme/protein	Specific Function
DNA pol I	Exonuclease activity removes RNA primer and replaces with newly synthesized DNA
DNA pol II	Repair function
DNA pol III	Main enzyme that adds nucleotides in the 5'-3' direction
Helicase	Opens the DNA helix by breaking hydrogen bonds between the nitrogenous bases
Ligase	Seals the gaps between the Okazaki fragments to create one continuous DNA strand
Primase	Synthesizes RNA primers needed to start replication
Sliding Clamp	Helps to hold the DNA polymerase in place when nucleotides are being added
Topoisomerase	Helps relieve the stress on DNA when unwinding by causing breaks and then resealing the DNA
Single-strand binding proteins (SSB)	Binds to single-stranded DNA to avoid DNA rewinding back.

Prokaryotic DNA Replication: Enzymes and Their Function: The enzymes involved in prokaryotic DNA replication and their functions are summarized on this table.

DNA Replication in Eukaryotes

DNA replication in eukaryotes occurs in three stages: initiation, elongation, and termination, which are aided by several enzymes.

Because eukaryotic genomes are quite complex, DNA replication is a very complicated process that involves several enzymes and other proteins. It occurs in three main stages: initiation, elongation, and termination.

Initiation

Eukaryotic DNA is bound to proteins known as histones to form structures called nucleosomes. During initiation, the DNA is made accessible to the proteins and enzymes involved in the replication process. There are specific chromosomal locations called origins of replication where replication begins. In some eukaryotes, like yeast, these locations are defined by having a specific sequence of basepairs to which the replication initiation proteins bind. In other eukaryotes, like humans, there does not appear to be a consensus sequence for their origins of replication. Instead, the replication initiation proteins might identify and bind to specific modifications to the nucleosomes in the origin region.

Certain proteins recognize and bind to the origin of replication and then allow the other proteins necessary for DNA replication to bind the same region. The first proteins to bind the DNA are said to "recruit" the other proteins. Two copies of an enzyme called helicase are among the proteins recruited to the origin. Each helicase unwinds and separates the DNA helix into single-stranded DNA. As the DNA opens up, Y-shaped structures called replication forks are formed. Because two helicases bind, two replication forks are formed at the origin of replication; these are extended in both directions as replication proceeds creating a replication bubble. There are multiple origins of replication on the eukaryotic chromosome which allow replication to occur simultaneously in hundreds to thousands of locations along each chromosome.

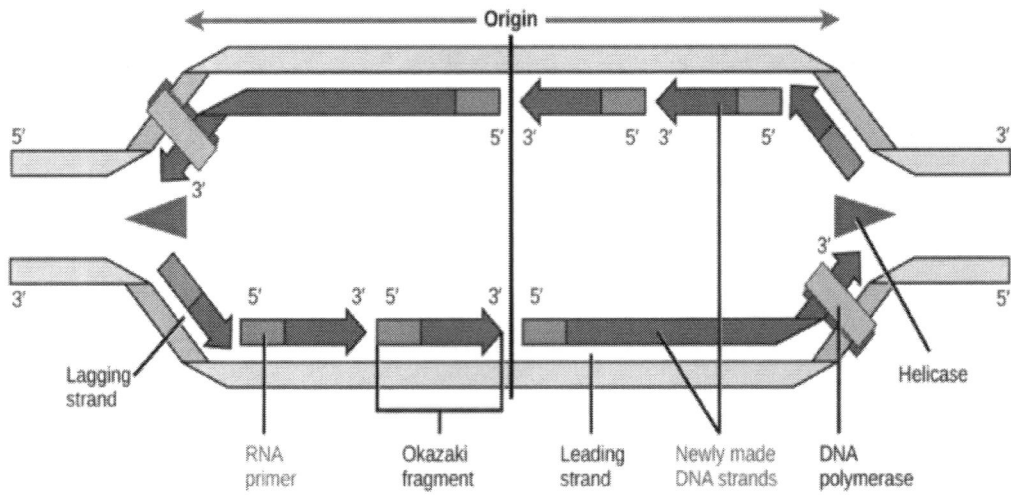

Fig.7.3. A replication fork is formed by the opening of the origin of replication

Replication Fork Formation: A replication fork is formed by the opening of the origin of replication; helicase separates the DNA strands. An RNA primer is synthesized by primase and is elongated by the DNA polymerase. On the leading strand, only a single RNA primer is needed, and DNA is synthesized continuously, whereas on the lagging strand, DNA is synthesized in short stretches, each of which must start with its own RNA primer. The DNA fragments are joined by DNA ligase (not shown).

Elongation

During elongation, an enzyme called DNA polymerase adds DNA nucleotides to the 3′ end of the newly synthesized polynucleotide strand. The template strand specifies which of the four DNA nucleotides (A, T, C, or G) is added at each position along the new chain. Only the nucleotide complementary to the template nucleotide at that position is added to the new strand.

DNA polymerase contains a groove that allows it to bind to a single-stranded template DNA and travel one nucleotide at at time. For example, when DNA polymerase meets an adenosine nucleotide on the template strand, it adds a thymidine to the 3′ end of the newly synthesized strand, and then moves to the next nucleotide on the template strand. This process will continue until the DNA polymerase reaches the end of the template strand.

DNA polymerase cannot initiate new strand synthesis; it only adds new nucleotides at the 3' end of an existing strand. All newly synthesized polynucleotide strands must be initiated by a specialized RNA polymerase called primase. Primase initiates polynucleotide synthesis and by creating a short RNA polynucleotide strand complementary to template DNA strand. This short stretch of RNA nucleotides is called the primer. Once RNA primer has been synthesized at the template DNA, primase exits, and DNA polymerase extends the new strand with nucleotides complementary to the template DNA.

Eventually, the RNA nucleotides in the primer are removed and replaced with DNA nucleotides. Once DNA replication is finished, the daughter molecules are made entirely of continuous DNA nucleotides, with no RNA portions.

The Leading and Lagging Strands

DNA polymerase can only synthesize new strands in the 5' to 3' direction. Therefore, the two newly-synthesized strands grow in opposite directions because the template strands at each replication fork are antiparallel. The "leading strand" is synthesized continuously toward the replication fork as helicase unwinds the template double-stranded DNA.

The "lagging strand" is synthesized in the direction away from the replication fork and away from the DNA helicase unwinds. This lagging strand is synthesized in pieces because the DNA polymerase can only synthesize in the 5' to 3' direction, and so it constantly encounters the previously-synthesized new strand. The pieces are called Okazaki fragments, and each fragment begins with its own RNA primer.

Termination

Eukaryotic chromosomes have multiple origins of replication, which initiate replication almost simultaneously. Each origin of replication forms a bubble of duplicated DNA on either side of the origin of replication. Eventually, the leading strand of one replication bubble reaches the lagging strand of another bubble, and the lagging strand will reach the 5' end of the previous Okazaki fragment in the same bubble.

DNA polymerase halts when it reaches a section of DNA template that has already been replicated. However, DNA polymerase cannot catalyze the formation of a phosphodiester bond between the two segments of the new DNA strand, and it drops off. These unattached sections of the sugar-phosphate backbone in an otherwise full-replicated DNA strand are called nicks.

Once all the template nucleotides have been replicated, the replication process is not yet over. RNA primers need to be replaced with DNA, and nicks in the sugar-phosphate backbone need to be connected.

The group of cellular enzymes that remove RNA primers include the proteins FEN1 (flap endonulcease 1) and RNase H. The enzymes FEN1 and RNase H remove RNA primers at the start of each leading strand and at the start of each Okazaki fragment, leaving gaps of unreplicated template DNA. Once the primers are removed, a free-floating DNA polymerase lands at the 3' end of the preceding DNA fragment and extends the DNA over the gap. However, this creates new nicks (unconnected sugar-phosphate backbone).

In the final stage of DNA replication, the enyzme ligase joins the sugar-phosphate backbones at each nick site. After ligase has connected all nicks, the new strand is one long continuous DNA strand, and the daughter DNA molecule is complete.

Telomere Replication

As DNA polymerase alone cannot replicate the ends of chromosomes, telomerase aids in their replication and prevents chromosome degradation.

The End Problem of Linear DNA Replication

Linear chromosomes have an end problem. After DNA replication, each newly synthesized DNA strand is shorter at its 5' end than at the parental DNA strand's 5' end. This produces a 3' overhang at one end (and one end only) of each daughter DNA strand, such that the two daughter DNAs have their 3' overhangs at opposite ends

Every RNA primer synthesized during replication can be removed and replaced with DNA strands except the RNA primer at the 5' end of the newly synthesized strand. This small section of RNA can only be removed, not replaced with DNA. Enzymes RNase H and FEN1 remove RNA primers, but DNA Polymerase will add new DNA only if the DNA Polymerase has an existing strand 5' to it ("behind" it) to extend. However, there is no more DNA in the 5' direction after the final RNA primer, so DNA polymerse cannot replace the RNA with DNA. Therefore, both daughter DNA strands have an incomplete 5' strand with 3' overhang.

In the absence of additional cellular processes, nucleases would digest these single-stranded 3' overhangs. Each daughter DNA would become shorter than the parental DNA, and eventually entire DNA would be lost. To prevent this shortening, the ends of linear eukaryotic chromosomes have special structures called telomeres.

Telomere Replication

The ends of the linear chromosomes are known as telomeres: repetitive sequences that code for no particular gene. These telomeres protect the important genes from being deleted as cells divide and as DNA strands shorten during replication.

In humans, a six base pair sequence, TTAGGG, is repeated 100 to 1000 times. After each round of DNA replication, some telomeric sequences are lost at the 5' end of the newly synthesized strand on each daughter DNA, but because these are noncoding sequences, their loss does not adversely affect the cell. However, even these sequences are not unlimited. After sufficient rounds of replication, all the telomeric repeats are lost, and the DNA risks losing coding sequences with subsequent rounds.

The discovery of the enzyme telomerase helped in the understanding of how chromosome ends are maintained. The telomerase enzyme attaches to the end of a chromosome and contains a catalytic part and a built-in RNA template. Telomerase adds complementary RNA bases to the 3' end of the DNA strand. Once the 3' end of the lagging strand template is sufficiently elongated, DNA polymerase adds the complementary nucleotides to the ends of the chromosomes; thus, the ends of the chromosomes are replicated.

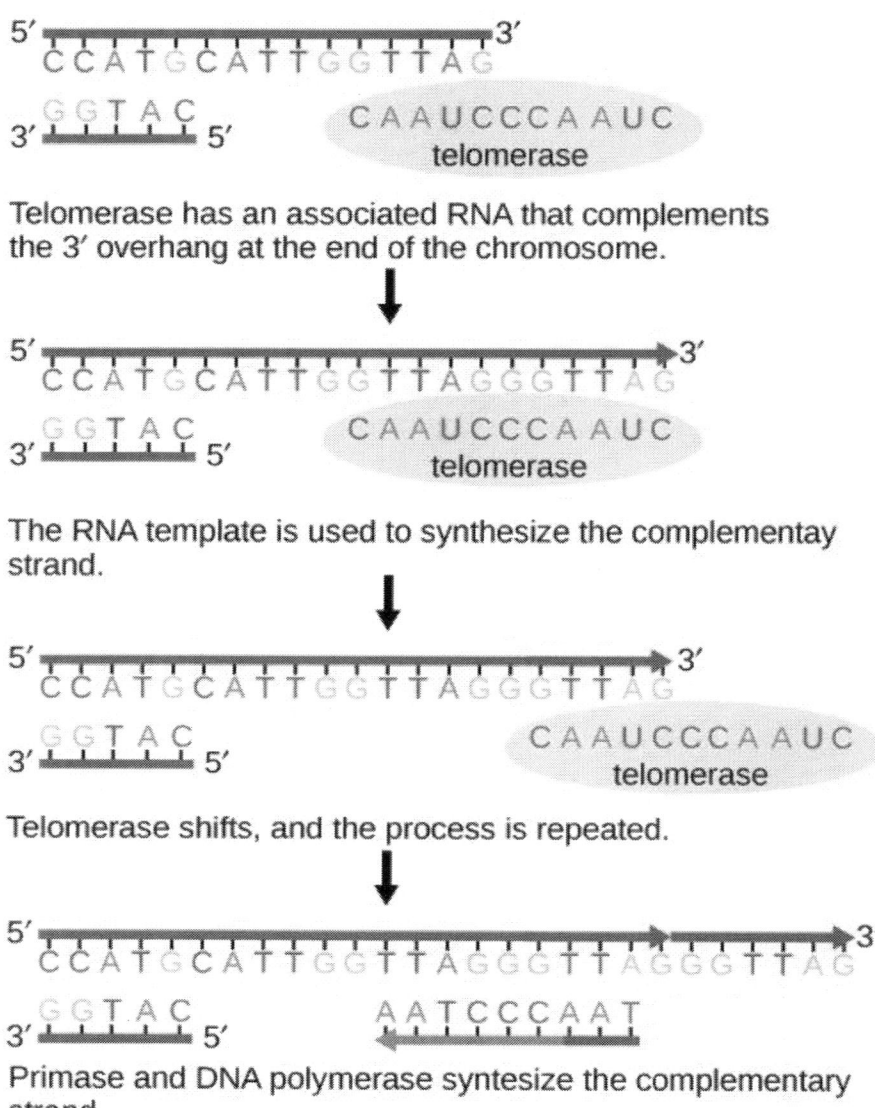

Fig.7.4. Telomere Replication

Telomerase is important for maintaining chromosome integrity: The ends of linear chromosomes are maintained by the action of the telomerase enzyme.

Telomerase and Aging

Telomerase is typically active in germ cells and adult stem cells, but is not active in adult somatic cells. As a result, telomerase does not protect the DNA of adult somatic cells and their telomeres continually shorten as they undergo rounds of cell division.

In 2010, scientists found that telomerase can reverse some age-related conditions in mice. These findings may contribute to the future of regenerative medicine. In the studies, the scientists used telomerase-deficient mice with tissue atrophy, stem cell depletion, organ failure, and impaired tissue injury responses. Telomerase reactivation in these mice caused extension of telomeres, reduced DNA damage, reversed neurodegeneration, and improved the function of the testes, spleen, and intestines. Thus, telomere reactivation may have potential for treating age-related diseases in humans.

2. Repair of DNA Damage and Replication Errors

One of the benefits of the double stranded DNA structure is that it lends itself to repair, when structural damage or replication errors occur. Several kinds of chemical change may cause damage to DNA:

- Spontaneous hydrolysis of a nucleoside removes the heterocyclic base component.
- Spontaneous hydrolysis of cytosine changes it to a uracil.
- Various toxic metabolites may oxidize or methylate heterocyclic base components.
- Ultraviolet light may dimerize adjacent cytosine or thymine bases.

All these transformations disrupt base pairing at the site of the change, and this produces a structural deformation in the double helix.. Inspection-repair enzymes detect such deformations, and use the undamaged nucleotide at that site as a template for replacing the damaged unit. These repairs reduce errors in DNA structure from about one in ten million to one per trillion.

RNA and Protein Synthesis

The genetic information stored in DNA molecules is used as a blueprint for making proteins. Why proteins? Because these macromolecules have diverse primary, secondary and tertiary structures that equip them to carry out the numerous functions necessary to maintain a living organism.

- Structural integrity (hair, horn, eye lenses etc.).

  - Molecular recognition and signaling (antibodies and hormones).

  - Catalysis of reactions (enzymes).

  - Molecular transport (hemoglobin transports oxygen).

  - Movement (pumps and motors).

The critical importance of proteins in life processes is demonstrated by numerous genetic diseases, in which small modifications in primary structure produce debilitating and often disastrous consequences. Such genetic diseases include Tay-Sachs, phenylketonuria (PKU), sickel cell anemia, achondroplasia, and Parkinson disease. The unavoidable conclusion is that proteins are of central importance in living cells, and that proteins must therefore be continuously prepared with high structural fidelity by appropriate cellular chemistry.

Early geneticists identified **genes** as hereditary units that determined the appearance and / or function of an organism (i.e. its phenotype). We now define genes as sequences of DNA that occupy specific locations on a chromosome. The original proposal that each gene controlled the formation of a single enzyme has since been modified as: **one gene = one polypeptide**. The intriguing question of how the information encoded in DNA is converted to the actual construction of a specific polypeptide has been the subject of numerous studies, which have created the modern field of **Molecular Biology**.

1. The Central Dogma and Transcription

Francis Crick proposed that information flows from DNA to RNA in a process called **transcription**, and is then used to synthesize polypeptides by a process called **translation**. Transcription takes place in a manner similar to DNA replication. A characteristic sequence of nucleotides marks the beginning of a gene on the DNA strand, and this region binds to a promoter protein that initiates RNA synthesis. The double stranded structure unwinds at the promoter site., and one of the strands serves as a template for RNA formation, as depicted in the following diagram. The RNA molecule thus formed is single stranded, and serves to carry information from DNA to the protein synthesis machinery called ribosomes. These RNA molecules are therefore called **messenger**-RNA (mRNA).

To summarize: a gene is a stretch of DNA that contains a pattern for the amino acid sequence of a protein. In order to actually make this protein, the relevant DNA segment is first copied into messenger-RNA. The cell then synthesizes the protein, using the mRNA as a template.

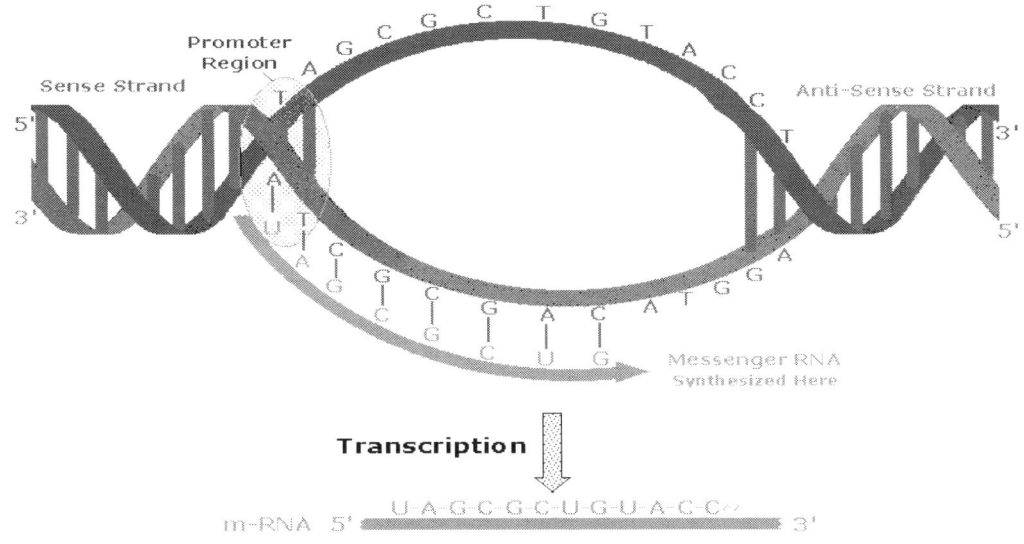

Fig.7.5. Transcription

An important distinction must be made here. One of the DNA strands in the double helix holds the genetic information used for protein synthesis. This is called the **sense strand**, or information strand (colored red above). The complementary strand that binds to the sense strand is called the **anti-sense strand** (colored green), and it serves as a template for generating a mRNA molecule that delivers a copy of the sense strand information to a ribosome. The promoter protein binds to a specific nucleotide sequence that identifies the sense strand, relative to the anti-sense strand. RNA synthesis is then initiated in the 3' direction, as nucleotide triphosphates bind to complementary bases on the template strand, and are joined by phosphate diester linkages. A characteristic "stop sequence" of nucleotides terminates the RNA synthesis. The messenger molecule (colored orange above) is released into the cytoplasm to find a ribosome, and the DNA then rewinds to its double helix structure.

In eucaryotic cells the initially transcribed m-RNA molecule is usually modified and shortened by an "editing" process that removes irrelevant material. The DNA of such organisms is often thousands of times larger and more complex than that composing the single chromosome of a procaryotic bacterial cell. This difference is due in part to repetitive nucleotide sequences (ca. 25% in the human genome). Furthermore, over 95% of human DNA is found in intervening sequences that separate genes and parts of genes. The informational DNA segments that make up genes are called **exons**, and the noncoding segments are called **introns**. Before the mRNA molecule leaves the nucleus, the nonsense bases that make up the introns are cut out, and the informationally useful exons are joined together in a step known as **RNA splicing**. In this fashion shorter mRNA molecules carrying the blueprint for a specific protein are sent on their way to the ribosome factories.

The **Central Dogma** of molecular biology, which at first was formulated as a simple linear progression of information from DNA to RNA to Protein, is summarized in the following illustration. The replication process on the left consists of passing information from a parent DNA molecule to daughter molecules. The middle transcription process copies this information to a mRNA molecule. Finally, this information is used by the chemical machinery of the ribosome to make polypeptides.

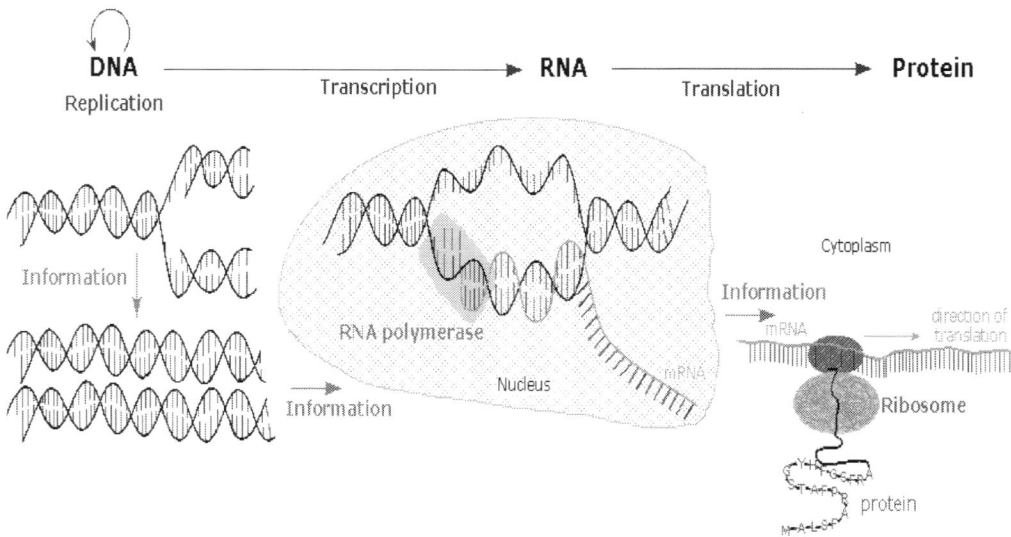

Fig.7.6. Central dogma

As more has been learned about these relationships, the central dogma has been refined to the representation displayed on the right. The dark blue arrows show the general, well demonstrated, information transfers noted above. It is now known that an RNA-dependent DNA polymerase enzyme, known as a reverse transcriptase, is able to transcribe a single-stranded RNA sequence into double-stranded DNA (magenta arrow). Such enzymes are found in all cells and are an essential component of retroviruses (e.g. HIV), which require RNA replication of their genomes (green arrow). Direct translation of DNA information into protein synthesis (orange arrow) has not yet been observed in a living organism. Finally, proteins appear to be an informational dead end, and do not provide a structural blueprint for either RNA or DNA.

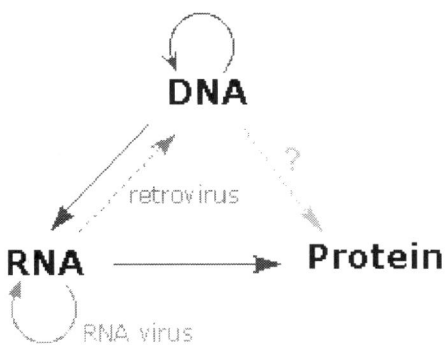

In the following section the last fundamental relationship, that of structural information translation from mRNA to protein, will be described

2. Translation

Translation is a more complex process than transcription. This would, of course, be expected. After all, the coded messages produced by the German Enigma machine could be copied easily, but required a considerable decoding effort before they could be read with understanding. In a similar sense, DNA replication is simply a complementary base pairing exercise, but the translation of the four letter (bases) alphabet code of RNA to the twenty letter (amino acids) alphabet of protein literature is far from trivial. Clearly, there could not be a direct one-to-one correlation of bases to amino acids, so the nucleotide letters must form short words or **codons** that define specific amino acids. Many questions pertaining to this genetic code were posed in the late 1950's:

> • **How many RNA nucleotide bases designate a specific amino acid?**
> If separate groups of nucleotides, called codons, serve this purpose, at least three are needed. There are $4^3 = 64$ different nucleotide triplets, compared with $4^2 = 16$ possible pairs.
> • **Are the codons linked separately or do they overlap?**
> Sequentially joined triplet codons will result in a nucleotide chain three times longer than the protein it describes. If overlapping codons are used then

fewer total nucleotides would be required.

• If triplet segments of mRNA designate specific amino acids in the protein, how are the codons identified?

For the sequence ~CUAGGU~ are the codons CUA & GGU or ~C, UAG & GU~ or ~CU, AGG & U~?

• Are all the codon words the same size?

In Morse code the most widely used letters are shorter than less common letters. Perhaps nature employs a similar scheme.

Physicists and mathematicians, as well as chemists and microbiologists all contributed to unravelling the genetic code. Although earlier proposals assumed efficient relationships that correlated the nucleotide codons uniquely with the twenty fundamental amino acids, it is now apparent that there is considerable redundancy in the code as it now operates. Furthermore, the code consists exclusively of non-overlapping triplet codons.

Clever experiments provided some of the earliest breaks in deciphering the genetic code. Marshall Nirenberg found that RNA from many different organisms could initiate specific protein synthesis when combined with broken E.coli cells (the enzymes remain active). A synthetic polyuridine RNA induced synthesis of poly-phenylalanine, so the UUU codon designated phenylalanine. Likewise an alternating ~CACA~ RNA led to synthesis of a ~His-Thr-His-Thr~ polypeptide.

The following table presents the present day interpretation of the genetic code. Note that this is the RNA alphabet, and an equivalent DNA codon table would have all the **U** nucleotides replaced by **T**. Methionine and tryptophan are uniquely represented by a single codon. At the other extreme, leucine is represented by eight codons. The average redundancy for the twenty amino acids is about three. Also, there are three **stop codons** that terminate polypeptide synthesis.

RNA Codons for Protein Synthesis

	Second Position				
First Position	**U**	**C**	**A**	**G**	**Third Position**
U	UUU Phe [F] UUC Phe [F] UUA Leu [L] UUG Leu [L]	UCU Ser [S] UCC Ser [S] UCA Ser [S] UCG Ser [S]	UAU Tyr [Y] UAC Tyr [Y] UAA Stop UAG Stop	UGU Cys [C] UGC Cys [C] UGA Stop UGG Trp [W]	U C A G
C	CUU Leu [L] CUC Leu [L] CUA Leu [L] CUG Leu [L]	CCU Pro [P] CCC Pro [P] CCA Pro [P] CCG Pro [P]	CAU His [H] CAC His [H] CAA Gln [Q] CAG Gln [Q]	CGU Arg [R] CGC Arg [R] CGA Arg [R] CGG Arg [R]	U C A G
A	AUU Ile [I] AUC Ile [I] AUA Ile [I] AUG Met [M]	ACU Thr [T] ACC Thr [T] ACA Thr [T] ACG Thr [T]	AAU Asn [N] AAC Asn [N] AAA Lys [K] AAG Lys [K]	AGU Ser [S] AGC Ser [S] AGA Arg [R] AGG Arg [R]	U C A G
G	GUU Val [V] GUC Val [V] GUA Val [V] GUG Val [V]	GCU Ala [A] GCC Ala [A] GCA Ala [A] GCG Ala [A]	GAU Asp [D] GAC Asp [D] GAA Glu [E] GAG Glu [E]	GGU Gly [G] GGC Gly [G] GGA Gly [G] GGG Gly [G]	U C A G

Transfer RNA Molecules

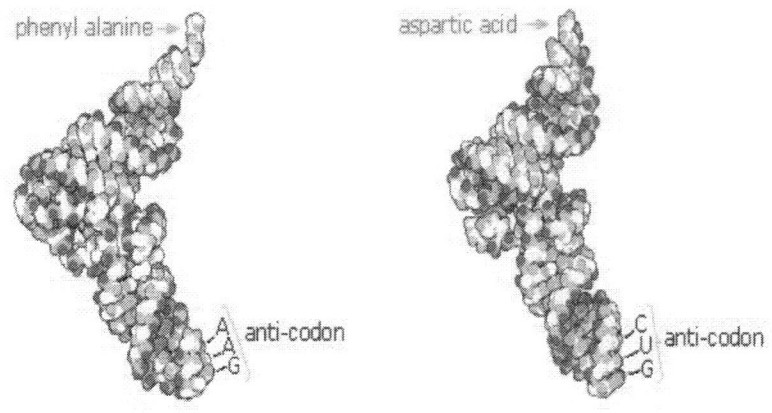

Transfer RNA Molecules

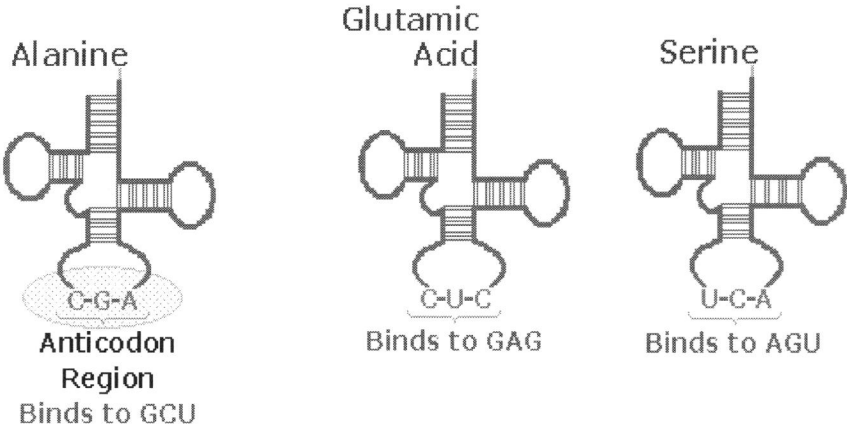

Fig.7.7.

The translation process is fundamentally straightforward. The mRNA strand bearing the transcribed code for synthesis of a protein interacts with relatively small RNA molecules (about 70-nucleotides) to which individual amino acids have been attached by an ester bond at the 3'-end. These **transfer RNA's** (tRNA) have distinctive three-dimensional structures consisting of loops of single-stranded RNA connected by double stranded segments. This cloverleaf secondary structure is further wrapped into an "L-shaped" assembly, having the amino acid at the end of one arm, and a characteristic **anti-codon** region at the other end. The anti-codon consists of a nucleotide triplet that is the complement of the amino acid's codon(s). Models of two such tRNA molecules are shown to the right. When read from the top to the bottom, the anti-codons depicted here should complement a codon in the previous table. Cloverleaf cartoons of three other tRNA molecules will be shown on the right by clicking on the diagram.

A cell's protein synthesis takes place in organelles called **ribosomes**. Ribosomes are complex structures made up of two distinct and separable subunits (one about twice the size of the other). Each subunit is composed of one or two RNA molecules (60-70%) associated with 20 to 40 small proteins (30-40%). The ribosome accepts a mRNA molecule, binding initially to a characteristic nucleotide sequence at the 5'-end (colored light blue in the following diagram). This unique binding assures that polypeptide synthesis starts at the right codon. A tRNA molecule with the appropriate anti-codon then attaches at the starting point and this is followed by a series of adjacent tRNA attachments, peptide bond formation and shifts of the ribosome along the mRNA chain to expose new codons to the ribosomal chemistry. The following diagram is designed as a slide show illustrating these steps. The outcome is synthesis of a polypeptide chain corresponding to the mRNA blueprint. A "stop codon" at a designated position on the mRNA terminates the synthesis by introduction of a "Release Factor".

Participating Species in Protein Synthesis

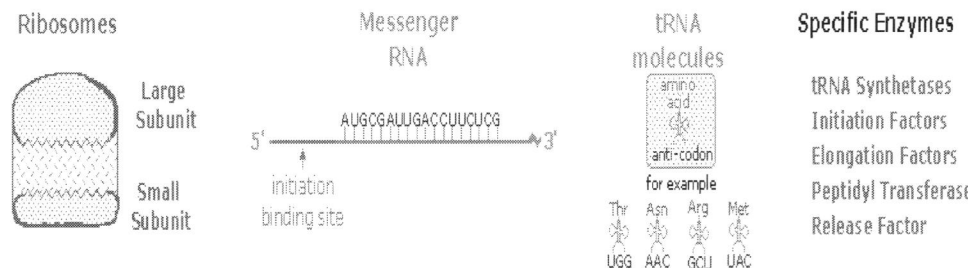

Ribosomes are composed of two major subunits, one larger than the other. Each of these is a complex assemblage of small proteins and rRNA molecules. The small subunit recognizes a characteristic base sequence at the 5' end of mRNA, and holds the mRNA chain in the ribosome so that synthesis may begin.

Click the Diagram

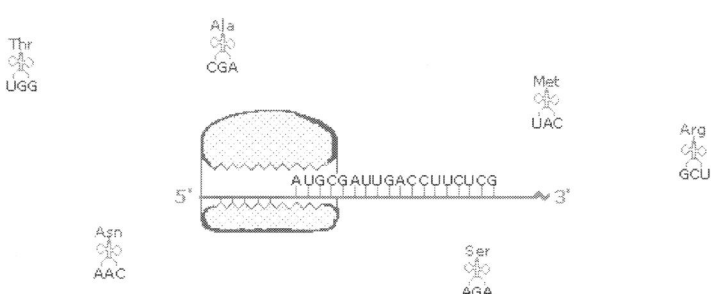

In this illustration the mRNA chain has docked with the ribosome. The cyan colored binding site at the 5' end of the chain is held by the small ribosomal unit. A mixture of tRNA molecules, each bearing a specific amino acid, are available for use when needed. Synthesis of these carrier molecules from their components is continuous.

Fig.7.8.

3. Post-translational Modification

Once a peptide or protein has been synthesized and released from the ribosome it often undergoes further chemical transformation. This **post-translational modification** may involve the attachment of other moieties such as acyl groups, alkyl groups, phosphates, sulfates, lipids and carbohydrates. Functional changes such as dehydration, amidation, hydrolysis and oxidation (e.g. disulfide bond formation) are also common. In this manner the limited array of twenty amino acids designated by the codons may be expanded in a variety of ways to enable proper functioning of the resulting protein. Since these post-translational reactions are generally catalyzed by enzymes, it may be said: *"Virtually every molecule in a cell is made by the ribosome or by enzymes made by the ribosome."*

Modifications, like phosphorylation and citrullination, are part of common mechanisms for controlling the behavior of a protein. As shown on the left below, citrullination is the post-translational modification of the amino acid arginine into the amino acid citrulline. Arginine is positively charged at a neutral pH, whereas citrulline is uncharged, so this change increases the hydrophobicity of a protein. Phosphorylation of serine, threonine or tyrosine residues renders them more hydrophilic, but such changes are usually transient, serving to regulate the biological activity of the protein. Other important functional changes include iodination of tyrosine residues in the peptide thyroglobulin by action of the enzyme thyroperoxidase. The monoiodotyrosine and diiodotyrosine formed in this manner are then linked to form the thyroid hormones T_3 and T_4, shown on the right below.

Amino acids may be enzymatically removed from the amino end of the protein. Because the "start" codon on mRNA codes for the amino acid methionine, this amino acid is usually removed from the resulting protein during post-translational modification. Peptide chains may also be cut in the middle to form shorter strands. Thus, insulin is initially synthesized as a 105 residue preprotein. The 24-amino acid signal peptide is removed, yielding a proinsulin peptide. This folds and forms disulfide bonds between cysteines 7 and 67 and between 19 and 80. Such dimeric cysteines, joined by a disulfide bond, are named **cystine**. A protease then cleaves the peptide at arg31 and arg60, with loss of the 32-60 sequence (chain C). Removal of arg31 yields mature insulin, with the A and B chains held together by disulfide bonds and a third cystine moiety in chain A. The following cartoon illustrates this chain of events.

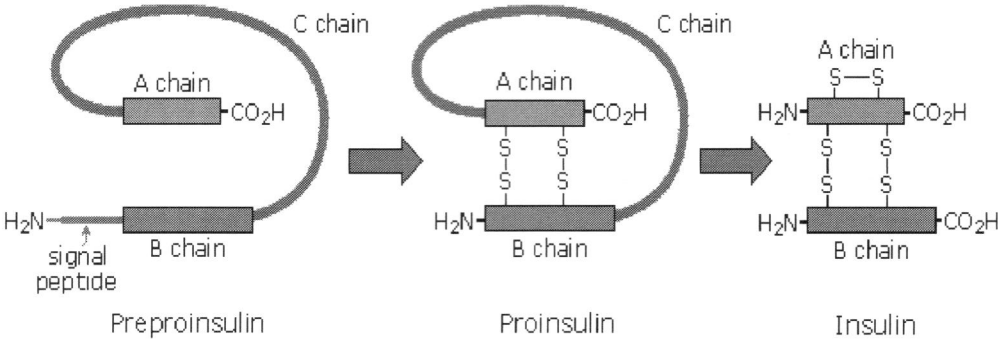

Preproinsulin → Proinsulin → Insulin

Nisin is a polypeptide (34 amino acids) made by the bacterium *Lactococcus lactis*. Nisin kills gram positive bacteria by binding to their membranes and targeting lipid II, an essential precursor of cell wall synthesis. Such antimicrobial peptides are a growing family of compounds which have received the name lantibiotics due to the presence of **lanthionine**, a nonproteinogenic amino acid with the chemical formula $HO_2C-CH(NH_2)-CH_2-S-CH_2-CH(NH_2)-CO_2H$. Lanthionine is composed of two alanine residues that are crosslinked on their β-carbon atoms by a thioether linkage (i.e. it is the monosulfide analog of the disulfide cystine). Lantibiotics are unique in that they are ribosomally synthesized as prepeptides, followed by post-translational processing of a number of amino acids (e.g. serine, threonine and cysteine) into dehydro residues and thioether crossbridges. Nisin is the only bacteriocin that is accepted as a food preservative. Several nisin subtypes that differ in amino acid composition and biological activity are known. A typical structure is drawn below, and a Jmol model will be presented by clicking on the diagram.

Nisin

Ile-Aca-Dal-Ile-Leu-Aaa-a-Cys-Abu-b-Cys-Lys-Abu-Pro-Gly-Gly-c-Gly-Ala-Leu-Met-Cys-Asn-Met-Lys-Abu-d-Cys-Asn-Ala-Abu-e-Cys-Ser-Ile-His-Val-Aaa-Lys

Dal = D-Alanine **Aaa** = 2-aminoacrylic acid (or **Dha** didehydroalanine) **Abu** = 2-aminobutyric acid (or **Dbb** = didehydrobutyrine) **Aca** = 2-aminocrotonic acid (or **Dhb** didehydroaminobutyric acid)

The bacterial cell wall is a cross-linked glycan polymer that surrounds bacterial cells, dictates their cell shape, and prevents them from breaking due to environmental changes in osmotic pressure. This wall consists mainly of peptidoglycan or murein, a three-dimensional polymer of sugars and amino acids located on the exterior of the cytoplasmic membrane. The monomer units are composed of two amino sugars, N-acetylglucosamine (NAG) and N-acetylmuramic acid (NAM), shown on the right. Transglycosidase enzymes join these units by glycoside bonds, and they are further interlinked to each other via peptide cross-links between the pentapeptide moieties that are attached to the NAM residues. Peptidoglycan subunits are assembled on the cytoplasmic side of the bacterial membrane from a polyisoprenoid anchor. Lipid II, a membrane-anchored cell-wall precursor that is essential for bacterial cell-wall biosynthesis, is one of the key components in the synthesis of peptidoglycan. Peptidoglycan synthesis via polymerization of Lipid II is illustrated in the following diagram. Cross-linking of the peptide side chains is then effected by transpeptidase enzymes. A model of Lipid II complexed with nisin may be examined as part of the previous Jmol display.

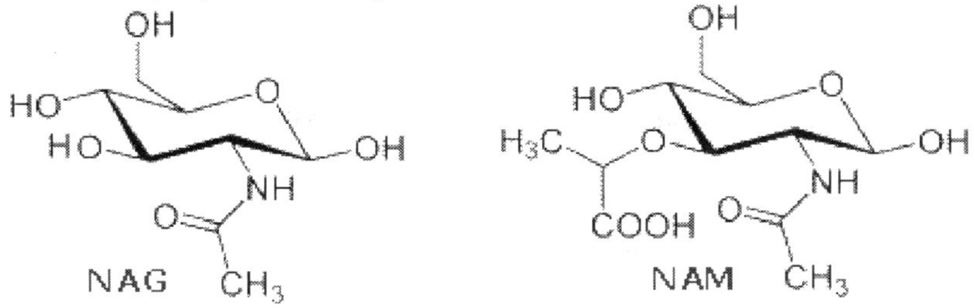

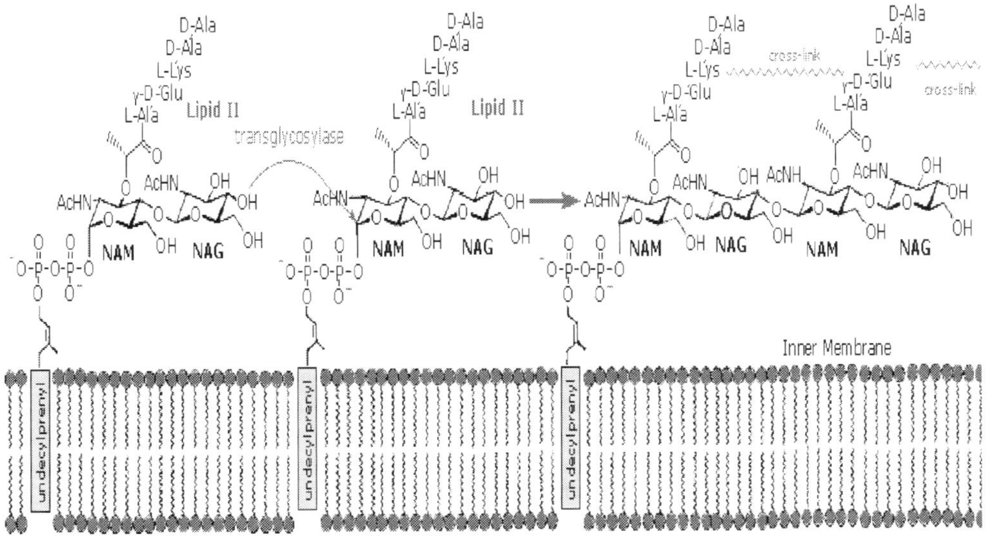

In order for bacteria to divide by binary fission and increase their size following division, links in the peptidoglycan must be broken, new peptidoglycan monomers must be inserted, and the peptide cross links must be resealed. Transglycosidase enzymes catalyze the formation of glycosidic bonds between the NAM and NAG of the peptidoglycan monomers and the NAG and NAM of the existing peptidoglycan. Finally, transpeptidase enzymes reform the peptide cross-links between the rows and layers of peptidoglycan making the wall strong. Many antibiotic drugs, including penicillin, target the chemistry of cell wall formation. The effectiveness of choosing Lipid II for an antibacterial strategy is highlighted by the fact that it is the target for at least four different classes of antibiotic, including the clinically important glycopeptide antibiotic vancomycin. The growing problem of bacterial resistance to many current drugs, including vancomycin, has led to increasing interest in the therapeutic potential of other classes of compound that target Lipid II. Lantibiotics such as nisin are part of this interest.

(https://www2.chemistry.msu.edu, https://courses.lumenlearning.com/boundless-biology/chapter/dna)

****

THE BACTERIA

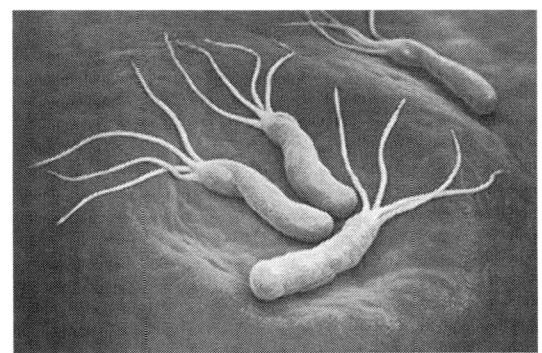

The first scientist who demonstrated the existence of micro-organisms was **Antony von Leeuwenhoek.** He was born at Delft, in Holland, in 1632, and enthusiastically pursued microscopy with primitive instruments. He corroborated **Harvey's discovery** of the circulation of the blood in the web of a frog's foot; he defined the red blood corpuscles of vertebrates, the fibres of the lens of the human eye, the scales of the skin, and the structure of hair. He was neither educated nor trained in science, but in the leisure time of his occupation as a linen-draper he learned the art of grinding lenses, in which he became so proficient that he was able to construct a microscope of greater power than had been previously manufactured. The compound microscope dates from 1590, and when Leeuwenhoek was about forty years old, Holland had already given to the world both microscope and telescope. **Robert Hooke** did for England what Hans Janssen had done for Holland, and established the same 2 conclusion that Leeuwenhoek arrived at independently, viz., that a simple globule of glass mounted between two metal plates and pierced with a minute aperture to allow rays of light to pass was a contrivance which would magnify more highly than the recognised microscopes of that day. It was with some such instrument as this that the first micro-organisms were observed in a drop of water. It was not until more than a hundred years later that these "animalcules," as they were termed, were thought to be anything more than accidental to any fluid or substance containing them. Plenciz, of Vienna, was one of the first to conceive the idea that decomposition could only take place in the presence of some of these "**animalcules.**" This was in the middle of the eighteenth century. Just about a century later, by a series of important discoveries, it was established beyond dispute that these micro-organisms had an intimate causal relation to fermentation, putrefaction, and infectious diseases. **Spallanzani, Pasteur, and Tyndall** are the three who more than others contributed to this discovery. **Spallanzani** was an Italian, who studied at Bologna, and was in 1754 appointed to the chair of logic at Reggio. But his inclinations led him into the realm of natural history. Amongst other things, his attention was directed to the doctrine of *spontaneous generation*, which had been propounded by Needham a few years previously. In 1768 **Spallanzani became Professor of Natural History at Pavia,** and whilst there he demonstrated that if infusions of vegetable matter were placed in flasks and hermetically sealed, and then brought to the boiling point, no living organisms could thereafter be detected, nor did the vegetable matter decompose. When, however, the flasks were very slightly cracked, and air gained admittance, then invariably both organisms and decomposition appeared. Schwann, the founder of the cell-theory, and Schulze, both showed that if the air gaining access to the flask were either passed through highly heated3 tubes or drawn through strong acid the result was the same as if no air entered at all, viz., no organisms and no decomposition. The result of these investigations was that scientific men began to believe that no form of life arose *de novo* (*abiogenesis*), but had its source in previous life (*biogenesis*). It remained to **Pasteur and Tyndall** to demonstrate this beyond dispute, and to put to rout the fresh arguments for spontaneous generation which Pouchet had advanced as late as 1859. Pasteur collected the floating dust of the air, and found by means of the microscope many organised particles, which he sowed on suitable infusions, and thus obtained rich crops of "**animalculæ.**" He also demonstrated that these organisms existed in different degrees in different atmospheres, few in the pure air of the **Mer de Glace**, more in the air of the plains, most in the air of towns. He further proved that it was not necessary to insist upon hermetic sealing or cotton filters to keep these living organisms in the air from gaining access to a flask of infusion. If the neck of the flask were drawn out into a long tube and turned downwards, and then a little upwards, even though the end be left open, no contamination gained access. Hence, if the infusion were boiled, no putrefaction would occur. The organisms which fell into the open end of the tube were arrested in the condensation water in the angle of the tube; but even if that were not so, the force of gravity acting upon them prevented them from passing up the long arm of the tube into the neck of the flask. A few years after Pasteur's first work on this subject Tyndall conceived a precise method of determining the absence or presence of dust particles in the air by passing a beam of sunlight through a glass box before and after its walls had been coated with glycerine. Into the floor of the box were fixed the mouths of flasks of infusion. These were boiled, after which they were allowed to cool, and might then be kept for weeks or months without putrefying or revealing the presence of germ life. Here all the con4ditions of the infusions were natural, except that in the air above them there was no dust.

The sum-total of result arising from all these investigations was to the effect that no spontaneous generation was possible, that the atmosphere contained unseen germs of life, that the smallest of organisms responded to the law of gravitation and adhered to moist surfaces, and that micro-organisms were in some way or other the cause of putrefaction.

The final refutation of the hypothesis of spontaneous generation was followed by an awakened interest in the unseen world of micro-organic life. Investigations into fermentation and putrefaction followed each other rapidly, and in **1863 Davaine claimed that Pollender's bacillus** of anthrax, which was found in the blood and body tissues of animals dead of anthrax, was the cause of that disease. From that time to this in every department of biology bacteria have been increasingly found to play an important part. They cause changes in milk, and flavour butter; they decompose animal matter, yet build up the broken-down elements into compounds suitable for use in nature's economy; they assist in the fixation of free nitrogen; they purify sewage; in certain well-established cases they are the cause of specific disease, and in many other cases they are the likely cause. No doubt the disposal of spontaneous generation did much to arouse interest in this branch of science. Yet it must not be forgotten that the advance of the microscope and bacteriological method and technique have played a large share in this development. The sterilisation of culture fluids by heat, the use of aniline dyes as staining agents, the introduction of solid culture media (like gelatine and agar), and Koch's "plate" method have all contributed not a little to the enormous strides of bacteriology. Owing to its relation to disease, physicians have entered keenly into the arena of bacteriological research. Hence, from a variety of causes, it has come about that the advance has been phenomenal.

General Characteristics
- Bacteria are prokaryotic organisms (Kingdom:Monera)
- They do not have cell defined organelles like mitochondria, Golgi bodies, Endoplasmic reticulum.,etc
- They are Microscopic, unicellular
- They may occur singly or in small groups to form colonies.

- They possess rigid cell wall. Cell wall is made up of peptidoglycan (Mureins) and Lipo polysaccharides.
- Absence of well-defined nucleus.i.e., DNA is not enclosed in a nuclear membrane.
- Ribosomes are scattered in the cytoplasmic matrix and are of 70S type.
- The plasma membrane is invaginated to form mesosomes.
- Most of the bacteria are heterotrophic. Some bacteria are autotrophic, possess bacteriochlorophyll,
- Motile bacteria possess one or more flagella.
- The common method of multiplication is binary fission.
- True sexual reproduction is lacking, but genetic recombination occurs by conjugation, transformation and transduction.

Bacteria Morphology:

It is the study of size, shape, structure, and arrangement of cells.

a) Size of Bacteria:

Bacteria are very small or minute, as single drop of water may contains about 50 billion of bacteria are usually measured in micrometer (μm) which is equivalent to 1/1000 mm.

1. Size may varies depending upon the species. Size generally ranges from 1 to 10 microns.

2. Most of the bacterial cells 0.5 to 1 μm in width or diameter.

3. Cylindrical or rod (Baclli) 2-3 μm length.

4. Spiral or helical (Spirlli) – 0.75 -1.25 μm.

b) Shape in Bacteria:

The shapes of bacterial cells are Spherical or Ellipsoidal or Oval – Cocci Cylindrical or rod- Bacilli Spiral or Helical – Spirilli

Some species have variety of shapes and thus termed as pleomorphic. E.g. *Arthrobacter.*

c) Arrangement of Bacterial Cells:

Bacterial cells are arranged in a characteristic manner of the particular species. The typical pattern of cell arrangement in different bacteria is an important characteristics used in identification of bacteria.

d) Cell grouping or Arrangement in Cocci:

i) Monococcus:

Single spherical bacterial cell.

ii) Diplococcus:

A coccus divides into two plane and cells remains in pairs.

iii) Streptococcus:

A coccus is arranged in a long chain e.g. *Streptococcus sp.*

iv) Tetrad:

A coccus divides in two plans, second division at a right angle to the first plane of division and forms a square of four cells. E.g. *Tetracoccus sp*

v) Sarcina:

A cube of eight coccus cell is formed by three divisions in alternate planes at right angle to each other. A cube of eight cells is known as *Sarcina*.

vi) Staphylococci:

A coccus cells divides in three planes in irregular pattern like cluster of cells or bunch of grapes. E.g. *Staphylococcus albus*

vii) Vibrio:

The short comma shaped cells are called as Vibrio. Short tightly coined rods are called spirillum. Very long cell with several cuvves and twiste are called "Spirochete".

e) Cell Grouping or Arrangement in Bacilli:

i) Monobacilli:

Single rod shaped bacterial cell. E.g. *Monobacillus.*

ii) Diplobacilli:

Bacilli are arranged in a pair of two cells. E.g. *Bacillus subtilis.*

iii) Streptobacilli:

Cells are arranged in a chain.E.g. *Lactobacillus bukgaricus.*

iv) Palisade:

Group of cells lined side by side like matchsticks in a match box called as palisade arrangement.

f) Flagellar Arrangement in Bacteria:

All types of bacteria do not have a flagella they are mostly present in Bacilli and Spirilli and rarely in cocci.

i) Atrichous:

A cell without flagella.

ii) Monotrichous:

Single polar flagellum at one end.

iii) Amphitrichous:

A cell with a single polar flagellum at both the ends. E.g *Spirillim*.

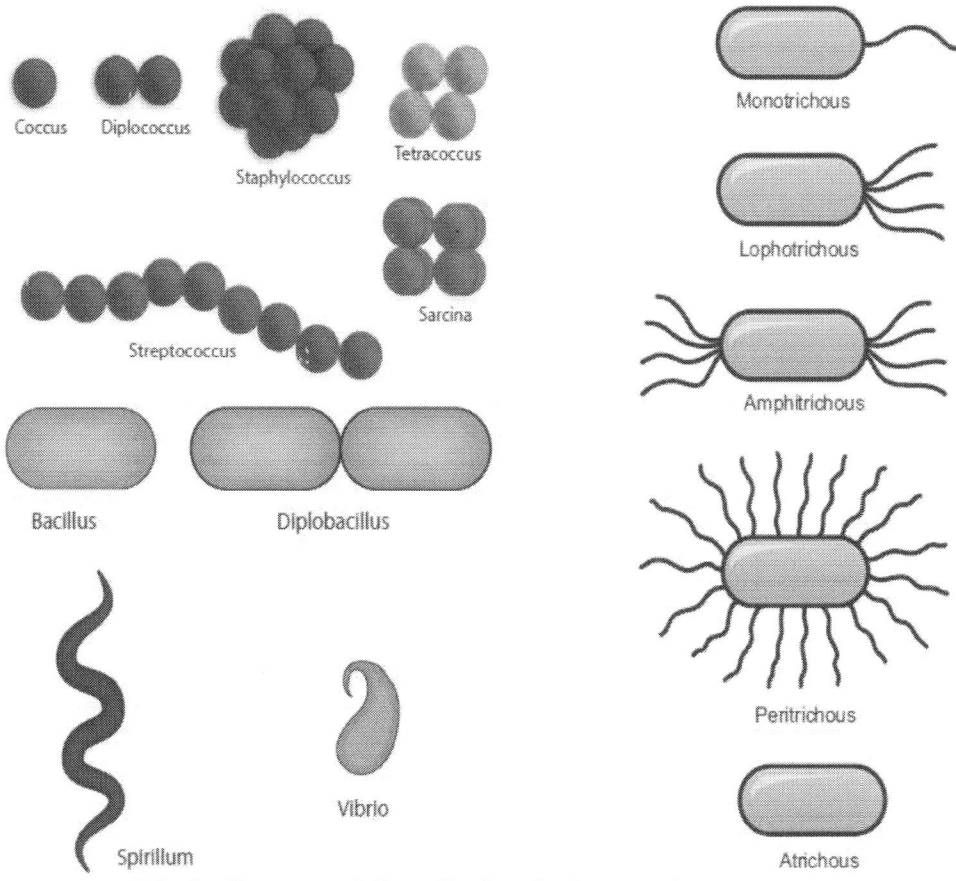

Fig.8. Shape and flagellation in bacteeria

iv) Lophotrichous:

A cell having tuft or bunch of flagella at both the ends. E.g. **Coli.**

v) Caphalotrichous:

A cell having a pair of flagella at one end.

Arrangement in Spirilli:

The Spirilli are predominantly unattached however they differ in frequency of turns and overall length. They are grouped in three types as:

a) Short tightly coiled rod called *spitillium*

b) Short incomplete spirals called as comma or *vibrios*.

c) They are long twisted with several curves calls "Spirochete".

Bacteria Cell Structure

They are as unrelated to human beings as living things can be, but bacteria are essential to human life and life on planet Earth. Although they are notorious for their role in causing human diseases, from tooth decay to the Black Plague, there are beneficial species that are essential to good health.

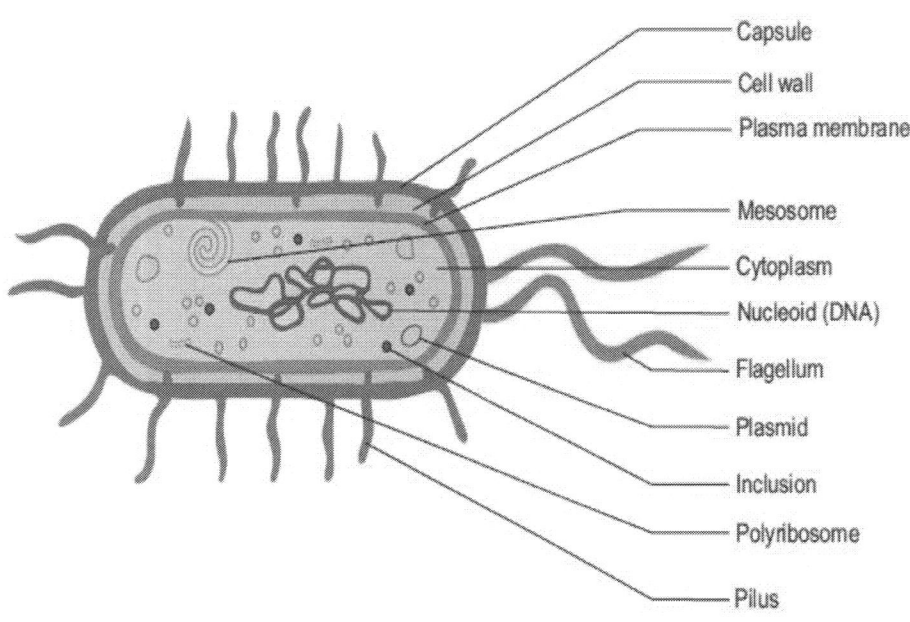

Fig.8.1. Ultra structure of a bacteria

For example, one species that lives symbiotically in the large intestine manufactures vitamin K, an essential blood clotting factor. Other species are beneficial indirectly. Bacteria give yogurt its tangy flavor and sourdough breads its sour taste. They make it possible for ruminant animals (cows, sheep, and goats) to digest plant cellulose and for some plants, (soybean, peas, and alfalfa) to convert nitrogen to a more usable form.

Bacteria are prokaryotes, lacking well-defined nuclei and membrane-bound organelles, and with chromosomes composed of a single closed DNA circle. They come in many shapes and sizes, from minute spheres, cylinders and spiral threads, to flagellated rods, and filamentous chains. They are found practically everywhere on Earth and live in some of the most unusual and seemingly inhospitable places.

Evidence shows that bacteria were in existence as long as 3.5 billion years ago, making them one of the oldest living organisms on the Earth. Even older than the bacteria are the archeans (also called archaebacteria) tiny prokaryotic organisms that live only in extreme environments: boiling water, super-salty pools, sulfur-spewing volcanic vents, acidic water, and deep in the Antarctic ice. Many scientists now believe that the archaea and bacteria developed separately from a common ancestor nearly four billion years ago. Millions of years later, the ancestors of today's eukaryotes split off from the archaea. Despite the superficial resemblance to bacteria, biochemically and genetically, the archea are as different from bacteria as bacteria are from humans.

In the late 1600s, **Antoni van Leeuwenhoek** became the first to study bacteria under the microscope. During the nineteenth century, the French scientist Louis Pasteur and the German physician Robert Koch demonstrated the role of bacteria as pathogens (causing disease). The twentieth century saw numerous advances in bacteriology, indicating their diversity, ancient lineage, and general importance. Most notably, a number of scientists around the world made contributions to the field of microbial ecology, showing that bacteria were essential to food webs and for the overall health of the Earth's ecosystems. The discovery that some bacteria produced compounds lethal to other bacteria led to the development of antibiotics, which revolutionized the field of medicine.

There are two different ways of grouping bacteria. They can be divided into three types based on their response to gaseous oxygen. Aerobic bacteria require oxygen for their health and existence and will die without it. Anerobic bacteria can't tolerate gaseous oxygen at all and die when exposed to it. Facultative aneraobes prefer oxygen, but can live without it.

The second way of grouping them is by how they obtain their energy. Bacteria that have to consume and break down complex organic compounds are heterotrophs. This includes species that are found in decaying material as well as those that utilize fermentation or respiration. Bacteria that create their own energy, fueled by light or through chemical reactions, are autotrophs.

- **Capsule** - Some species of bacteria have a third protective covering, a capsule made up of polysaccharides (complex carbohydrates). Capsules play a number of roles, but the most important are to keep the bacterium from drying out and to protect it from phagocytosis (engulfing) by larger microorganisms. The capsule is a major virulence factor in the major disease-causing bacteria, such as ***Escherichia coli*** and ***Streptococcus pneumoniae***. Nonencapsulated mutants of these organisms are avirulent, i.e. they don't cause disease.
- **Cell Envelope** - The cell envelope is made up of two to three layers: the interior cytoplasmic membrane, the cell wall, and -- in some species of bacteria -- an outer capsule.

Bacteria: Cell Walls

It is important to note that not all bacteria have a **cell wall**. Having said that though, it is also important to note that **most** bacteria (about 90%) have a cell wall and they typically have one of two types: a **gram positive** cell wall or a **gram negative** cell wall.

The two different cell wall types can be identified in the lab by a differential stain known as the **Gram stain**. Developed in 1884, it's been in use ever since. Originally, it was not known why the Gram stain allowed for such reliable separation of bacterial into two groups. Once the electron microscope was invented in the 1940s, it was found that the staining difference correlated with differences in the cell walls. Here is a website that shows the actual steps of the Gram stain. After this stain technique is applied the gram positive bacteria will stain purple, while the gram negative bacteria will stain pink.

Gram + Bacteria   Gram - Bacteria

Overview of Bacterial Cell Walls

A cell wall, not just of bacteria but for all organisms, is found outside of the cell membrane. It's an additional layer that typically provides some strength that the cell membrane lacks, by having a semi-rigid structure.

Both gram positive and gram negative cell walls contain an ingredient known as **peptidoglycan** (also known as **murein**). This particular substance hasn't been found anywhere else on Earth, other than the cell walls of bacteria. But both bacterial cell wall types contain additional ingredients as well, making the bacterial cell wall a complex structure overall, particularly when compared with the cell walls of eukaryotic microbes. The cell walls of eukaryotic microbes are typically composed of a single ingredient, like the cellulose found in algal cell walls or the chitin in fungal cell walls.

The bacterial cell wall performs several functions as well, in addition to providing overall strength to the cell. It also helps maintain the cell shape, which is important for how the cell will grow, reproduce, obtain nutrients, and move. It protects the cell from **osmotic lysis**, as the cell moves from one environment to another or transports in nutrients from its surroundings. Since water can freely move across both the cell membrane and the cell wall, the cell is at risk for an osmotic imbalance, which could put pressure on the relatively weak plasma membrane. Studies have actually shown that the internal pressure of a cell is similar to the pressure found inside a fully inflated car tire. That is a lot of pressure for the plasma membrane to withstand! The cell wall can keep out certain molecules, such as toxins, particularly for gram negative bacteria. And lastly, the bacterial cell wall can contribute to the pathogenicity or disease –causing ability of the cell for certain bacterial pathogens.

Structure of Peptidoglycan

Let us start with peptidoglycan, since it is an ingredient that both bacterial cell walls have in common.

Peptidoglycan is a polysaccharide made of two glucose derivatives, **N-acetylglucosamine (NAG)** and **N-acetylmuramic acid (NAM)**, alternated in long chains. The chains are cross-linked to one another by a **tetrapeptide** that extends off the NAM sugar unit, allowing a lattice-like structure to form. The four amino acids that compose the tetrapeptide are: **L-alanine, D-glutamine, L-lysine or** *meso*-**diaminopimelic acid (DPA),** and **D-alanine**. Typically only the L-isomeric form of amino acids are utilized by cells but the use of the mirror image D-amino acids provides protection from proteases that might compromise the integrity of the cell wall by attacking the peptidoglycan. The tetrapeptides can be **directly cross-linked** to one another, with the D-alanine on one tetrapeptide binding to the L-lysine/ DPA on another tetrapeptide. In many gram positive bacteria there is a cross-bridge of five amino acids such as glycine (**peptide interbridge**) that serves to connect one tetrapeptide to another. In either case the cross-linking serves to increase the strength of the overall structure, with more strength derived from **complete cross-linking**, where every tetrapeptide is bound in some way to a tetrapeptide on another NAG-NAM chain.

While much is still unknown about peptidoglycan, research in the past ten years suggests that peptidoglycan is synthesized as a cylinder with a coiled substructure, where each coil is cross-linked to the coil next to it, creating an even stronger structure overall.

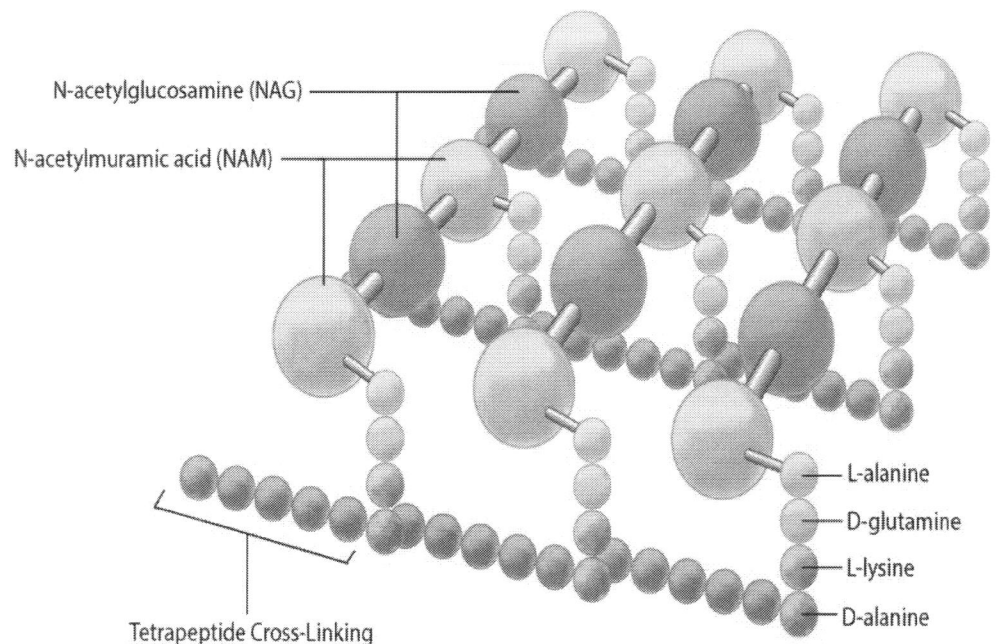

Fig.8.2. Peptidoglycan Structure.

Gram Positive Cell walls

The cell walls of gram positive bacteria are composed predominantly of peptidoglycan. In fact, peptidoglycan can represent up to 90% of the cell wall, with layer after layer forming around the cell membrane. The NAM tetrapeptides are typically cross-linked with a peptide interbridge and complete cross-linking is common. All of this combines together to create an incredibly strong cell wall.

The additional component in a gram positive cell wall is **teichoic acid**, a glycopolymer, which is embedded within the peptidoglycan layers. Teichoic acid is believed to play several important roles for the cell, such as generation of the net negative charge of the cell, which is essential for development of a proton motive force. Teichoic acid contributes to the overall rigidity of the cell wall, which is important for the maintenance of the cell shape, particularly in rod-shaped organisms. There is also evidence that teichoic acids participate in cell division, by interacting with the peptidoglycan biosynthesis machinery. Lastly, teichoic acids appear to play a role in resistance to adverse conditions such as high temperatures and high salt concentrations, as well as to β-lactam antibiotics. Teichoic acids can either be covalently linked to peptidoglycan (**wall teichoic acids or WTA**) or connected to the cell membrane via a lipid anchor, in which case it is referred to as **lipoteichoic acid**.

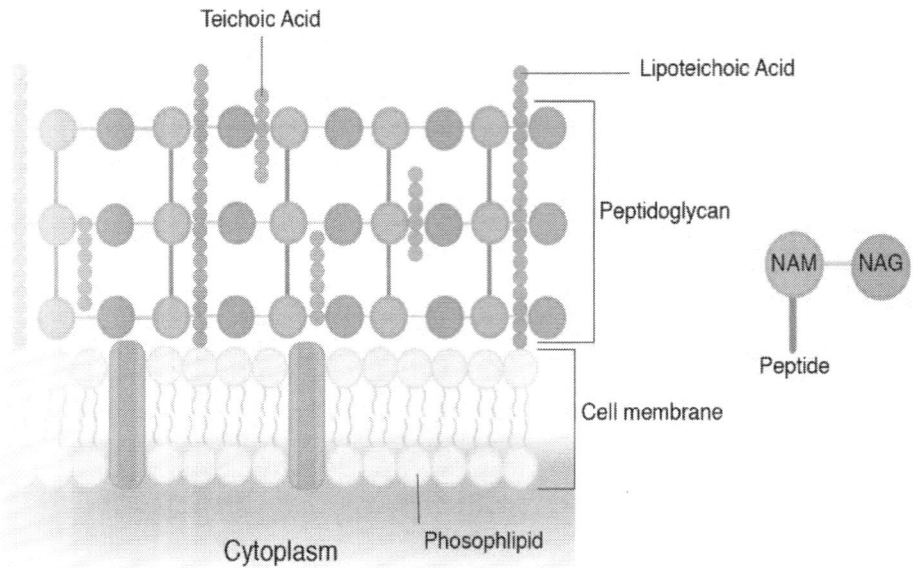

Fig.8.3

Since peptidoglycan is relatively porous, most substances can pass through the gram positive cell wall with little difficulty. But some nutrients are too large, requiring the cell to rely on the use of **exoenzymes**. These extracellular enzymes are made within the cell's cytoplasm and then secreted past the cell membrane, through the cell wall, where they function outside of the cell to break down large macromolecules into smaller components.

Gram Negative Cell Walls

The cell walls of gram negative bacteria are more complex than that of gram positive bacteria, with more ingredients overall. They do contain peptidoglycan as well, although only a couple of layers, representing 5-10% of the total cell wall. What is most notable about the gram negative cell wall is the presence of a plasma membrane located outside of the peptidoglycan layers, known as the **outer membrane**. This makes up the bulk of the gram negative cell wall. The outer membrane is composed of a lipid bilayer, very similar in composition to the cell membrane with polar heads, fatty acid tails, and integral proteins. It differs from the cell membrane by the presence of large molecules known as **lipopolysaccharide (LPS)**, which are anchored into the outer membrane and project from the cell into the environment. LPS is made up of three different components: 1) the **O-antigen or O-polysaccharide**, which represents the outermost part of the structure, 2) the **core polysaccharide**, and 3) **lipid A**, which anchors the LPS into the outer membrane. LPS is known to serve many different functions for the cell, such as contributing to the net negative charge for the cell, helping to stabilize the outer membrane, and providing protection from certain chemical substances by physically blocking access to other parts of the cell wall. In addition, LPS plays a role in the host response to pathogenic gram negative bacteria. The O-antigen triggers an immune response in an infected host, causing the generation of antibodies specific to that part of LPS (think of *E. coli* **O**157). Lipid A acts as a toxin, specifically an **endotoxin**, causing general symptoms of illness such as fever and diarrhea. A large amount of lipid A released into the bloodstream can trigger endotoxic shock, a body-wide inflammatory response which can be life-threatening.

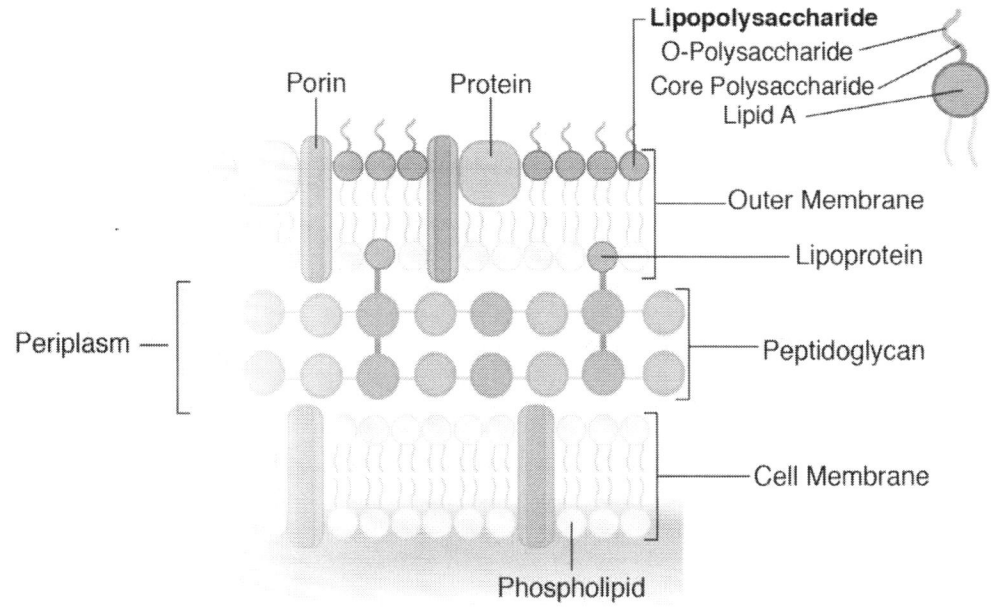

Gram Negative Bacteria Cell Wall

Fig.8.4

The outer membrane does present an obstacle for the cell. While there are certain molecules it would like to keep out, such as antibiotics and toxic chemicals, there are nutrients that it would like to let in and the additional lipid bilayer presents a formidable barrier. Large molecules are broken down by enzymes, in order to allow them to get past the LPS. Instead of exoenzymes (like the gram positive bacteria), the gram negative bacteria utilize **periplasmic enzymes** that are stored in the **periplasm**. Where is the periplasm, you ask? It is the space located between the outer surface of the cell membrane and the inner surface of the outer membrane, and it contains the gram negative peptidoglycan. Once the periplasmic enzymes have broken nutrients down to smaller molecules that can get past the LPS, they still need to be transported across the outer membrane, specifically the lipid bilayer. Gram negative cells utilize **porins**, which are transmembrane proteins composed of a trimer of three subunits, which form a pore across the membrane. Some porins are non-specific and transport any molecule that fits, while some porins are specific and only transport substances that they recognize by use of a binding site. Once across the outer membrane and in the periplasm, molecules work their way through the porous peptidoglycan layers before being transported by integral proteins across the cell membrane.

The peptidoglycan layers are linked to the outer membrane by the use of a lipoprotein known as **Braun's lipoprotein** (good ol' Dr. Braun). At one end the lipoprotein is covalently bound to the peptidoglycan while the other end is embedded into the outer membrane via its polar head. This linkage between the two layers provides additional structural integrity and strength.

Unusual and Wall-less Bacteria

Having emphasized the important of a cell wall and the ingredient peptidoglycan to both the gram positive and the gram negative bacteria, it does seem important to point out a few exceptions as well. Bacteria belonging to the phylum **Chlamydiae** appear to lack peptidoglycan, although their cell walls have a gram negative structure in all other regards (i.e. outer membrane, LPS, porin, etc). It has been suggested that they might be using a protein layer that functions in much the same way as peptidoglycan. This has an advantage to the cell in providing resistance to β-lactam antibiotics (such as penicillin), which attack peptidoglycan.

Bacteria belonging to the phylum **Tenericutes** lack a cell wall altogether, which makes them extremely susceptible to osmotic changes. They often strengthen their cell membrane somewhat by the addition of **sterols**, a substance usually associated with eukaryotic cell membranes. Many members of this phylum are pathogens, choosing to hide out within the protective environment of a host.

- **Cytoplasm** - The cytoplasm, or protoplasm, of bacterial cells is where the functions for cell growth, metabolism, and replication are carried out. It is a gel-like matrix composed of water, enzymes, nutrients, wastes, and gases and contains cell structures such as ribosomes, a chromosome, and plasmids. The cell envelope encases the cytoplasm and all its components. Unlike the eukaryotic (true) cells, bacteria do not have a membrane enclosed nucleus. The chromosome, a single, continuous strand of DNA, is localized, but not contained, in a region of the cell called the nucleoid. All the other cellular components are scattered throughout the cytoplasm.

One of those components, plasmid, are small, extrachromosomal genetic structures carried by many strains of bacteria. Like the chromosome, plasmids are made of a circular piece of DNA. Unlike the chromosome, they are not involved in reproduction. Only the chromosome has the genetic instructions for initiating and carrying out cell division, or binary fission, the primary means of reproduction in bacteria. Plasmids replicate independently of the chromosome and, while not essential for survival, appear to give bacteria a selective advantage.

Plasmids are passed on to other bacteria through two means. For most plasmid types, copies in the cytoplasm are passed on to daughter cells during binary fission. Other types of plasmids, however, form a tube like structure at the surface called a pilus that passes copies of the plasmid to other bacteria during conjugation, a process by which bacteria exchange genetic information. Plasmids have been shown to be instrumental in the transmission of special properties, such as antibiotic drug resistance, resistance to heavy metals, and virulence factors necessary for infection of animal or plant hosts. The ability to insert specific genes into plasmids have made them extremely useful tools in the fields of molecular biology and genetics, specifically in the area of genetic engineering.

- **Cytoplasmic Membrane** - A layer of phospholipids and proteins, called the cytoplasmic membrane, encloses the interior of the bacterium, regulating the flow of materials in and out of the cell. This is a structural trait bacteria share with all other living cells; a barrier that allows them to selectively interact with their environment. Membranes are highly organized and asymmetric having two sides, each side with a different surface and different functions. Membranes are also dynamic, constantly adapting to different conditions.

- **Flagella** - Flagella (singular, flagellum) are hair like structures that provide a means of locomotion for those bacteria that have them. They can be found at either or both ends of a bacterium or all over its surface. The flagella beat in a propeller-like motion to help the bacterium move toward nutrients; away from toxic chemicals; or, in the case of the photosynthetic cyanobacteria; toward the light.
- **Nucleoid** - The nucleoid is a region of cytoplasm where the chromosomal DNA is located. It is not a membrane bound nucleus, but simply an area of the cytoplasm where the strands of DNA are found. Most bacteria have a single, circular chromosome that is responsible for replication, although a few species do have two or more. Smaller circular auxiliary DNA strands, called plasmids, are also found in the cytoplasm.
- **Pili** - Many species of bacteria have pili (singular, pilus), small hairlike projections emerging from the outside cell surface. These outgrowths assist the bacteria in attaching to other cells and surfaces, such as teeth, intestines, and rocks. Without pili, many disease-causing bacteria lose their ability to infect because they're unable to attach to host tissue. Specialized pili are used for conjugation, during which two bacteria exchange fragments of plasmid DNA.
- **Ribosomes** - Ribosomes are microscopic "factories" found in all cells, including bacteria. They translate the genetic code from the molecular language of nucleic acid to that of amino acids—the building blocks of proteins. Proteins are the molecules that perform all the functions of cells and living organisms. Bacterial ribosomes are similar to those of eukaryotes, but are smaller and have a slightly different composition and molecular structure. Bacterial ribosomes are never bound to other organelles as they sometimes are (bound to the endoplasmic reticulum) in eukaryotes, but are free-standing structures distributed throughout the cytoplasm. There are sufficient differences between bacterial ribosomes and eukaryotic ribosomes that some antibiotics will inhibit the functioning of bacterial ribosomes, but not a eukaryote's, thus killing bacteria but not the eukaryotic organisms they are infecting.

Reproduction in Bacteria

1. Binary Fission:

In binary fission, single cell divides into two equal cells (Fig.8.5). Initially the bacterial cell reaches a critical mass in its structure and cellular constituents.

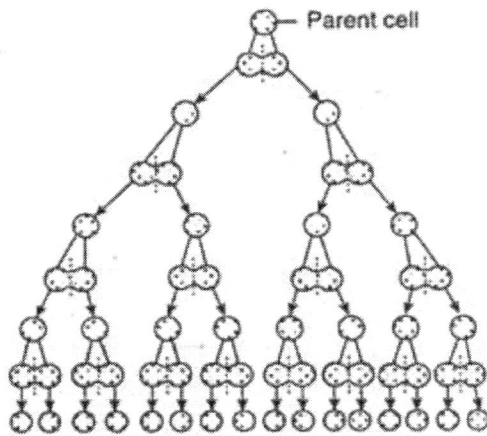

Fig. 8.5. Binary Fission

The circular double stranded DNA of bacteria undergoes replication, where both the strands separate and new complementary strands are formed on the original strands — results in the formation of two identical double stranded DNA (Fig.8.6).

The new double stranded DNA molecule i.e., incipient nuclei, are then distributed into two poles of the dividing cell (no spindle formation takes place like mitotic division). A transverse septum develops in the middle region of the cell, which separates the two daughter cells.

The binary fission is a rapid process and cell undergoes division at an interval of 20-30 minutes. The division becomes gradually slow after certain time due to accumulation of toxic substance and exhaustion of nutrients.

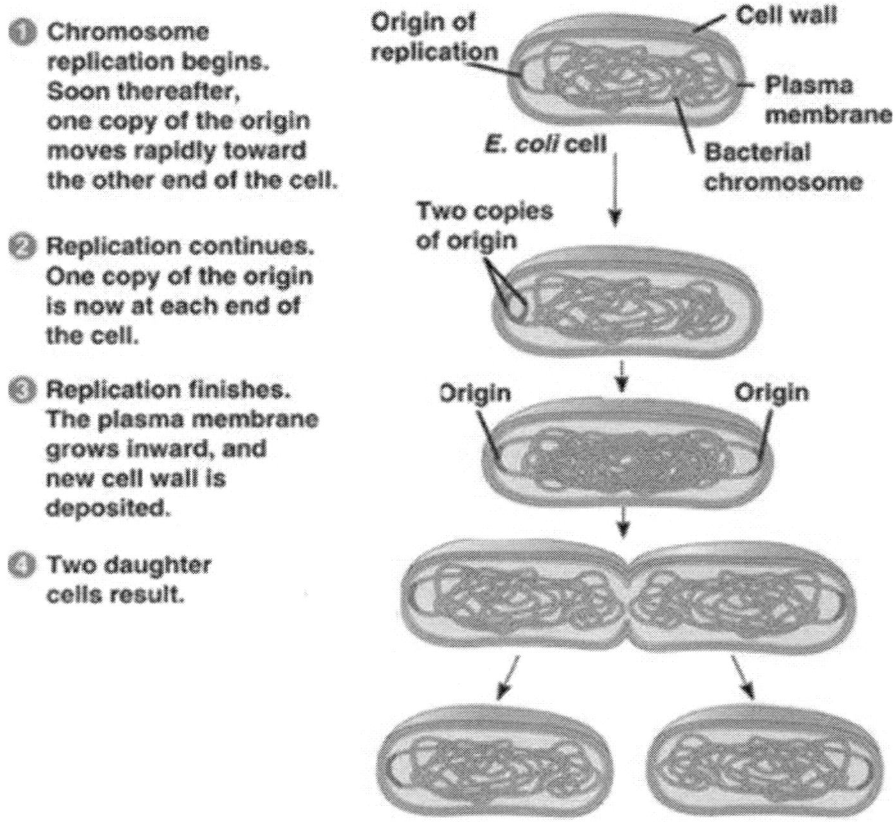
1 Chromosome replication begins. Soon thereafter, one copy of the origin moves rapidly toward the other end of the cell.
2 Replication continues. One copy of the origin is now at each end of the cell.
3 Replication finishes. The plasma membrane grows inward, and new cell wall is deposited.
4 Two daughter cells result.
Origin of replication
Cell wall
Plasma membrane
E. coli cell
Bacterial chromosome
Two copies of origin
Origin
Origin

Fig.8.6. Binary fission of a bacterium

2. Conidia:

Conidia formation takes place in filamentous bacteria like Streptomyces etc., by the formation of a transverse septum at the apex of the filament (Fig. **8.7.**). The part of this filament which bears conidia is called conidiophore. After detachment from the mother and getting contact with suitable substratum, the conidium germinates and gives rise to new mycelium.

3. By endospore. Endospores are formed in Bacillus (rod-shaped bacteria), Clostridium and Sporosarcina. In these forms, some amount of protoplast containing genophore shrinks, becomes spherical and develops a thick cell wall. It is called endospore. Endospore develops singly, and thus, the mother cell is known as sporangium. Endospore wall is three layered and strong. It is the most resistant stage of bacteria in which adverse environment has no effect. During favourable condition, endospore comes out by breaking the sporangial wall and forms a new bacterial cell.

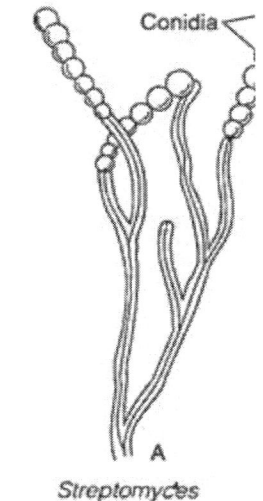

Fig.8.7. **Conidia formation**

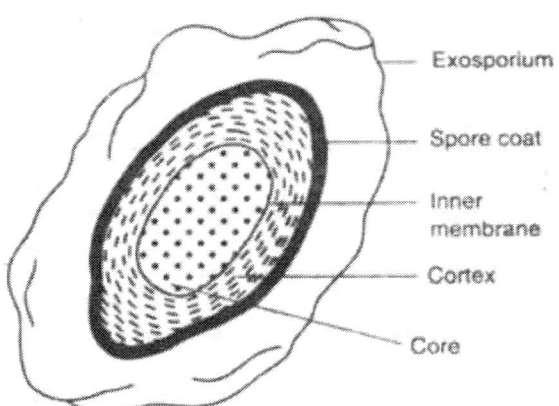

Fig.8.8. **Bacterial endosperm**

Sexual Reproduction

Typical sexual reproduction involving the formation and fusion of gametes is absent in bacteria. However gene recombination can occur in bacteria by three different methods they are

1. Conjugation

2. Transformation

3. Transduction

1. Conjugation

J. Lederberg and Edward L. Tatum demon-strated conjugation in *E. coli.* in the year 1946. In this method of gene transfer the donor cell gets attached to the recipient cell with the help of pili. The pilus grows in size and forms the conjugation tube. The plas-mid of donor cell which has the F+ (fer-tility factor) undergoes replication. Only one strand of DNA is transferred to the re-cipient cell through conjugation tube. The recipient completes the structure of double
stranded DNA by synthesizing the strand that complements the strand acquired from the donor (Figure 6.12.).

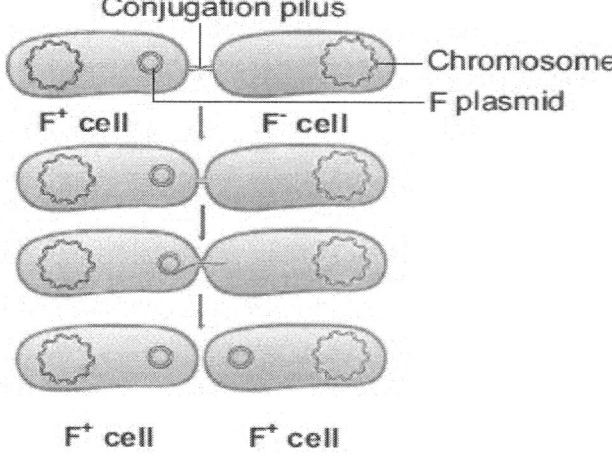

Fig.8.9. Conjugation

2. Transformation

Transfer of DNA from one bacterium to another is called transformation (Figure 6.13.). In 1928 the bacteriologist **Frederick Griffith** demonstrated transformation in Mice using ***Diplococcus pneumoniae***. Two strains of this bacterium are present. One strain produces smooth colonies and are virulent in nature (S-type). In addition another strain produce rough colonies and are avirulent (R-type). When S-type of cells were injected into the mouse, the mouse died. When R-type of cells were injected, the mouse survived.

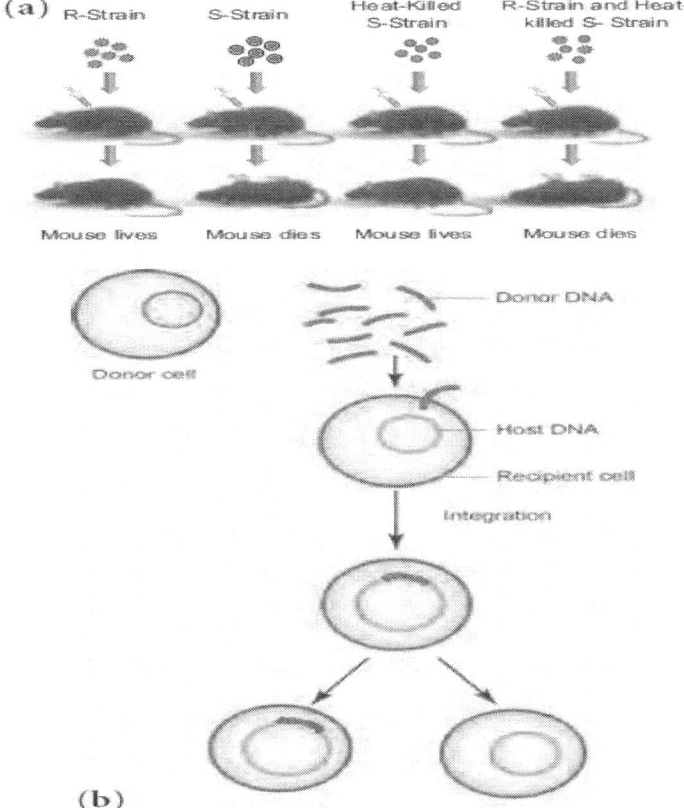

Fig.8.10. Transformation
    a.   **Griffith of experiment transformation**
    b.   **Mechanism of Transformation**

He injected heat killed S-type cells into the mouse the mouse did not die. When the mixture of heat killed S-type cells and R-type cells were injected into the mouse. The mouse died. The avirulent rough strain of ***Diplococcus*** had been transformed into S-type cells. The hereditary material of heat killed S-type cells had transformed R-type cell into virulent smooth strains. Thus the phenomenon of changing the character of one strain by transferring the DNA of another strain into the former is called Transformation.

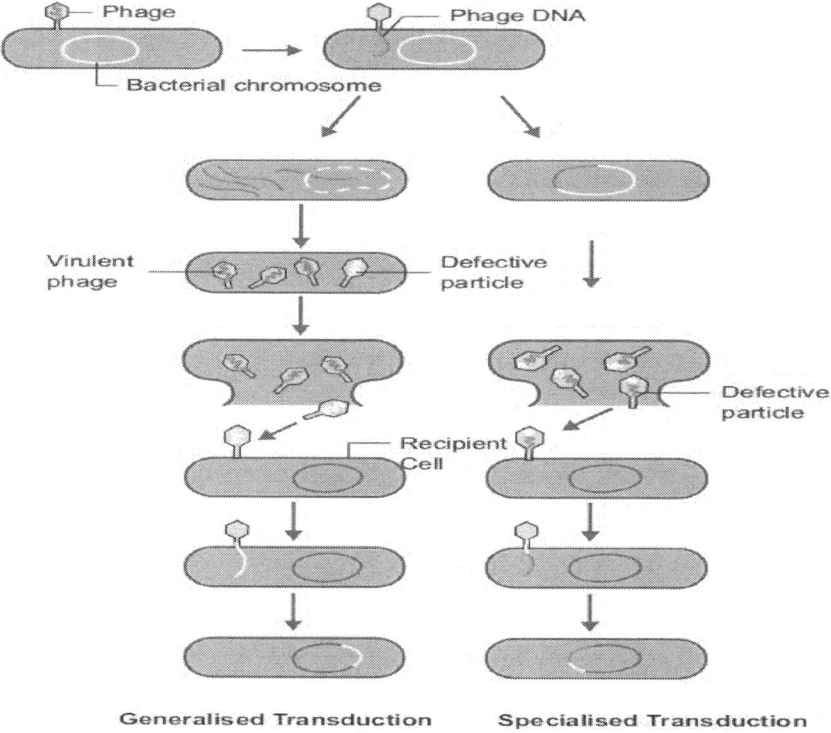

Fig.8.11. Transduction

3. Transduction

Zinder and Lederberg (1952) discovered Transduction in ***Salmonella typhimurum***. Phage mediated DNA transfer is called Transduction (Figure 8.11.).

Transduction is of two types

(i) Generalized Transduction (ii) Specialized or Restricted Transduction

(i) Generalized Transduction

The ability of a bacteriophage to carry genetic material of any region of bacterial DNA is called generalized transduction.

(ii) Specialized or Restricted Transduction

The ability of the bacteriophage to carry only a specific region of the bacterial DNA is called specialized or restricted transduction.

****

9.
BACTERIAL GENETICS

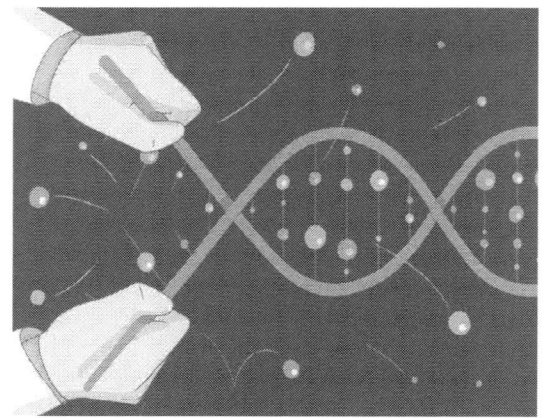

Genetics is the study of genes including the structure of genetic materials, what information is stored in the genes, how the genes are expressed and how the genetic information is transferred. Genetics is also the study of heredity and variation. The arrangement of genes within organisms is its genotype and the physical characteristics an organism based on its genotype and the interaction with its environment, make up its phenotype. The order of DNA bases constitutes the bacterium's genotype. A particular organism may possess alternate forms of some genes. Such alternate forms of genes are referred to as alleles. The cell's genome is stored in chromosomes, which are chains of double stranded DNA. Genes are sequences of nucleotides within DNA that code for functional proteins. The genetic material of bacteria and plasmids is DNA. The two essential functions of genetic material are replication and expression.

Structure of DNA the DNA molecule is composed of two chains of nucleotides wound around each other in the form of "double helix". Double-stranded DNA is helical, and the two strands in the helix are antiparallel. The backbone of each strand comprises of repeating units of deoxyribose and phosphate residue. Attached to the deoxyribose is purine (AG) or pyrimidine (CT) base. Nucleic acids are large polymers consisting of repeating nucleotide units. Each nucleotide contains one phosphate group, one deoxyribose sugar, and one purine or pyrimidine base. In DNA the sugar is deoxyribose; in RNA the sugar is ribose. The double helix is stabilized by hydrogen bonds between purine and pyrimidine bases on the opposite strands. A on one strand pairs by two hydrogen bonds with T on the opposite strand, or G pairs by three hydrogen bonds with C. The two strands of double-helical DNA are, therefore complementary. Because of complementarity, double-stranded DNA contains equimolar amounts of purines (A + G) and pyrimidines (T + C), with A equal to T and

G equal to C, but the mole fraction of G + C in DNA varies widely among different bacteria. One of the differences between DNA and RNA is that RNA contains uracil instead of the base thymine. Structure of chromosome In contrast to the linear chromosomes found in eukaryotic cells, most bacteria have single, covalently closed, circular chromosomes. Not all bacteria have a single circular chromosome: some bacteria have multiple circular chromosomes, and many bacteria have linear chromosomes and linear plasmids. Multiple chromosomes have also been found in many other bacteria, including Brucella, Leptospira interrogans, Burkholderia and Vibrio cholerae. Borrelia and Streptomyces have linear chromosomes and most strains contain both linear and circular plasmids. The chromosome of E coli has a length of approximately 1.35 mm, several hundred times longer than the bacterial cell, but the circular DNA is then looped and supercoiled to allow the chromosome to fit into the small space inside the cell. Codon A set of three base pairs constitutes a codon, which codes for a single amino acid. The "triplet code" is said to be degenerate or redundant because more than codon may exist for the same amino acid. For example, the codons AGA, AGG, CGU, CGC, CGA and CGG all code for arginine. There are 64 codons, of which 3 (UAA, UAG and UGA) are nonsense codons. They don't code for any amino acid, but act as stop codons. There are specific codons which code for start and stop sequences. The start codon (AUG) indicates the beginning of the sequence to be translated, and the stop codons (UAA, UGA, UAG) terminate the protein synthesis. With the exception of methionine, all amino acids are coded for by more than one codon. The DNA in a gene that are expressed into the protein product are called exons and the non-coding DNA segments are called introns. There are no introns in bacterial chromosome. A segment of DNA carrying codons specifying a particular polypeptide is called a cistron or a gene.

Flow of genetic information

The central dogma of molecular biology is that DNA carries all genetic information. The flow of genetic information includes the replication of DNA to make more DNA, the transcription of the DNA into mRNA and the translation of mRNA into proteins. Replication of DNA first involves the separation of the two strands of DNA followed by synthesis of new identical DNA strand by enzymes called DNA polymerases. The RNA strand is synthesized by enzymes called RNA polymerases. The RNA sequence will be complementary to the DNA sequence. The mRNA strands are then guided to the ribosomes for protein translation. Amino acid residues are brought to the mRNA strand on the ribosomes by transfer RNA (tRNA).

GENETIC ORGANIZATION

Very less non coding region are present in bacterial genome. Characteristic feature of bacterial genome is operon. Operon concept does not hold for Archaean, all genes are not biochemically related. An operon consists of a Structural gene encoding protein or RNA and Regulatory gene encoding for proteins regulating gene expression. Cis acting regulate only of the gene or genes physically connected to it (Eg: operator, promoter, and terminator). Trans acting regulate other genes also (Eg: transcription factors). Control may be negative control and positive control. Negative control suppress protein expression and positive control induce protein expression. Regulatory proteins are trans acting factors that recognize cis acting elements upstream of the gene.

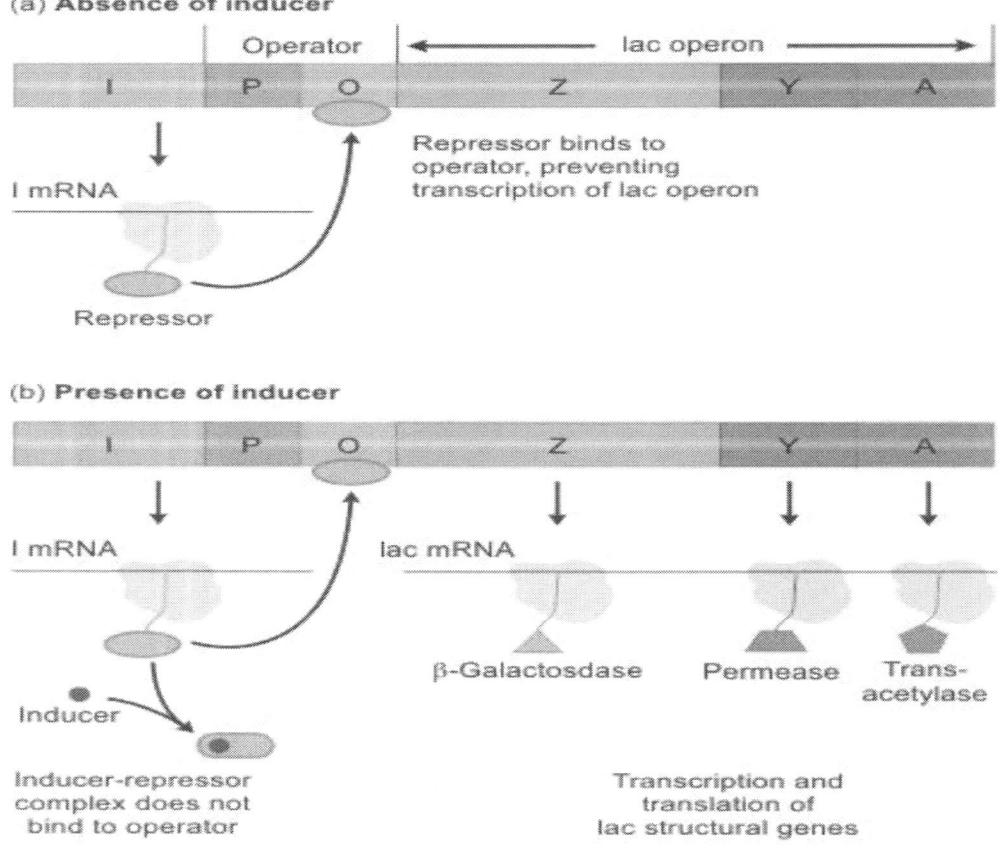

Fig.9: Lac operon model

Lac I – repressor (monomer), forms tetramer binds to operator of lac, blocks transcription, lactose binds to repressor, releases transcription. Lac Z – β-galactosidase Lac Y – Permease Lac A – transacetylase

Transposable elements are DNA molecules that move from one site to another. Movement of DNA elements from one site to another is called transposition. The mobile segments transposons, usually contains easily recognizable genes example, antibiotic resistance. They are designated as Tn followed by a number. Sometimes indicate the phenotype Tn 1 (AmpR). First discovered transposons do not have any phenotypic character like insertion.

elements Insertion sequences are present in the bacterial genome. This is proved by the following experiment. A gene which were highly polar mutated was selected for the study. Mapped first gene of an operon where mutation was seen and downstream proteins were not synthesized. No reversion possible by base analog /frameshift. If plasmid was inserted similar mutation observed and need not be the same site. Mutated gene was larger than original. Hence concluded that a segment of DNA had been inserted.

Termini of each IS elements have inverted repeat sequences of 10-40bp. Insertion can be either left to right/right to left. Different IS elements have different number of bases. This IS elements contain atleast two apparent coding sequences initiated by an AUG and terminated with an in phase stop codon. There are two types of transposons:

- Composite transposons – carrying anterior flanked by two identical / nearly identical copies of an IS element.
- Tn 3 transposon family – three genes
- Transposable phages – Mu and D108

Mutations

Bacteria grow and multiply fast and can reach large numbers. When bacteria multiply, one cell divides into two cells. Before the bacterium can divide, it needs to make two identical copies of the DNA in its chromosome; one for each cell. Every time the bacterium goes through this process there is a chance (or risk, depending on the end result) that errors occur; so-called mutations. These mutations are random and can be located anywhere in the DNA. Mutations can also form due to external factors like radiation or harmful chemicals.

Natural selection

While some mutations are harmful to the bacteria, others can provide an advantage given the right circumstances. Here, Darwin's theory of natural selection comes in. If a mutation gives the bacterium an advantage in a particular environment, this bacterium will grow better than its neighbors and can increase in numbers – it is selected for.

Mutations can provide resistance to antibiotics

Mutations are one way for bacteria to become resistant to antibiotics. Some spontaneous mutations (or genes that have been acquired from other bacteria through horizontal gene transfer) may make the bacterium resistant to an antibiotic (See: Resistance mechanisms for information about how bacteria resist antibiotic action). If we were to treat the bacterial population with that specific antibiotic, only the resistant bacteria will be able to multiply; the antibiotic selects for them. These bacteria can now increase in numbers and the end result is a population of mainly resistant bacteria.

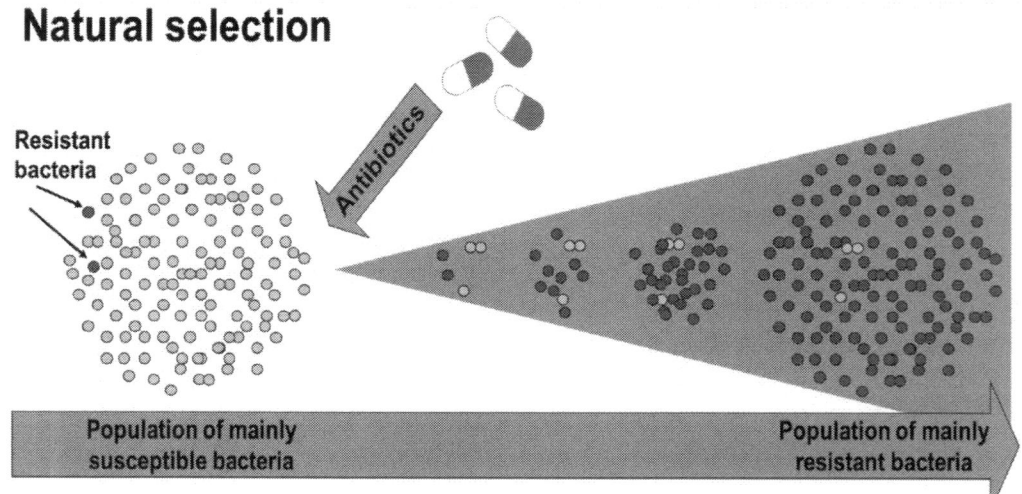

Fig. 9.1. Natural selection of antibiotic resistant bacteria. The starting point in this example is a large bacterial population mainly consisting of bacteria that are susceptible to antibiotics and a couple of bacteria that are antibiotic-resistant by chance. A bactericidal antibiotic is added, which kills most of the susceptible bacteria in the population, while the resistant bacteria survives. Only the resistant bacteria will continue to proliferate in the presence of the antibiotic and increases in number over time. The end result is a population of mainly resistant bacteria.

It is important to understand that selection of antibiotic resistant bacteria can occur anywhere an antibiotic is present at a selective concentration. When we treat an infection, selection can occur at any site in the body to which the antibiotic reaches. Thus, the antibiotic can select for resistance genes and mechanisms in both pathogenic bacteria and in commensal bacteria living in the body that have nothing to do with the infection in question. By using narrow-spectrum antibiotics (when possible), the risk of selecting for antibiotic resistance in the commensal flora decreases.

Types of Mutations

There are three types of DNA Mutations: base substitutions, deletions and insertions.

1. Base Substitutions

Single base **substitutions** are called point mutations, recall the point mutation Glu -----> Val which causes sickle-cell disease. Point mutations are the most common type of mutation and there are two types.

Transition: this occurs when a purine is substituted with another purine or when a pyrimidine is substituted with another pyrimidine.

Transversion: when a purine is substituted for a pyrimidine or a pyrimidine replaces a purine.

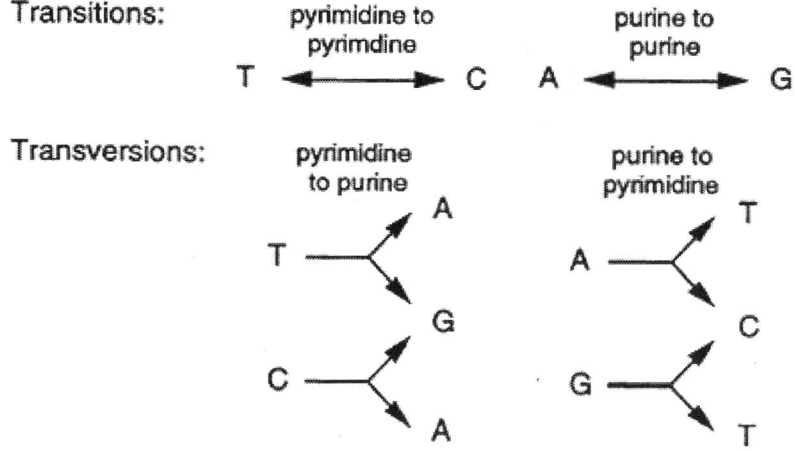

Point mutations that occur in DNA sequences encoding proteins are either silent, missense or nonsense.

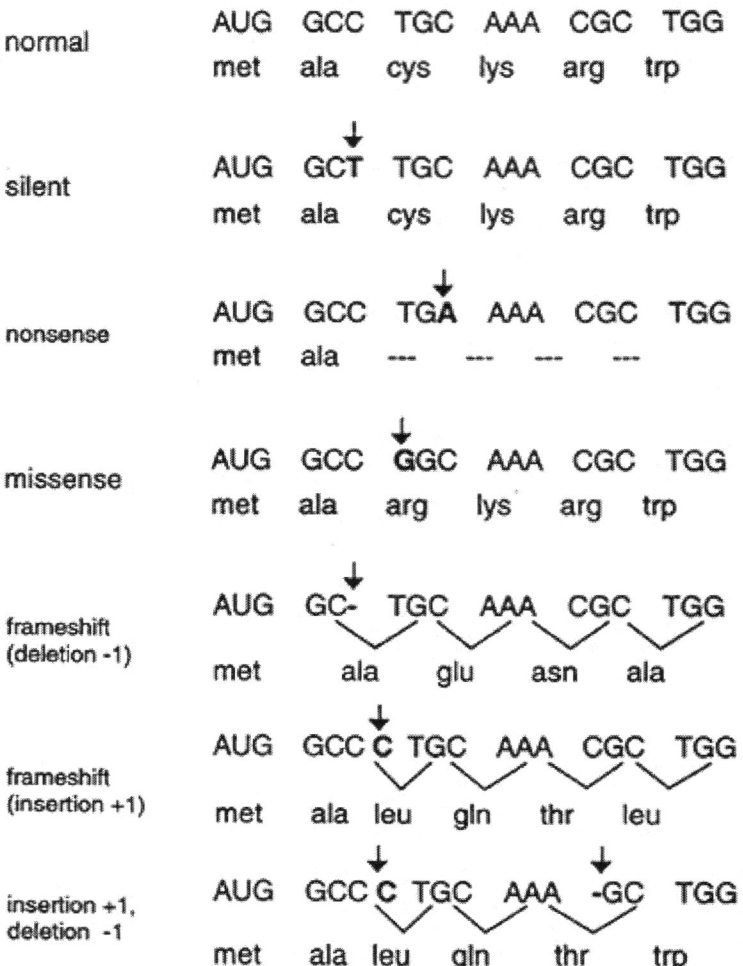

Silent: If a base **substitution** occurs in the third position of the codon there is a good chance that a synonymous codon will be generated. Thus the amino acid sequence encoded by the gene is not changed and the mutation is said to be silent.

Missence: When base **substitution** results in the generation of a codon that specifies a different amino acid and hence leads to a different polypeptide sequence. Depending on the type of amino acid substitution the missense mutation is either conservative or nonconservative. For example if the structure and properties of the substituted amino acid are very similar to the original amino acid the mutation is said to be conservative and will most likely have little effect on the resultant proteins structure / function. If the substitution leads to an amino acid with very different structure and properties the mutation is nonconservative and will probably be deleterious (bad) for the resultant proteins structure / function (i.e. the sickle cell point mutation).

Nonsense: When a base **substitution** results in a stop codon ultimately truncating translation and most likely leading to a nonfunctional protein.

2. Deletions

A deletion, resulting in a frameshift, results when one or more base pairs are lost from the DNA (see Figure above). If one or two bases are deleted the translational frame is altered resulting in a garbled message and nonfunctional product. A deletion of three or more bases leave the reading frame intact. A deletion of one or more codons results in a protein missing one or more amino acids. This may be deleterious or not.

3. Insertions

The insertion of additional base pairs may lead to frameshifts depending on whether or not multiples of three base pairs are inserted. Combinations of insertions and deletions leading to a variety of outcomes are also possible.

Causes of Mutations

Errors in DNA Replication

On very, very rare occasions DNA polymerase will incorporate a non-complementary base into the daughter strand. During the next round of replication the missincorporated base would lead to a mutation. This, however, is very rare as the exonuclease functions as a proofreading mechanism recognizing mismatched base pairs and excising them.

Errors in DNA Recombination

DNA often rearranges itself by a process called recombination which proceeds via a variety of mechanisms. Occasionally DNA is lost during replication leading to a mutation.

Chemical Damage to DNA

Many chemical mutagens, some exogenous, some man-made, some environmental, are capable of damaging DNA. Many chemotherapeutic drugs and intercalating agent drugs function by damaging DNA.

Radiation

Gamma rays, X-rays, even UV light can interact with compounds in the cell generating free radicals which cause chemical damage to DNA

Transfer of genetic material

Sometimes when two pieces of DNA come into contact with each other, sections of each DNA strand will be exchanged. This is usually done through a process called crossing over in which the DNA breaks and is attached on the other DNA strand leading to the transfer of genes and possibly the formation of new genes. Genetic recombination is the transfer of DNA from one organism to another. The transferred donor DNA may then be integrated into the recipient's nucleoid by various mechanisms. In the case of homologous recombination, homologous DNA sequences having nearly the same nucleotide sequences are exchanged by means of breakage and reunion of paired DNA segments. Genetic information can be transferred from organism to organism through vertical transfer (from a parent to offspring) or through horizontal transfer methods such as conjugation, transformation or transduction. Bacterial genes are usually transferred to members of the same species but occasionally transfer to other species can also occur.

Plasmids:

Plasmids are extra chromosomal elements found inside a bacterium. These are not essential for the survival of the bacterium but they confer certain extra advantages to the cell.

Number and size: A bacterium can have no plasmids at all or have many plasmids (20-30) or multiple copies of a plasmid. Usually they are closed circular molecules; however they occur as linear molecule in *Borrelia burgdorferi*. Their size can vary from 1 Kb to 400 Kb.

Multiplication: Plasmids multiply independently of the chromosome and are inherited regularly by the daughter cells.

Types of plasmids: R factor, Col factor, RTF and F factor.

F factor: This is also known as fertility factor or sex factor. Most plasmids are unable to mediate their own transfer to other cells. Vertical (inheritance) or horizontal (transfer) transmissions maintain plasmids. F factor is a plasmid that codes for sex pili and its transfer to other cells. Those bacteria that possess transfer factor are called F+, such bacteria have sex pili on their surface. Those cells lacking this factor are designated F-. The F factor plasmid transferred to other cells through conjugation. An F- cell will become F+ when it receives the fertility factor from another F+ cell.

R factor: Those plasmids that code for the transmissible drug resistance are called R factor. These plasmids contain genes that code for resistance to many antibiotics.

R factors may be transferred by conjugation and its transfer to other bacteria is independent of the F factor. Bacteria possessing such plasmids are resistant to many antibiotics and this drug resistance is transferred to closely related species. R factors may simultaneously confer resistance to five antibiotics. They are usually transferred to related species along with RTF.

Significance of plasmids:

1. Codes for resistance to several antibiotics. Gram-negative bacteria carry plasmids that give resistance to antibiotics such as neomycin, kanamycin, streptomycin, chloramphenicol, tetracycline, penicillins and sulfonamides.

2. Codes for the production of bacteriocines.

3. Codes for the production of toxins (such as Enterotoxins by Escherichia coli, Vibrio cholerae, exfoliative toxin by Staphylococcus aureus and neurotoxin of *Clostridium tetani*).

4. Codes for resistance to heavy metals (such as Hg, Ag, Cd, Pb etc.).

5. Plasmids carry virulence determinant genes. Eg, the plasmid Col V of *Escherichia coli* contains genes for iron sequestering compounds.

6. Codes resistance to uv light (DNA repair enzymes are coded in the plasmid).

7. Codes for colonization factors that is necessary for their attachment. Eg, as produced by the plasmids of *Yersinia enterocolitica, Shigella flexneri*, Enteroinvasive *Escherichia coli.*

8. Contains genes coding for enzymes that allow bacteria unique or unusual materials for carbon or energy sources. Some strains are used for clearing oil spillage.

Application of plasmids: 1. Used in genetic engineering as vectors.

2. Plasmid profiling is a useful genotyping method.

Episomes: Jacob and Wollman coined the term episome. Previously, it was considered synonymous with plasmids. F factors are those plasmids that can code for self-transfer to other bacteria. Occasionally such plasmids get spontaneously integrated into chromosome. Plasmids with this capability are called episomes and such bacterial cells are called Hfr cells i.e. high frequency of recombination.

Transposable genetic elements:

Transposable genetic elements are segments of DNA that have the capacity to move from one location to another (i.e. jumping genes). Properties of Transposable Genetic Elements:

1. Random movement: Transposable genetic elements can move from any DNA molecule to any DNA other molecule or even to another location on the same molecule. The movement is not totally random; there are preferred sites in a DNA molecule at which the transposable genetic element will insert.

2. Not capable of self-replication: The transposable genetic elements do not exist autonomously and thus, to be replicated they must be a part of some other replicon.

3. Transposition mediated by site-specific recombination: Transposition requires little or no homology between the current location and the new site. The transposition event is mediated by an enzyme transposase that is coded by the transposable genetic element. Recombination that does not require homology between the recombining molecules is called illegitimate or nonhomologous recombination.

4. Transposition can be accompanied by duplication: In many instances transposition of the transposable genetic element results in removal of the element from the original site and insertion at a new site. However, in some cases the transposition event is accompanied by the duplication of the transposable genetic element. One copy remains at the original site and the other is transposed to the new site.

Types of Transposable Genetic Elements:A.

Insertion sequences (IS): Insertion sequences are transposable genetic elements that carry no known genes except those that are required for transposition. Insertion sequences are small stretches of DNA that have at their ends repeated sequences, which are involved in transposition. In between the terminal repeated sequences there are genes involved in transposition and sequences that can control the expression of the genes but no other nonessential genes are present.

Importance of IS:

    i)   Mutation - The introduction of an insertion sequence into a bacterial gene will result in the inactivation of the gene.

ii) The sites at which plasmids insert into the bacterial chromosome are at or near insertion sequence in the chromosome.

iii) Phase Variation: In Salmonella there are two genes, which code for two antigenically different flagellar antigens. The expression of these genes is regulated by an insertion sequences.

B. Transposons: Transposons are transposable genetic elements that carry one or more other genes in addition to those, which are essential for transposition. The structure of a transposon is similar to that of an insertion sequence. The extra genes are located between the terminal repeated sequences.

Importance of transposons: Many antibiotic resistance genes are located on transposons. Since transposons can jump from one DNA molecule to another, these antibiotic resistance transposons are a major factor in the development of plasmids, which can confer multiple drug resistance on a bacterium harboring such a plasmid. These multiple drug resistance plasmids have become a major medical problem. (www.microrao.com)

****

10.

NUTRITION AND GROWTH OF BACTERIA

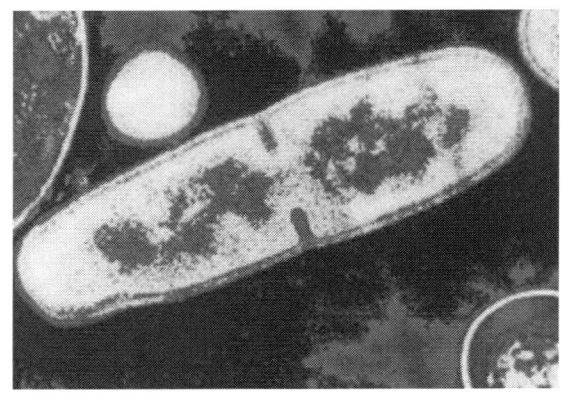

Nutritional Requirements of Cells

Every organism must find in its environment all of the substances required for energy generation and cellular biosynthesis. The chemicals and elements of this environment that are utilized for bacterial growth are referred to as **nutrients** or **nutritional requirements**. Many bacteria can be grown the laboratory in **culture media** which are designed to provide all the essential nutrients in solution for bacterial growth. Bacteria that are symbionts or obligate intracellular parasites of other cells, usually eucaryotic cells, are (not unexpectedly) difficult to grow outside of their natural host cells. Whether the microbe is a mutualist or parasite, the host cell must ultimately provide the nutritional requirements of its resident.

Many bacteria can be identified in the environment by inspection or using genetic techniques, but attempts to isolate and grow them in artificial culture has been unsuccessful. This, in part, is the basis of the estimate that we may know less than one percent of all procaryotes that exist.

The Major Elements

At an elementary level, the nutritional requirements of a bacterium such as *E. coli* are revealed by the cell's elemental composition, which consists of C, H, O, N, S. P, K, Mg, Fe, Ca, Mn, and traces of Zn, Co, Cu, and Mo. These elements are found in the form of water, inorganic ions, small molecules, and macromolecules which serve either a structural or functional role in the cells. The general physiological functions of the elements are outlined in Table 1.

Table 1. Major elements, their sources and functions in bacterial cells.

Element	% of dry weight	Source	Function
Carbon	50	organic compounds or CO_2	Main constituent of cellular material
Oxygen	20	H_2O, organic compounds, CO_2, and O_2	Constituent of cell material and cell water; O_2 is electron acceptor in aerobic respiration
Nitrogen	14	NH_3, NO_3, organic compounds, N_2	Constituent of amino acids, nucleic acids

			nucleotides, and coenzymes
Hydrogen	8	H$_2$O, organic compounds, H$_2$	Main constituent of organic compounds and cell water
Phosphorus	3	inorganic phosphates (PO$_4$)	Constituent of nucleic acids, nucleotides, phospholipids, LPS, teichoic acids
Sulfur	1	SO$_4$, H$_2$S, S^o, organic sulfur compounds	Constituent of cysteine, methionine, glutathione, several coenzymes
Potassium	1	Potassium salts	Main cellular inorganic cation and cofactor for certain enzymes
Magnesium	0.5	Magnesium salts	Inorganic cellular cation, cofactor for certain enzymatic reactions
Calcium	0.5	Calcium salts	Inorganic cellular cation, cofactor for certain enzymes and a

			component of endospores
Iron	0.2	Iron salts	Component of cytochromes and certain nonheme iron-proteins and a cofactor for some enzymatic reactions

Trace Elements

Table 1 ignores the occurrence of trace elements in bacterial nutrition. **Trace elements** are metal ions required by certain cells in such small amounts that it is difficult to detect (measure) them, and it is not necessary to add them to culture media as nutrients. Trace elements are required in such small amounts that they are present as "contaminants" of the water or other media components. As metal ions, the trace elements usually act as cofactors for essential enzymatic reactions in the cell. One organism's trace element may be another's required element and vice-versa, but the usual cations that qualify as trace elements in bacterial nutrition are Mn, Co, Zn, Cu, and Mo.

Carbon and Energy Sources for Bacterial Growth

In order to grow in nature or in the laboratory, a bacterium must have an energy source, a source of carbon and other required nutrients, and a permissive range of physical conditions such as O_2 concentration, temperature, and pH. Sometimes bacteria are referred to as individuals or groups based on their patterns of growth under various chemical (nutritional) or physical conditions. For example, phototrophs are organisms that use light as an energy source; anaerobes are organisms that grow without oxygen; thermophiles are organisms that grow at high temperatures.

All living organisms require a source of energy. Organisms that use radiant energy (light) are called **phototrophs**. Organisms that use (oxidize) an organic form of carbon are called **heterotrophs** or **(chemo)heterotrophs**. Organisms that oxidize inorganic compounds are called **lithotrophs**.

The carbon requirements of organisms must be met by organic carbon (a chemical compound with a carbon-hydrogen bond) or by CO_2. Organisms that use organic carbon are **heterotrophs** and organisms that use CO_2 as a sole source of carbon for growth are called **autotrophs**.

Thus, on the basis of carbon and energy sources for growth four major nutritional types of procaryotes may be defined (Table 2).

Table 2. Major nutritional types of procaryotes

Nutritional Type	Energy Source	Carbon Source	Examples
Photoautotrophs	Light	CO_2	Cyanobacteria, some Purple and Green Bacteria
Photoheterotrophs	Light	Organic compounds	Some Purple and Green Bacteria
Chemoautotrophs or Lithotrophs (Lithoautotrophs)	Inorganic compounds, e.g. H_2,	CO_2	A few Bacteria and many Archaea

		NH_3, NO_2, H_2S		
Chemoheterotrophs or Heterotrophs		Organic compounds	Organic compounds	Most Bacteria, some Archaea

Almost all eucaryotes are either photoautotrophic (e.g. plants and algae) or heterotrophic (e.g. animals, protozoa, fungi). Lithotrophy is unique to procaryotes and photoheterotrophy, common in the Purple and Green Bacteria, occurs only in a very few eucaryotic algae. Phototrophy has not been found in the Archaea, except for nonphotosynthetic light-driven ATP synthesis in the extreme halophiles.

Growth Factors

This simplified scheme for use of carbon, either organic carbon or CO_2, ignores the possibility that an organism, whether it is an autotroph or a heterotroph, may require small amounts of certain organic compounds for growth because they are essential substances that the organism is unable to synthesize from available nutrients. Such compounds are called **growth factors**.

Growth factors are required in small amounts by cells because they fulfill specific roles in biosynthesis. The need for a growth factor results from either a blocked or missing metabolic pathway in the cells. Growth factors are organized into three categories.

1. **Purines and pyrimidines**: required for synthesis of nucleic acids (DNA and RNA)

2. **Amino acids**: required for the synthesis of proteins

3. **Vitamins**: needed as coenzymes and functional groups of certain enzymes

Some bacteria (e.g. *E. coli*) do not require any growth factors: they can synthesize all essential purines, pyrimidines, amino acids and vitamins, starting with their carbon source, as part of their own intermediary metabolism. Certain other bacteria (e.g. *Lactobacillus*) require purines, pyrimidines, vitamins and several amino acids in order to grow. These compounds must be added in advance to culture media that are used to grow these bacteria. The growth factors are not metabolized directly as sources of carbon or energy, rather they are assimilated by cells to fulfill their specific role in metabolism. Mutant strains of bacteria that require some growth factor not needed by the wild type (parent) strain are referred to as **auxotrophs**.

Thus, a strain of *E. coli* that requires the amino acid tryptophan in order to grow would be called a tryptophan auxotroph and would be designated *E. coli* **trp-**

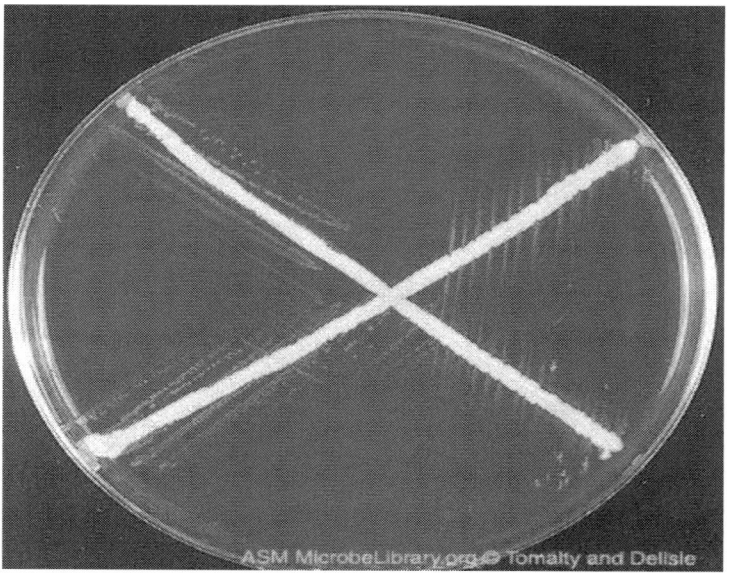

Figure 10. Cross-feeding between *Staphylococcus aureus* and *Haemophilus influenzae* growing on blood agar. © Gloria J. Delisle and Lewis Tomalty, Queens University, Kingston, Ontario, Canada. Licensed for use by ASM Microbe

Library http://www.microbelibrary.org. *Haemophilus influenzae* was first streaked on to the blood agar plate followed by a cross streak with *Staphylococcus aureus*. *H. influenzae* is a fastidious bacterium which requires both hemin and NAD for growth. There is sufficient hemin in blood for growth of *Haemophilus*, but the medium is insufficient in NAD. *S. aureus* produces NAD in excess of its own needs and secretes it into the medium, which supports the growth of *Haemophilus* as satellite colonies.

Some vitamins that are frequently required by certain bacteria as growth factors are listed in Table 3. The function(s) of these vitamins in essential enzymatic reactions gives a clue why, if the cell cannot make the vitamin, it must be provided exogenously in order for growth to occur.

Table 3. Common vitamins required in the nutrition of certain bacteria.

Vitamin	Coenzyme form	Function
p-Aminobenzoic acid (PABA)	-	Precursor for the biosynthesis of folic acid
Folic acid	Tetrahydrofolate	Transfer of one-carbon units and required for synthesis of thymine, purine bases, serine, methionine and pantothenate
Biotin	Biotin	Biosynthetic reactions that require CO_2 fixation
Lipoic acid	Lipoamide	Transfer of acyl groups in oxidation of keto acids
Mercaptoethane-sulfonic acid	Coenzyme M	CH_4 production by methanogens

Nicotinic acid	NAD (nicotinamide adenine dinucleotide) and NADP	Electron carrier in dehydrogenation reactions
Pantothenic acid	Coenzyme A and the Acyl Carrier Protein (ACP)	Oxidation of keto acids and acyl group carriers in metabolism
Pyridoxine (B_6)	Pyridoxal phosphate	Transamination, deamination, decarboxylation and racemation of amino acids
Riboflavin (B_2)	FMN (flavin mononucleotide) and FAD (flavin adenine dinucleotide)	Oxidoreduction reactions
Thiamine (B_1)	Thiamine pyrophosphate (TPP)	Decarboxylation of keto acids and transaminase reactions
Vitamin B_{12}	Cobalamine coupled to adenine nucleoside	Transfer of methyl groups
Vitamin K	Quinones and napthoquinones	Electron transport processes

Culture Media for the Growth of Bacteria

For any bacterium to be propagated for any purpose it is necessary to provide the appropriate biochemical and biophysical environment. The biochemical (nutritional) environment is made available as a **culture medium**, and depending upon the special needs of particular bacteria (as well as particular investigators) a large variety and types of culture media have been developed with different purposes and uses. Culture media are employed in the isolation and maintenance of pure cultures of bacteria and are also used for identification of bacteria according to their biochemical and physiological properties.

The manner in which bacteria are cultivated, and the purpose of culture media, varies widely. **Liquid media** are used for growth of pure batch cultures, while solidified media are used widely for the isolation of pure cultures, for estimating viable bacterial populations, and a variety of other purposes. The usual gelling agent for solid or **semisolid medium** is **agar**, a hydrocolloid derived from red algae. Agar is used because of its unique physical properties (it melts at 100°C and remains liquid until cooled to 40°C, the temperature at which it gels) and because it cannot be metabolized by most bacteria. Hence as a medium component it is relatively inert; it simply holds (gels) nutrients that are in aqueous solution.

Physical and Environmental Requirements for Microbial Growth

The procaryotes exist in nature under an enormous range of physical conditions such as O_2 concentration, Hydrogen ion concentration (pH) and temperature. The exclusion limits of life on the planet, with regard to environmental parameters, are always set by some microorganism, most often a procaryote, and frequently an Archaeon. Applied to all microorganisms is a vocabulary of terms used to describe their growth (ability to grow) within a range of physical conditions. A thermophile grows at high temperatures, an acidiphile grows at low pH, an osmophile grows at high solute concentration, and so on. This nomenclature will be employed in this section to describe the response of the procaryotes to a variety of physical conditions.

The Effect of Oxygen

Oxygen is a universal component of cells and is always provided in large amounts by H_2O. However, procaryotes display a wide range of responses to molecular oxygen O_2 (Table 4).

Obligate aerobes require O_2 for growth; they use O_2 as a final electron acceptor in aerobic respiration.

Obligate anaerobes (occasionally called **aerophobes**) do not need or use O_2 as a nutrient. In fact, O_2 is a toxic substance, which either kills or inhibits their growth. Obligate anaerobic procaryotes may live by fermentation, anaerobic respiration, bacterial photosynthesis, or the novel process of methanogenesis.

Facultative anaerobes (or **facultative aerobes**) are organisms that can switch between aerobic and anaerobic types of metabolism. Under anaerobic conditions (no O_2) they grow by fermentation or anaerobic respiration, but in the presence of O_2 they switch to aerobic respiration.

Aerotolerant anaerobes are bacteria with an exclusively anaerobic (fermentative) type of metabolism but they are insensitive to the presence of O_2. They live by fermentation alone whether or not O_2 is present in their environment

Table 4. Terms used to describe O_2 Relations of Microorganisms.

Group	Environmet		O_2 Effect
	Aerobic	**Anaerobic**	
Obligate Aerobe	Growth	No growth	Required (utilized for aerobic respiration)

Microaerophile	Growth if level not too high	No growth	Required but at levels below 0.2 atm
Obligate Anaerobe	No growth	Growth Toxic	
Facultative Anaerobe (Facultative Aerobe)	Growth	Growth	Not required for growth but utilized when available
Aerotolerant Anaerobe	Growth	Growth	Not required and not utilized

The response of an organism to O_2 in its environment depends upon the occurrence and distribution of various enzymes which react with O_2 and various oxygen radicals that are invariably generated by cells in the presence of O_2. All cells contain enzymes capable of reacting with O_2. For example, oxidations of flavoproteins by O_2 invariably result in the formation of H_2O_2 (peroxide) as one major product and small quantities of an even more toxic free radical, superoxide or O_2^-. Also, chlorophyll and other pigments in cells can react with O_2 in the presence of light and generate singlet oxygen, another radical form of oxygen which is a potent oxidizing agent in biological systems.

In aerobes and aerotolerant anaerobes the potential for lethal accumulation of superoxide is prevented by the enzyme superoxide dismutase (Figure 10). All organisms which can live in the presence of O_2 (whether or not they utilize it in their metabolism) contain superoxide dismutase. Nearly all organisms contain the enzyme catalase, which decomposes H_2O_2. Even though certain aerotolerant bacteria such as the lactic acid bacteria lack catalase, they decompose H_2O_2 by means of peroxidase enzymes which derive electrons from $NADH_2$ to reduce peroxide to H_2O. Obligate anaerobes lack superoxide dismutase and catalase and/or peroxidase, and therefore undergo lethal oxidations by various oxygen radicals when they are exposed to O_2. See Figure 10.1. below.

All photosynthetic (and some nonphotosynthetic) organisms are protected from lethal oxidations of singlet oxygen by their possession of carotenoid pigments which physically react with the singlet oxygen radical and lower it to its nontoxic "ground" (triplet) state. Carotenoids are said to "quench" singlet oxygen radicals.

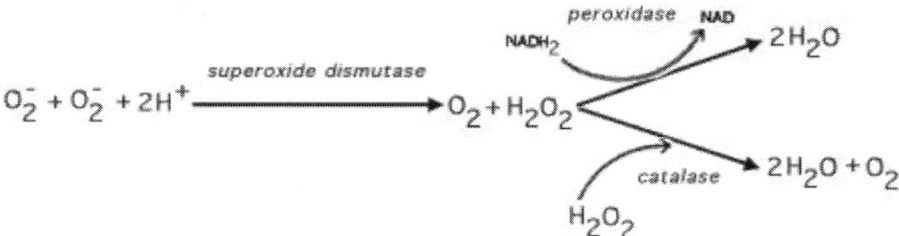

Figure.10.1. The action of superoxide dismutase, catalase and peroxidase. These enzymes detoxify oxygen radicals that are inevitably generated by living systems in the presence of O_2. The distribution of these enzymes in cells determines their ability to exist in the presence of O_2

Table 5. Distribution of superoxide dismutase, catalase and peroxidase in procaryotes with different O_2 tolerances.

Group	Superoxide dismutase	Catalase	Peroxidase
Obligate aerobes and most facultative anaerobes (e.g. Enterics)	+	+	-

Most aerotolerant anaerobes (e.g. Streptococci)	+	-	+
Obligate anaerobes (e.g. Clostridia, Methanogens, Bacteroides)	-	-	-

The Effect of pH on Growth

The pH, or hydrogen ion concentration, [H$^+$], of natural environments varies from about 0.5 in the most acidic soils to about 10.5 in the most alkaline lakes. Appreciating that pH is measured on a logarithmic scale, the [H$^+$] of natural environments varies over a billion-fold and some microorganisms are living at the extremes, as well as every point between the extremes! Most free-living procaryotes can grow over a range of 3 pH units, about a thousand fold change in [H$^+$]. The range of pH over which an organism grows is defined by **three cardinal points**: the **minimum pH**, below which the organism cannot grow, the **maximum pH**, above which the organism cannot grow, and the **optimum pH**, at which the organism grows best. For most bacteria there is an orderly increase in growth rate between the minimum and the optimum and a corresponding orderly decrease in growth rate between the optimum and the maximum pH, reflecting the general effect of changing [H$^+$] on the rates of enzymatic reaction (Figure 10.2.).

Microorganisms which grow at an optimum pH well below neutrality (7.0) are called **acidophiles**. Those which grow best at neutral pH are called **neutrophiles** and those that grow best under alkaline conditions are called **alkaliphiles**. Obligate acidophiles, such as some *Thiobacillus* species, actually require a low pH for growth since their membranes dissolve and the cells lyse at neutrality. Several genera of Archaea, including *Sulfolobus* and *Thermoplasma*, are obligate acidophiles. Among eukaryotes, many fungi are acidophiles, but the champion of growth at low pH is the eucaryotic alga *Cyanidium* which can grow at a pH of 0.

In the construction and use of culture media, one must always consider the optimum pH for growth of a desired organism and incorporate **buffers** in order to maintain the pH of the medium in the changing milieu of bacterial waste products that accumulate during growth. Many pathogenic bacteria exhibit a relatively narrow range of pH over which they will grow. Most diagnostic media for the growth and identification of human pathogens have a pH near 7.

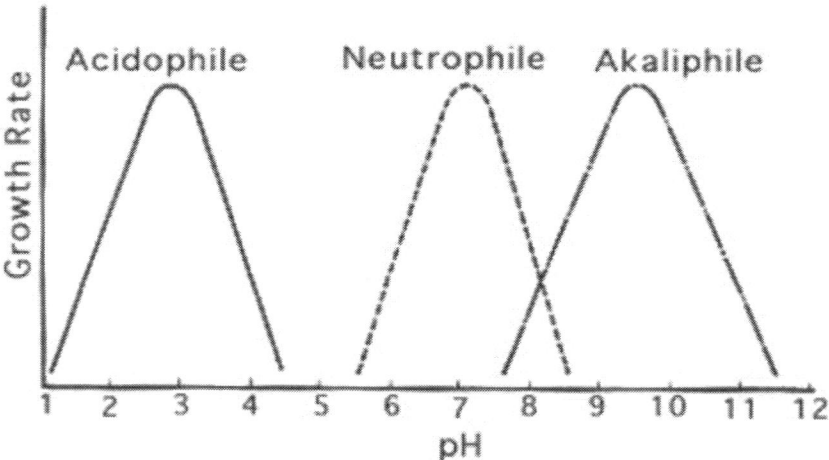

Figure 10.2. Growth rate vs pH for three environmental classes of procaryotes. Most free-living bacteria grow over a pH range of about three units. Note the symmetry of the curves below and above the optimum pH for growth.

Table 6. Minimum, maximum and optimum pH for growth of certain procaryotes.

Organism	Minimum pH	Optimum pH	Maximum pH
Thiobacillus thiooxidans	0.5	2.0-2.8	4.0-6.0
Sulfolobus acidocaldarius	1.0	2.0-3.0	5.0
Bacillus acidocaldarius	2.0	4.0	6.0
Zymomonas lindneri	3.5	5.5-6.0	7.5
Lactobacillus acidophilus	4.0-4.6	5.8-6.6	6.8
Staphylococcus aureus	4.2	7.0-7.5	9.3
Escherichia coli	4.4	6.0-7.0	9.0
Clostridium sporogenes	5.0-5.8	6.0-7.6	8.5-9.0
Erwinia caratovora	5.6	7.1	9.3
Pseudomonas aeruginosa	5.6	6.6-7.0	8.0
Thiobacillus novellus	5.7	7.0	9.0
Streptococcus pneumoniae	6.5	7.8	8.3
Nitrobacter sp	6.6	7.6-8.6	10.0

The Effect of Temperature on Growth

Microorganisms have been found growing in virtually all environments where there is liquid water, regardless of its temperature. In 1966, Professor Thomas D. Brock, then at Indiana University, made the amazing discovery in boiling hot springs of Yellowstone National Park that bacteria were not just surviving there, they were growing and flourishing. Brock's discovery of thermophilic bacteria, archaea and other "extremophiles" in Yellowstone is summarized for the general public in an article at this web site. See Life at High Temperatures.

Subsequently, procaryotes have been detected growing around black smokers and hydrothermal vents in the deep sea at temperatures at least as high as 120 degrees. Microorganisms have been found growing at very low temperatures as well. In supercooled solutions of H_2O as low as -20 degrees, certain organisms can extract water for growth, and many forms of life flourish in the icy waters of the Antarctic, as well as household refrigerators, near 0 degrees.

A particular microorganism will exhibit a range of temperature over which it can grow, defined by three cardinal points in the same manner as pH. Considering the total span of temperature where liquid water exists, the procaryotes may be subdivided into several subclasses on the basis of one or another of their cardinal points for growth. For example, organisms with an optimum temperature near 37 degrees (the body temperature of warm-blooded animals) are called **mesophiles**. Organisms with an optimum T between about 45 degrees and 70 degrees are **thermophiles**. Some Archaea with an optimum T of 80 degrees or higher and a maximum T as high as 115 degrees, are now referred to as **extreme thermophiles** or **hyperthermophiles**. The cold-loving organisms are **psychrophiles** defined by their ability to grow at 0 degrees. A variant of a psychrophile (which usually has an optimum T of 10-15 degrees) is a **psychrotroph**, which grows at 0 degrees but displays an optimum T in the mesophile range, nearer room temperature. Psychrotrophs are the scourge of food storage in refrigerators since they are invariably brought in from their mesophilic habitats and continue to grow in the refrigerated environment where they spoil the food. Of course, they grow slower at 2 degrees than at 25 degrees. Think how fast milk spoils on the counter top versus in the refrigerator.

Psychrophilic bacteria are adapted to their cool environment by having largely unsaturated fatty acids in their plasma membranes. Some psychrophiles, particularly those from the Antarctic have been found to contain polyunsaturated fatty acids, which generally do not occur in procaryotes. The degree of unsaturation of a fatty acid correlates with its solidification T or thermal transition stage (i.e., the temperature at which the lipid melts or solidifies); unsaturated fatty acids remain liquid at low T but are also denatured at moderate T; saturated fatty acids, as in the membranes of thermophilic bacteria, are stable at high temperatures, but they also solidify at relatively high T. Thus, saturated fatty acids (like butter) are solid at room temperature while unsaturated fatty acids (like safflower oil) remain liquid in the refrigerator. Whether fatty acids in a membrane are in a liquid or a solid phase affects the fluidity of the membrane, which directly affects its ability to function. Psychrophiles also have enzymes that continue to function, albeit at a reduced rate, at temperatures at or near 0 degrees. Usually, psychrophile proteins and/or membranes, which adapt them to low temperatures, do not function at the body temperatures of warm-blooded animals (37 degrees) so that they are unable to grow at even moderate temperatures.

Thermophiles are adapted to temperatures above 60 degrees in a variety of ways. Often thermophiles have a high G + C content in their DNA such that the melting point of the DNA (the temperature at which the strands of the double helix separate) is at least as high as the organism's maximum T for growth. But this is not always the case, and the correlation is far from perfect, so thermophile DNA must be stabilized in these cells by other means. The membrane fatty acids of thermophilic bacteria are highly saturated allowing their membranes to remain stable and functional at high temperatures. The membranes of hyperthermophiles, virtually all of which are Archaea, are not composed of fatty acids but of repeating subunits of the C5 compound, phytane, a branched, saturated, "isoprenoid" substance, which contributes heavily to the ability of these bacteria to live in superheated environments. The structural proteins (e.g. ribosomal proteins, transport proteins (permeases) and enzymes of thermophiles and hyperthermophiles are very heat stable compared with their mesophilic counterparts. The proteins are modified in a number of ways including dehydration and through slight changes in their primary structure, which accounts for their thermal stability.

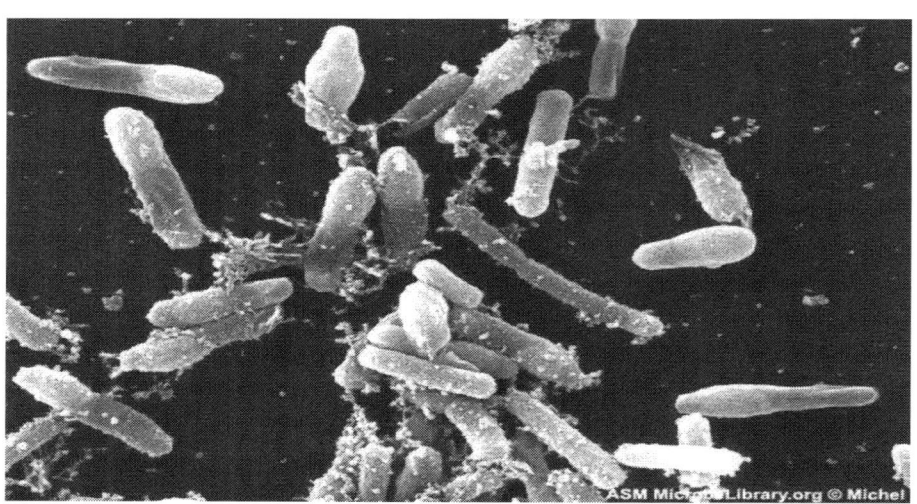

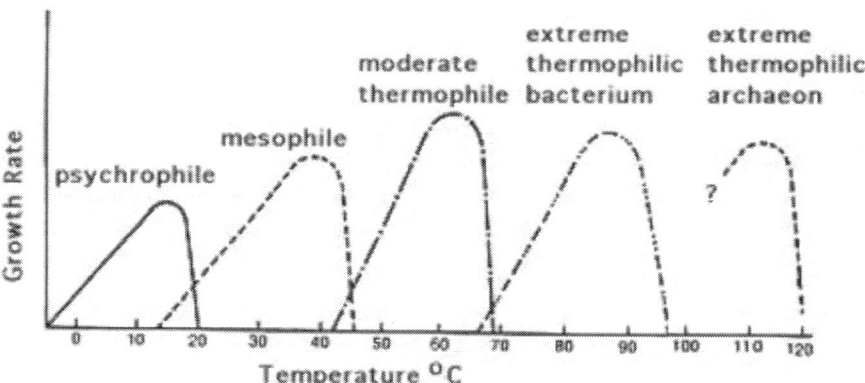

Figure 10.3. SEM of a thermophilic *Bacillus* species isolated from a compost pile at 55º C. © Frederick C. Michel. The Ohio State University -OARDC, Wooster, Ohio. Licensed for use by ASM Microbe Library http://www.microbelibrary.org. The rods are 3-5 microns in length and 0.5 to 1 micron in width with terminal endospores in a slightly-swollen sporangium.

Figure 10.4. (Below). Growth rate vs temperature for five environmental classes of procaryotes. Most procaryotes will grow over a temperature range of about 30 degrees. The curves exhibit three cardinal points: minimum, optimum and maximum temperatures for growth. There is a steady increase in growth rate between the minimum and optimum temperatures, but slightly past the optimum a critical thermolabile cellular event occurs, and the growth rates plunge rapidly as the maximum T is approached. As expected and as predicted by T.D. Brock, life on earth, with regard to temperature, exists wherever water remains in a liquid state. Thus, psychrophiles grow in solution wherever water is supercooled below 0 degrees; and extreme thermophilic archaea (hyperthermophiles) have been identified growing near deep-sea thermal vents at temperatures up to 120 degrees. Theoretically, the bar can be pushed to even higher temperatures.

Table 7. Terms used to describe microorganisms in relation to temperature requirements for growth.

Temperature for growth (degrees C)

Group	Minimum	Optimum	Maximum	Comments
Psychrophile	Below 0	10-15	Below 20	Grow best at relatively low T
Psychrotroph	0	15-30	Above 25	Able to grow at low T but prefer moderate T
Mesophile	10-15	30-40	Below 45	Most bacteria esp. those living in association

				with warm-blooded animals
Thermophile*	45	50-85	Above 100 (boiling)	Among all thermophiles is wide variation in optimum and maximum T

* For "degrees" of thermophily see text and graphs above

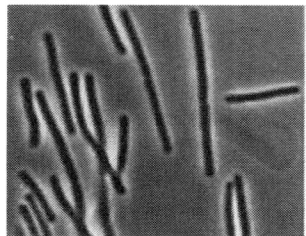 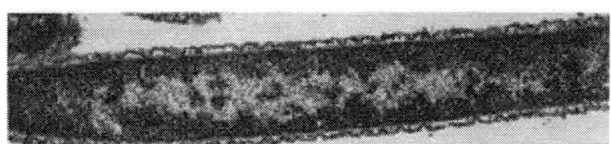

Figure 10.5. *Thermus aquaticus, the* thermophilic bacterium that is the source of taq polymerase. L wet mount; R electron micrograph. T.D. Brock. Life at High Temperatures.

Table 8a. Minimum, maximum and optimum temperature for growth of certain bacteria and archaea.

	Temperature for growth (degrees C)		
Bacterium	Minimum	Optimum	Maximum

Genus and species			
Listeria monocytogenes	1	30-37	45
Vibrio marinus	4	15	30
Pseudomonas maltophilia	4	35	41
Thiobacillus novellus	5	25-30	42
Staphylococcus aureus	10	30-37	45
Escherichia coli	10	37	45
Clostridium kluyveri	19	35	37
Streptococcus pyogenes	20	37	40
Streptococcus pneumoniae	25	37	42
Bacillus flavothermus	30	60	72
Thermus aquaticus	40	70-72	79
Methanococcus jannaschii	60	85	90
Sulfolobus acidocaldarius	70	75-85	90
Pyrobacterium brockii	80	102-105	115

Table 8b. Optimum growth temperature of some procaryotes.

Genus and species	Optimal growth temp (degrees C)
Vibrio cholerae	18-37
Photobacterium phosphoreum	20
Rhizobium leguminosarum	20
Streptomyces griseus	25
Rhodobacter sphaeroides	25-30
Pseudomonas fluorescens	25-30
Erwinia amylovora	27-30

Staphylococcus aureus	30-37
Escherichia coli	37
Mycobacterium tuberculosis	37
Pseudomonas aeruginosa	37
Streptococcus pyogenes	37
Treponema pallidum	37
Thermoplasma acidophilum	59
Thermus aquaticus	70
Bacillus caldolyticus	72
Pyrococcus furiosus	100

Table.8c.Hyperthermophilic Archaea.

Temperature for growth (degrees C)

Genus	Minimum	Optimum	Maximum	Optimum pH
Sulfolobus	55	75-85	87	2-3
Desulfurococcus	60	85	93	6
Methanothermus	60	83	88	6-7
Pyrodictium	82	105	113	6
Methanopyrus	85	100	110	7

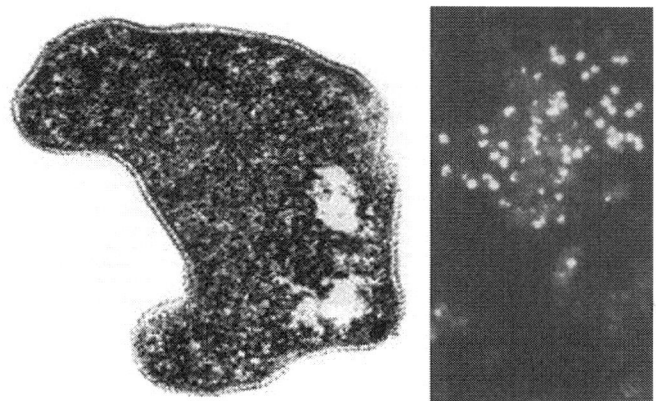

Figure .10.6. *Sulfolobus acidocaldarius* is an extreme thermophile and an acidophile found in geothermally-heated acid springs, mud pots and surface soils with temperatures from 60 to 95 degrees C, and a pH of 1 to 5. Left: Electron micrograph of a thin section (85,000X). Under the electron microscope the organism appears as irregular spheres which are often lobed. Right: Fluorescent photomicrograph of cells attached to a sulfur crystal. Fimbrial-like appendages have been observed on the cells attached to solid surfaces such as sulfur crystals. T.D. Brock. Life at High Temperatures.

Water Availability

Water is the solvent in which the molecules of life are dissolved, and the availability of water is therefore a critical factor that affects the growth of all cells. The availability of water for a cell depends upon its presence in the atmosphere (relative humidity) or its presence in solution or a substance (**water activity**). The water activity (A_w) of pure H_2O is 1.0 (100% water). Water activity is affected by the presence of solutes such as salts or sugars, that are dissolved in the water. The higher the solute concentration of a substance, the lower is the water activity and vice-versa. Microorganisms live over a range of A_w from 1.0 to 0.7. The A_w of human blood is 0.99; seawater = 0.98; maple syrup = 0.90; Great Salt Lake = 0.75. Water activities in agricultural soils range between 0.9 and 1.0.

The only common solute in nature that occurs over a wide concentration range is salt [NaCl], and some microorganisms are named based on their growth response to salt. Microorganisms that require some NaCl for growth are **halophiles**. **Mild halophiles** require 1-6% salt, **moderate halophiles** require 6-15% salt; **extreme halophiles** that require 15-30% NaCl for growth are found among the archaea. Bacteria that are able to grow at moderate salt concentrations, even though they grow best in the absence of NaCl, are called **halotolerant**. Although halophiles are "osmophiles" (and halotolerant organisms are "osmotolerant") the term **osmophiles** is usually reserved for organisms that are able to live in environments high in sugar. Organisms which live in dry environments (made dry by lack of water) are called **xerophiles**.

The concept of lowering water activity in order to prevent bacterial growth is the basis for preservation of foods by drying (in sunlight or by evaporation) or by addition of high concentrations of salt or sugar.

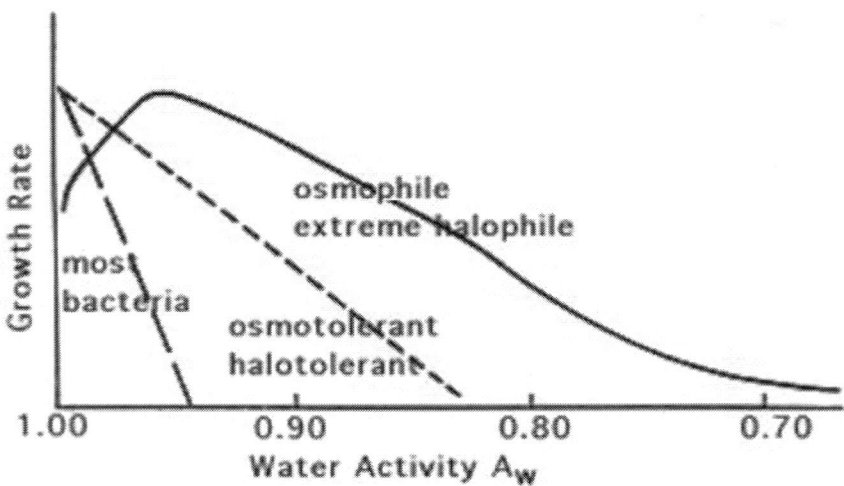

Figure 10.7.. Growth rate vs osmolarity for different classes of procaryotes. Osmolarity is determined by solute concentration in the environment. Osmolarity is inversely related to water activity (A_w), which is more like a measure of the concentration of water (H_2O) in a solution. Increased solute

concentration means increased osmolarity and decreased A_w. From left to right the graph shows the growth rate of a normal (nonhalophile) such as *E. coli* or *Pseudomonas,* the growth rate of a halotolerant bacterium such as *Staphylococcus aureus, and* the growth rate of an extreme halophile such as the archaean *Halococcus. Note* that a true halophile grows best at salt concentrations where most bacteria are inhibited.

Table 9. Limiting water activities (A_w) for growth of certain procaryotes.

Organism	Minimum A_w for growth
Caulobacter	1.00
Spirillum	1.00
Pseudomonas	.91
Salmonella/E. coli	.91
Lactobacillus	.90
Bacillus	.90
Staphylococcus	.85
Halococcus	.75

(www.textbookofbacteriology.net)

****

11. METHODS IN BACTERIOLOGY

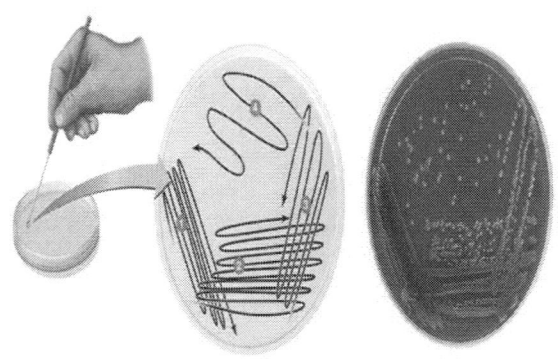

Meaning of Culture Medium:

A culture medium is a solid or liquid preparation of nutrient or combination of nutrients used to grow, transport, and store microorganisms. Much of the study of microbiology depends on the ability to grow and maintain microorganisms in the laboratory, and this is possible only if suitable culture media are available. To be effective, the medium must contain all the nutrients the microorganism requires for growth.

Although all microorganisms require energy sources, carbon, nitrogen, phosphorus, sulphur and various minerals, the precise composition of a satisfactory medium depends on the microbial species one is trying to grow because nutritional requirements vary so greatly. Knowledge about microorganism's natural habitat often is useful in selecting an appropriate culture medium because its nutritional requirements reflect its natural surroundings.

In preparing a culture medium for any microorganism, the primary goal is to provide a balanced mixture of the required nutrients at concentrations that will permit good growth. Additionally, the culturing of microorganisms requires careful control of various environmental factors which normally are maintained within narrow limits.

Microbiological culture media, however, consist of various nutrient substances supporting the growth of particular types of microorganisms. Some media contain solutions of inorganic salts and may be supplemented with one or more organic compounds.

Other media are prepared from complex ingredients such as extracts or digests of plant and animal tissues. Culture media would, thereinafter, be called 'media' (sing, medium).

Types of Culture Medium:

Following are some important types of culture medium:

1. Non-Synthetic Media (or Complex Media):

A medium in which the exact chemical composition of each of the constituents is not known with certainty is referred to as non-synthetic medium, undefined medium, or complex medium.

Potato-Dextrose-Agar (GM-25), Soil-Extract-Agar (SM-1), Oatmeal-Agar (GM-24), Malt-Extract-Agar (GM-19b), Waksmans medium (GM- 40) are some of the most widely used non-synthetic media. For convenience, the undefined chemical composition medium (complex medium) used to grow either Escherichia coli or Leuconostoc mesenteroides is as follows.

Glucose	**15g**
Yeast extract	**5g**
Peptone	**5g**
KH_2PO_4	**2g**
Distilled water	**1000 ml**
pH	**7**

Non-synthetic media often employ digests of caesin (milk protein), soybeans, beef, yeast cells, or any of a number of highly nutritious but chemically undefined substances. Such digests are available commercially in powdered form and can be weighed out rapidly and dissolved in distilled water to prepare a medium.

2. Synthetic Media (or Defined Media):

A medium in which only pure chemicals in definite concentrations are used is called synthetic medium or defined medium. On account of their known chemical compositions these media are useful for nutritional and metabolic studies. Czakek's Dox medium (GM-9) and Richard's solution (GM-27) are the most widely used synthetic media.

For example, a defined medium used for Escherichia coli is as follows:

K_2HPO_4 .... 7 g
KH_2PO_4 .... 2 g
$(NH_4)_2SO_4$ .... 1 g
$MgSO_4$ .... 0.1 g
$CaCl_2$ .... 0.02 g
Glucose .... 4-10 g
Trace elements (Fe, Co, Mn, Zn, Cu, Ni, Mo) .... 2-10 µm each
Distilled water .... 1000 ml
pH 7

3. Solid Media:

Solid media may either always remain solid (e.g., potato slices, coagulated blood serum, coagulated egg) or be liquefied (e.g., nutrient-agar medium, nutrient-gelatin medium, potato dextrose-agar medium). Liquefiable solid media are prepared by adding suitable amount of gelatin or agar to the liquid medium to remain solid when cooled but become liquid when warmed or vice-versa.

Agar is a complex polysaccharide (carbohydrate) consisting of 3, 6-anhydro-L-galactose and D-galactopyranose, free of nitrogen, produced from various red algae belonging to genera Gelidium, Gracilaria, Gigartina and Pterocladia. It liquefies on heating to 96°C and hardens into a jelly on cooling to 40-45°C.

The solidified medium kept in a Petri dish provides an artificial environment suitable for a rapid growth of fungi. While in liquefied state, solid media can be taken in test tubes, which are either allowed to cool and harden in a slanted position producing agar slants or allowed to harden in the upright position producing agar deep tubes.

4. Liquid Media:

Liquid media remain in liquid form and are called liquid broth (i.e., media lacking agar). Bacteria, in contrast to fungi, are often cultured in liquid broth. Nutrient broth, glucose broth, beef extract, skimmed milk, peptone solution are examples of liquid media.

The most commonly used liquid media in bacteriological laboratory are beef extract (a beef derivative which is a source of organic carbon, nitrogen, vitamins and inorganic salts) and peptone solution (a semidigested protein).

These may be modified in a variety of ways by adding some specific chemicals or materials to provide a medium suitable for cultivation or demonstration of a reaction for specific types or groups of bacteria.

5. Semi-Solid Media:

Semi-solid media contain a smaller amount (0.5% or less) of agar which imparts a "custard consistency".

Example:

Cystine trypticase-agar medium.

6. Special Media:

(i) Enrichment Media:

Enrichment culture is that in which the growth of a particular microorganism is favoured as against a mixed population by adjusting the nutritional requirements and environmental factors. The so grown microbial population in the medium is called enrichment culture.

(ii) Selective Media:

A selective medium is one which prevents or retards the growth of unwanted microorganisms while permits and promotes the growth of wanted microorganisms to form distinctive colonies.

The selective action of the medium is due to the addition of certain chemicals to the medium. For example, addition of crystal violet dye in the culture medium selectively inhibits the growth of gram-positive bacteria and permits and promotes the growth of gram-negative bacteria.

Important examples of selective media are MacConkey-Agar for E. coli, Deoxycholate-Citrate-Agar (DCA) for Samonella and Shigella, Wilson and Blair's medium for Salmonella, and Mannitol-Salt-Agar medium for pathogenic staphylococci.

Mannitol-Salt-Agar medium contains high concentration of sodium chloride which inhibits the growth of bacteria other than pathogenic staphylococci. The fermenting ability of the staphylococci colonies induced by a yellow halo can be detected by mannitol present in the medium. Mannitol is metabolized and fermented by staphylococci and the acid produced can be detected by phenol-red, a pH indicator.

The following is the composition for the preparation of Mannitol-Salt-Agar medium:

Mannitol	10.0 g
Beef extract	1.0 g
Peptone	10.0 g
Sodium chloride	75.0 g
Phenol red	0.025 g
Agar	15.0 g
Distilled water	100 ml

(iii) Transport Media:

Transport media have wide applicability in medical field. It is used specifically to maintain the pathogenic microorganisms during transportation to laboratory from hospital or from a distant area.

When a patient is far from the pathological laboratory, the delicate pathogenic microorganism (e.g., Neisseria gonorrhoeae that causes gonorrhoea) may not survive or the normal microorganisms (e.g., Escherichia coli) may overgrow pathogenic microorganisms (e.g., Salmonella, Shigella, Vibrio cholerae) even before the transportation of clinical sample to the testing laboratory. To avoid this, culture media have been devised to maintain the viability of the pathogen.

Some of the best examples of the transport media are the following:

Stuart's medium:

Stuart's medium used to maintain the viability of gonococci bacteria.

Pike's medium:

Pike's medium used to preserve Streptococcus pyogenes.

Glycerol-saline medium:

Glycerol-saline medium used to prevent normal intestinal microflora from overgrowing the enteric fever bacilli.

Bile-Peptone medium:

Bile-Peptone medium used to maintain the viability of cholera causing bacteria.

(iv) Differential (Indicator) Media:

A differential (indicator) medium is one which causes a visible change between different groups of bacteria growing in the medium and even permits tentative identification of microorganisms based on their biological characteristics. Blood-Agar medium, MacConkey-Agar medium and Christensen's medium are good examples of differential (indicator) medium.

Blood-Agar medium is used to differentiate between hemolytic and non-hemolytic bacteria. Hemolytic bacteria (e.g., many streptococci and staphylococci isolated from throats) produced clear zones around their colonies because of red blood cell destruction.

Such a clear zone is not formed around non-hemolytic bacterial colonies. MacConkey-Agar medium contains lactose and neutral red. Lactose fermenting bacteria after growth on this medium produce acid and in acidic pH the neutral red becomes red in colour.

Thus E. coli, which is lactose fermenter, produces red or pink colonies on the medium. Christensen's medium possesses urea and phenol red. When urease producing bacteria (e.g., Proteus, Klebsiella) grow on this medium, urea is broken into ammonia and CO_2. Ammonia turns the medium alkaline and in alkaline pH the medium becomes pink in colour because phenol red becomes red in colour in alkaline medium.

However, the constituents of Blood-Agar medium are the following:

Infusion from beef heart.........500.0 g
Sodium chloride......................5.0 g
Tryptose...............................5.0 g
Agar...................................15.0 g
Distilled water......................100 ml

(v) Enriched Media:

Enriched media are prepared to meet the nutritional requirements of metabolically fastidious microorganisms by addition of specific growth substances such as blood, serum and egg to a basal medium.

Important examples of enriched media are blood agar for isolation of Streptococcus, chocolate agar for isolation of Neisseria and Haemophilus, Bordet-Gengou for isolation of Bordetella, and Loeffer's serum slope for the isolation of Corynebacterium dephtheriae.

Pouring a plate

1. Collect one bottle of sterile molten agar from the water bath.

2. Hold the bottle in the right hand; remove the cap with the little finger of the left hand.

3. Flame the neck of the bottle.

4. Lift the lid of the Petri dish slightly with the left hand and pour the sterile molten agar into the Petri dish and replace the lid.

5. Flame the neck of the bottle and replace the cap.

6. Gently rotate the dish to ensure that the medium covers the plate evenly.

7. Allow the plate to solidify.

The base of the plate must be covered, agar must not touch the lid of the plate and the surface must be smooth with no bubbles. The plates should be used as soon as possible after pouring. If they are not going to be used straight away they need to be stored inside sealed plastic bags to prevent the agar from drying out.

Sterilisation vs disinfection

Sterilisation means the complete destruction of all the micro-organisms including spores, from an object or environment. It is usually achieved by heat or filtration but chemicals or radiation can be used. Disinfection is the destruction, inhibition or removal of microbes that may cause disease or other problems, e.g. spoilage. It is usually achieved by the use of chemicals.

Sterilisation using the autoclave/pressure cooker

The principle of sterilisation in an autoclave or pressure cooker is that steam under pressure is used to produce a temperature of 121 °C which if held for 15 minutes will kill all micro-organisms, including bacterial endospores.

Sterilisation of equipment and materials

1. Wire loop

Heat to redness in Bunsen burner flame.

2. Empty . (not plastic!) pipettes and Petri dishes

Either: hot air oven, wrapped in either greaseproof paper or aluminium and held at 160 °C for 2 hours, allowing additional time for items to come to temperature (and cool down!). Or: autoclave/pressure cooker. Note: plastic Petri dishes are supplied in already sterilised packs; packs of sterile plastic pipettes are also available but cost may be a consideration.

3. Culture media and solutions

Autoclave/pressure cooker.

4. Glass spreaders and metal forceps

Flaming in alcohol (70 % IDA).

Choice, preparation and use of disinfectants Specific disinfectants at specified working strengths are used for specific purposes. The choice is now much morestraightforward as the range available from suppliers has decreased. Commonly available disinfectants and their uses

Disinfectant	Use	Working strength
VirKon	Work surfaces, discard pots for pipettes and slides, skin disinfection	1% (w/v)
	Spillages	Powder
Hypochlorite (sodium chlorate I)	Discard pots for pipettes and slides	2,500 p.p.m. (0·25%, v/v) available chlorine
Alcohol	Skin disinfection	70% (v/v) industrial denatured alcohol (IDA)

When preparing working strength solutions from stock for class use and dealing with powder form, wear eye protectionand gloves to avoid irritant or harmful effects. Disinfectants for use at working strength should be freshly prepared from full strength stock or powder form. Activityof VirKon solution may remain for up to a week (as indicated by retention of pink colour) but less, e.g. 1 day, after use. Use working strength hypochlorite on day of preparation.

Inoculation and other aseptic procedures

There are several essential precautions that must be taken during inoculation procedures to control the opportunities for the contamination of cultures, people or the environment.

- Operations must not be started until all requirements are within immediate reach and must be completed as quickly as possible.
- Vessels must be open for the minimum amount of time possible and while they are open all work must be done close to the Bunsen burner flame where air currents are drawn upwards.

- On being opened, the neck of a test tube or bottle must be immediately warmed by flaming so that any air movement is outwards and the vessel held as near as possible to the horizontal.
- During manipulations involving a Petri dish, exposure of the sterile inner surfaces to contamination from the air must be limited to the absolute minimum.
- The parts of sterile pipettes that will be put into cultures or sterile vessels must not be touched or allowed to come in contact with other non-sterile surfaces, e.g. clothing, the surface of the working area, the outside of test tubes/bottles.

Using a wire loop

Wire loops are sterilised using red heat in a Bunsen flame before and after use. They must be heated to red hot to make sure that any contaminating bacterial spores are destroyed. The handle of the wire loop is held close to the top, as you would a pen, at an angle that is almost vertical. This leaves the little finger free to take hold of the cotton wool plug/screw cap of a test tube/bottle. Flaming procedure The flaming procedure is designed to heat the end of the loop gradually because after use it will contain culture, which may 'splutter' on rapid heating with the possibility of releasing small particles of culture and aerosol formation.

1. Position the handle end of the wire in the light blue cone of the flame. This is the cool area of the flame.
2. Draw the rest of the wire upwards slowly up into the hottest region of the flame, (immediately above the light blue cone).
3. Hold there until it is red hot.
4. Ensure the full length of the wire receives adequate heating.
5. Allow to cool then use immediately.
6. Do not put the loop down or wave it around.
7. Re-sterilise the loop immediately after use.

Storage of culture media

- Dehydrated culture media and dry ingredients should be stored at an even temperature in a cool dry place away from direct light.

• Plates of culture media, and additives like serum, blood and antimicrobials in solid form require storage at 2-8 OC.

• Antimicrobials in solution form should be stored at –20 OC.

• All culture media and additives should be labeled with the name and date of preparation.

Inoculation of culture media

When inoculating culture media, an aseptic technique must be used to prevent contamination of specimens and culture media, and laboratory worker and the environment.

Aseptic technique during inoculation of culture media

• Decontaminate the workbench before and after the work of the day.

• Use facemask and gloves during handling highly infectious specimens.

• Flame sterilize wire loops, straight wires, and metal forceps before and after use.

• Flame the neck of specimen and culture bottles, and tubes after removing and before replacing caps and plugs.

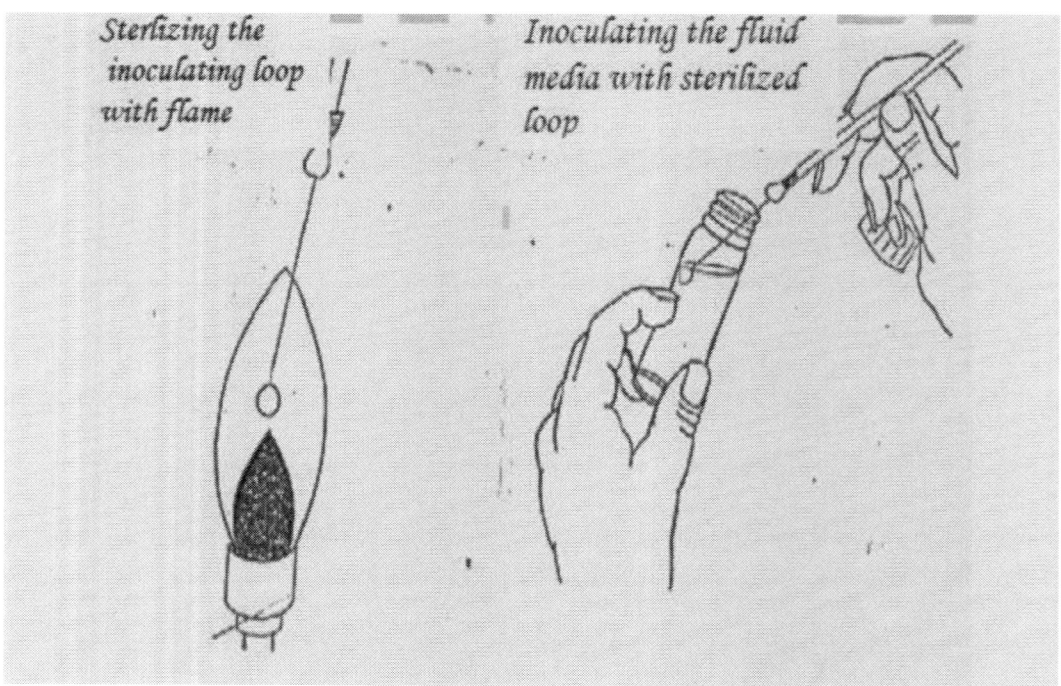

Fig. 11. Aseptic inoculation technique

Inoculation of media in petridishes

The inoculation of media in petridishes is named as 'plating out' or 'looping out'.

Before inoculating a plate of culture media, dry the surface of the media by incubating at 37 OC for 30 minutes. To inoculate a plate, apply the inoculum to a small area of the plate ('the well') using sterile wire loop and then spread and thin out the inoculum to ensure single colony growth.

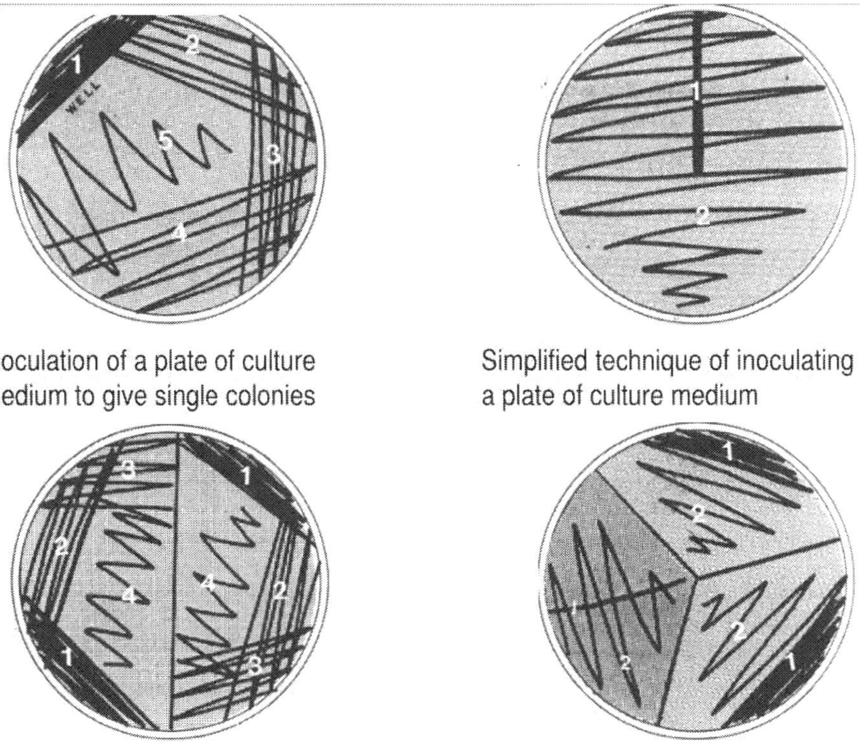

Inoculation of a plate of culture medium to give single colonies

Simplified technique of inoculating a plate of culture medium

Inoculation of half a plate of culture medium

Different ways of inoculating a third of a plate of culture medium

Fig. 11.1. methods of inoculating solid culture media in petridishes

Inoculation of butt and slant media

To inoculate butt and slant media, use a sterile straight wire to stab into the butt and then streak the slant in a zigzag pattern.

Inoculation of slant media

To inoculate slant media, use a straight wire to streak the inoculum down the center of the slant and then spread the inoculum in a zigzag pattern.

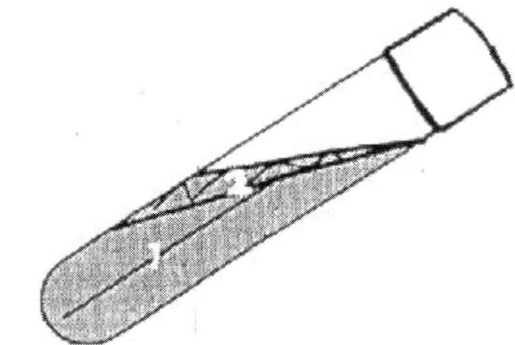

Fig. 11.2. Inoculation of slant and bath media

Inoculation of stab media

To inoculate stab media, use a straight wire to stab through the center of the medium and withdraw the wire along the line of inoculum.

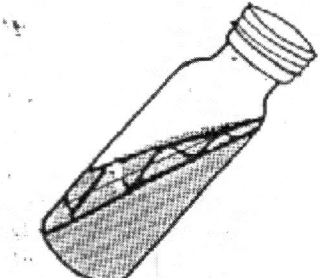

Fig. 11.3. Inoculation of slant media

Inoculation of fluid media
To inoculate fluid media, use straight wire or wire loops.
Incubation of cultures
Inoculated media should be incubated as soon as possible.

Optimal temperature, humidity and gaseous atmosphere should be provided for microorganisms to grow best. The temperature selected for routine culturing is 35-37 OC. Some pathogens require CO_2-enriched atmosphere to grow in culture media, and the simplest way to provide CO_2-enriched atmosphere is to enclose a lighted candle in an airtight jar which

provides 3-5% CO2 by the time the candle is extinguished. Anaerobic atmosphere is essential for the growth of strict anaerobes, and the techniques for obtaining anaerobic conditions
are the following:
. Anaerobic jar with a gas generating kit.
. Reducing agents in culture media.

Fig. 11.4. Co2-enriched atmosphere

Staining of bacteria

Bacterial staining is the process of coloring of colorless bacterial structural components using stains (dyes).The principle of taining is to identify microorganisms selectively by using dyes,

fluorescence and radioisotope emission. Staining reactions are made possible because of the physical phenomena of capillary osmosis, solubility, adsorption, and absorption of stains or dyes by cells of microorganisms. Individual variation in the cell wall constituents among different groups of bacteria will consequently produce variations in colors during microscopic examination. Nucleus is acidic in character and hence, it has greater affinity for basic dyes. Whereas, cytoplasm is basic in character and has greater affinity for acidic dyes.

There are many types of affinity explaining this attraction force:
1. hydrophobic bonding
2. reagent-cell interaction
3. reagent-reagent interaction
4. ionic bonding
5. hydrogen bonding
6. covalent bonding

Why are stains not taken up by every microorganism?

Factors controlling selectivity of microbial cells are:

1. number and affinity of binding sites
2. rate of reagent uptake
3. rate of reaction
4. rate of reagent loss (differentiation or regressive staining)

Properties of dyes

Why dyes color microbial cells?

Because dyes absorb radiation energy in visible region of electromagnetic spectrum i.e., "light"(wave length 400-650). And absorption is anything outside this range it is colorless. E.g., acid fuschin absorbs blue green and transmit red.

General methods of staining

1. Direct staining. Is the process by which microorganisms are stained with simple dyes. E.g., methylene blue

2. Indirect staining – is the process which needs mordants. A mordant is the substance which, when taken up by the microbial cells helps make dye in return, serving as a link or bridge to make the staining recline possible. It combines with a dye to form a colored "lake", which in turn combines with the microbial cell to form a " cell-mordant-dyecomplex". It is an integral part of the staining reaction itself, without which no staining could possibly occur. E.g., iodine. A mordant may be applied before the stain or it may be included as part of the staining technique, or it may be added to the dye solution itself. An accentuator, on the other hand is not essential to the chemical union of the microbial cells and the dye. It does not participate in the staining reaction, but merely accelerate or hasten the speed of the staining reaction by increasing the staining power and selectivity of the dye.

Progressive staining - is the process whereby microbial cells are stained in a definite sequence, in order that a satisfactory differential coloration of the cell may be achieved at the end of the

correct time with the staining solution.

Regressive staining - with this technique, the microbial cell is first over stained to obliteratethe cellulare desires, and the excess stain is removed or decolorized from unwanted part.

Differentiation (decolorization) - is the selective removal of excess stain from the tissue from microbial cells during regressive staining in order that a specific substance may be stained differentiallyh from the surrounding cell.

Differentiation is usually controlled visually by examination under the microscope

Uses

1. To observe the morphology, size, and arrangement of bacteria.
2. To differentiate one group of bacteria from the other group.

Biological stains are dyes used to stain micro-organisms.

Types of microbiological stains

- . Basic stains
- . Acidic stains
- . Neutral stains

Basic stains are stains in which the coloring substance is contained in the base part of the stain. The acidic part is colorless. Eg.

Acidic stains are stains in which the coloring substance is contained in the acidic part of the stain. The base part is colorless. It is not commonly used in microbiology laboratory.

Eg. Eosin stain

Neutral stains are stains in which the acidic and basic components of stain are colored. Neutral dyes stain both nucleic acid and cytoplasm. Eg. Giemsa stain

Types of staining methods

1. Simple staining method
2. Differential staining method
3. Special staining method

1. Simple staining method

It is type of staining method in which only a single dye is used. Usually used to demonstrate bacterial morphology and arrangement.

Two kinds of simple stains

1. Positive staining: The bacteria or its parts are stained by the dye.

Eg. Carbol fuchsin stain

Methylene blue stain

Crystal violet stain

Procedure:

- . Make a smear and label it.
- . Allow the smear to dry in air.
- . Fix the smear over a flame.
- .Apply a few drops of positive simple stain like 1% methylene blue,

1% carbolfuchsin or

1% gentian violet for 1 minute.

- . Wash off the stain with water.
- . Air-dry and examine under the oil immersion objective.

2. Negative staining: The dye stains the background and the bacteria remain unstained. Eg. Indian ink stain Negrosin stain

2. Differential staining method

Multiple stains are used in differential staining method to distinguish different cell structures and/or cell types. Eg. Gram stain and Ziehl-Neelson stain

A. Gram staining method

Developed by Christian Gram. Most bacteria are differentiated by their gram reaction due to differences in their cell wall structure. Gram-positive bacteria are bacteria that stain purple with crystal violet after decolorizing with acetone-alcohol. Gram-negative bacteria are bacteria that stain pink with the counter stain (safranin) after losing the primary stain (crystal violet) when treated with acetone-alcohol.

Required reagents:
- ❖ . Gram's Iodine
- ❖ . Acetone-Alcohol
- ❖ . Safranin

Procedure:
1. Prepare the smear from the culture or from the specimen.
2. Allow the smear to air-dry completely.
3. Rapidly pass the slide (smear upper most) three times through the flame.
4. Cover the fixed smear with crystal violet for 1 minute and wash with distilled water.
5. Tip off the water and cover the smear with gram's iodine for 1 minute.
6. Wash off the iodine with clean water.
7. Decolorize rapidly with acetone-alcohol for 30 seconds.
8. Wash off the acetone-alcohol with clean water.
9. Cover the smear with safranin for 1 minute.
10. Wash off the stain wipe the back of the slide. Let the smear to air-dry.
11. Examine the smear with oil immersion objective to look for bacteria.

Interpretation:
- ➢ . Gram-positive bacterium ……………Purple
- ➢ . Gram-negative bacterium …………..Pink

B. Ziehl-Neelson staining method

Developed by Paul Ehrlichin 1882, and modified by Ziehl and Neelson. Ziehl-Neelson stain (Acid-fast stain) is used for staining Mycobacteria which are hardly stained by gram staining method. Once the Mycobacteria is stained with primary stain it cannot be decolorized with acid, so named as acid-fast bacteria.

Reagents required:
- ❖ . Carbol-fuchsin
- ❖ . Acid-Alcohol
- ❖ . Methylene blue/Malachite green

Procedure for Ziehl-Neelson staining method

1. Prepare the smear from the primary specimen and fix it by passing through the flame and label clearly
2. Place fixed slide on a staining rack and cover each slide with concentrated carbol fuchsin solution.
3. Heat the slide from underneath with sprit lamp until vapour rises (do not boil it) and wait for 3-5 minutes.
4. Wash off the stain with clean water.
5. Cover the smear with 3% acid-alcohol solution until all color is removed (two minutes).
6. Wash off the stain and cover the slide with 1% methylene blue.for one minute.
7. Wash off the stain with clean water and let it air-dry.
8. Examine the smear under the oil immersion objective to look for acid fast bailli.

Interpretation:

Acid fast bacilli…………..Red
Back ground……………….Blue
Reporting system
0 AFB/100 field …………………..No AFB seen
1-2 AFB/ 300 field……………….. Scanty
1-10 AFB/100 field………………1+
11-100AFB/100 field…………….2+
1-10 AFB/field…… ……………3+
>10 AFB/field……………… ….4+
NB: AFB means number of acid fast bacilli seen.

3. Special stains
a. Spore staining method
b. Capsule staining method

a. Spore staining method
Procedure:
1. Prepare smear of the spore-forming bacteria and fix in flame.

2. Cover the smear with 5% malachite green solution and heat over steaming water bath for 2 3 minutes.

3. Wash with clean water.
4. Apply 1% safranin for 30 seconds.

5. Wash with clean water.
6. Dry and examine under the oil immersion objective.

b. Capsule staining method: Welch method

Procedure:
1. Prepare smear of capsulated bacteria.
2. Allow smear to air-dry; do not fix the smear.
3. Cover the smear with 1% aqueous crystal violet for 1 minute over steaming water bath.
4. Wash with 20% copper sulfate solution. Do not use water.
5. Dry and examine under the oil immersion objective.

CULTIVATION OF BACTERIA IN CULTURE MEDIA

Culture media

It is the media containing the required nutrients for bacterial growth.

Uses: . Isolation and identification of micro-organisms
. Performing anti-microbial sensitivity tests

Common ingredients of culture media
- . Peptone
- . Meat extract
- . Yeast extract
- . Mineral salts
- . Carbohydrates
- . Agar
- .Water

Peptone: Hydrolyzed product of animal and plant proteins: Free amino acids, peptides and proteoses(large sized peptides). It provides nitrogen; as well carbohydrates, nucleic acid fractions, minerals and vitamins.

Meat extract: supply amino acids, vitamins and mineral salts.

Yeast extract: It is bacterial growth stimulants.

Mineral salts: these are: Sulfates as a source of sulfur.
- . Phosphates as a source of phosphorus.

- . Sodium chloride
- . Other elements

Carbohydrates: Simple and complex sugars are a source of carbon and energy.

.Assist in the differentiation of bacteria.

Eg. Sucrose in TCBS agar differentiates vibro species.

Lactose in MacConkey agar differentiates enterobacteria.

Agar: It is an inert polysaccharide of seaweed.

It is not metabolized by micro-organism.

Property
- . It has . high gelling strength
- . high melting temperature(90-95 oc)
- . low gelling temperature
- . It forms firm gel at 1.5% W/V concentration.
- . It forms semisolid gel at 0.4-0.5% W/V concentration.

Uses:
- . Solidify culture media

May provide calcium and organic ions to inoculated bacteria.

Water

Deionized or distilled water must be used in the preparation of culture media.

****

12. ACTINOMYCETES

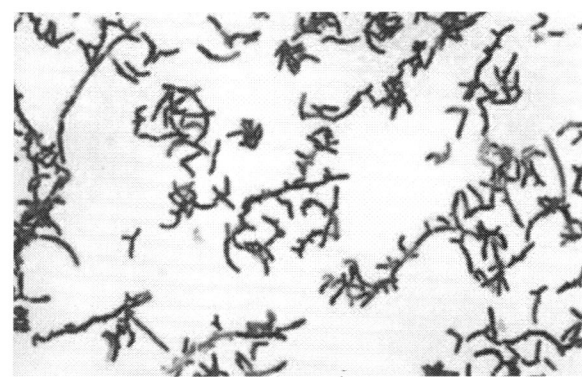

Introduction

Actinomycetes belong to the group of prokaryotic, gram-positive bacteria which are having a filamentous structure. Its filamentous structure resembles the fungal mycelium, which consists of a highly dense and **filamentous** network. Due to the filamentous structure of Actinomycetes, it also refers as "**Thread** or **Ray bacteria**". The cell wall and the internal structure of the actinomycetes are similar to the group of bacteria. Thus, Actinomycetes also refers as filamentous Actinobacteria and acts as a **connecting link** between the bacteria and fungi as it shows resemblance with both. Actinomycetes are "**True bacteria**", not fungus and therefore these are placed in the kingdom "Bacteria" and a class "**Actinobacteria**". These are ubiquitous and commonly found in soil and are soil microorganisms.

Actinomycetes also act as "**Decomposers**" which carry out the decomposition of organic compounds like chitin, complex sugars, hemicellulose etc. In addition to soil, these are also very common in marine habitat and considered as a treasure house of **secondary metabolites**. Its filamentous forms are predominantly aerobic, and few are anaerobic.

The early exploratory studies by McCormack (1935) and Alexopoulos and Herrick (1938-1942) were followed by the intensive studies by Professor S. A. Waksman and his students (1943-1951) which culminated in the discovery of streptomycin and other new and potentially useful chemotherapeutic agents.

Nearly 100 antibiotic substances have been reported in the literature as metabolites of the Actinomycetes. A few of these have been isolated in pure form and their chemistry studied in detail, while others have been described only as concentrates or in a preliminary way.

Actinomycetes can define as the **prokaryotic** or unicellular organisms, which are having a **gram-positive** cell wall. The morphology of actinomycetes is similar to fungi as it produces a filamentous, dense, branched and raised colony over the substrate. Most of its features are common to the bacteria than that of fungi and thus placed in the group of bacteria which includes members like Mycobacterium, Corynebacterium, Streptomyces, and Actinomyces etc.

A diameter of the Actinomycete is much smaller (1-2 µm) than the branches of fungi which range from 5µm-10µm. The filamentous forms of Actinomycetes are aerobic and may produce spore singly or in chains. Its colony appears as a powdery mass and are pigmented by the formation of aerial spores.

Classification:	
Domain:	**Bacteria**
Phylum:	**Actinobacteria**
Class:	**Actinobacteria**
Order:	**Actinomycetales**
Family:	**Actinomycetaceae**

Characteristics

The following characteristics of Actinomycetes are given below:

1. The size of Actinomycetes is **1-2 μm** in diameter.
2. These are usually rod shaped with a filamentous and branched structure. The filaments contain **mumaric acid**.
3. Most of the species are **aerobic**, but few are anaerobic to facultative aerobes.
4. Cell wall and internal structures are similar to the bacteria. The cell wall of Actinomycetes consists of **mycolic acid**.
5. The growth or reproduction of Actinomycetes is slower than the bacteria and fungi and hence also refers as "**Slow growers**".
6. These are having **60-78%** of G+C content.
7. Actinomycetes are most abundant in **soil** (10^6-10^8g) and **marine habitat**.
8. These are most usually non-motile, non-capsulated and non-acid fast.
9. The growth of actinomycetes is optimal at **alkaline pH**.

Life Cycle

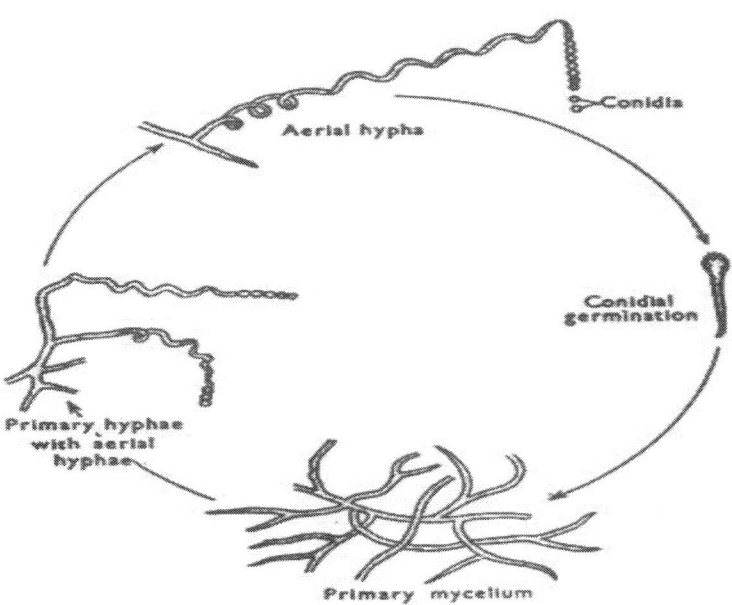

Fig. 12. Life cycle of actinomycetes

Most species reproduce by conidia which are developed in chains from the aerial hyphae. The chains may be straight, flexuous (wavy) or coiled to various degrees. The conidia bearing filaments are often spirally twisted. Sometimes the whole length of the aerial hypha, sometimes only its upper part is transformed into conidia. Each conidium has a roundish nucleus and is surrounded by a firm outer wall. The conidial wall may be smooth, warty, spiny, or hairy. The conidia can persist in the dry state for many years. Even the vegetative forms of the Actinomycetes are quite hardy and are able to adapt themselves to the changing soil conditions. The conidia appear as a fine powdery coat on the surface of cultures. When the conidia have been scattered on the ground and conditions are favourable they germinate producing one to three or even occasionally four little germ tubes which give rise to mycelioid condition (Fig. 12). The primary mycelium in some species commonly breaks up into small fragments called arthrospores, which often look like bacterial cells and which might easily be mistaken for the latter.

Classification

Based on hyphal and reproductive structures, Actinomycetes divides into seven families:

Streptomycetaceae: Members of this family consist of non-segmented hyphae and consists of 5-50 conidial spores per chain of aerial hyphae. Examples: *Streptomyces*, *Microdlobaspone* and *Sporoctilhya*.

Nocardiaceae: Members of this family consist of typical non-segmented hyphae. Examples: *Nocardia*, *Pseudonocardia*.

Micromonosporaceae: Members of this family consist of typical non-segmented conidia which occur singly, pairs or in chains. Examples: *Micromonospora*, *Thermonospora*, *Thermoactinomycetes*, *Actinobifida*.

Actinoplanaceae: Members of this family consist sporangiospores and the diameter of hyphae ranges from 0.2-2.0 μm. Examples: *Streptosporangium*, *Actinoplanes*, *Plasmobispora* and *Dactylosporangium*.

Dermatophilaceae: Members of this family consist of hyphae which undergo fragmentation to produce a large number of motile structures. Examples: *Geodermatophilus*.

Frankiaceae: Members of this family are strictly associated with the roots of a non-leguminous plant and helps in nitrogen fixation by forming root nodules. Example: *Frankia*.

Actinomycetaceae: The members of this family do not contain true mycelium and are facultative anaerobes. Examples: *Actinomyces*.

Importance of actinomycetes

Practical utilization:

The practical utilization of antibiotics has also much to disqualify it as basis for a classification, since relatively few antibiotics produced by actinomycetes, have found practical application.

Enzymes :

For many years, actinomycetes were known as the source of large numbers of antibiotics. More recently they have been found to be a promising source of a wide range of important enzymes. Some of these actinomycete enzymes have already been produced on an industrial scale, but many other remain to be harnessed. At present, enzymes of microbial origin are widely used in food processing, detergent manufacturing, the textile and pharmaceutical industries, medical therapy, bioorganic chemistry and molecular biology.

Actinomycetes in Agriculture and Forestry:

It is well known that actinomycetes carry out numerous activities in soil; in degradation of organic matter; inhibition or stimulation of other microorganisms and plants; transformation of chemical compounds such as herbicides, and other agriculturally useful compounds (Lechevalier, 1981; Krause et al., 1985). Lignocellulose is converted to humus by Streptomyces imparting desirable properties to soil (Kaluskar, 1995). Actinomycetes, with activity against plant pathogen and antagonistic actinomycetes have been detected in the rhizosphere of diseased plants. (Kundu and Nandi, 1985). Plant growth has been enhanced by free living actinomycetes. Actinomycetes isolated from rhizosphere of pine, produced B vitamin; since mycorrhizae require these vitamins and are known to enhance the growth of plants. It was proposed that actinomycetes may thus indirectly contribute to plant growth stimulation (Strzelczyk and Leniarska; 1985).

Industrial importance of

Actinomycetes are fungi-like bacteria. They are widespread in nature, and can be found in the soil, water and in compost. Actinomycetes form branching hyphae or mycelium (which is typical of fungi), and this is why they are often called fungi-like bacteria. Actinomycetes are chemoorganotrophic organisms that derive their carbon from organic molecules or substrates. They degrade chitin, agar cellulose, paraffin, rubber and keratin. Actinomycetes also produce antibiotics. Actinomycetes are Gram positive bacteria, and they are spore forming bacteria.

The bacteria genera in the group of organisms known as actinomycetes include: *Nocardia, Actinomycetes, Corynebacteria, Streptomyces* and *Micromonospora*. Not all members of actinomycetes are known to produce mycelium. Actinomycetes are filamentous bacteria because they form mycelium. Actinomycetes are aerobic bacteria that produce asexual spores. Actinomycetes have mycelia morphology (that makes them have similar resemblance to fungi); and majority of bacteria in this group of actinomycetes are known for their antibiotic production, especially those in the genus *Streptomyces*.

INDUSTRIAL SIGNIFICANCE OF ACTINOMYCETES

- Actinomycetes are well recognized because they produce primary and secondary metabolites that are of economic significance.
- Actinomycetes produce enzymes such as lipase, cellulases and amylase which are used in industrial fermentation processes.
- They produce some valuable antibiotics including, amphotericin, neomycin, vancomycin, gentamicin, tetracycline, erythromycin, nystatin, novobiocin and chloramphenicol.
- Actinomycetes are also used as plant growth promoting agents.
- They produce biopesticide agents used to control pests in farmlands.
- Actinomycetes also have application in bioremediation.
- Actinomycetes produce protease enzyme which is used as anti-inflammation agents and also in cancer treatment.
- Actinomycetes produce enzymes that have application in wine production.

ACTINOMYCOSIS

Members of the genus Actinomyces are normal residents of the mouth, throat, and intestinal tract. But they are capable of causing infections both in humans and in cattle if they are able to enter other regions. This can occur as the result of an accident such as a cut or abrasion. An infection known as Actinomycosis is characterized by the formation of an abscess—a process "walling off" the site of infection as the body responds to the infection—and by swelling. Pus can also be present. The pus, which is composed of dead bacteria, is granular, because of the presence of granules of sulfur that are made by the bacteria.

The diagnosis of an Actinomyces infection can be challenging, as the symptoms and appearance of the infection is reminiscent of a tumor or of a tuberculosis lesion. A wellestablished infection can produce a great deal of tissue damage.

Additionally, the slow growth of the bacteria can make the treatment of infection with antibiotics very difficult, because antibiotics rely on bacterial growth in order to exert their lethal effect. The culturing of Actinomyces in the laboratory is also challenging. The bacteria do not grow on nonselective media, but instead require the use of specialized and nutritionally complex selective media. Furthermore, incubation needs to be in the absence of oxygen. The growth of the bacteria is quite slow. Solid growth medium may need to be incubated for up to 14 days to achieve visible growth. In contrast, a bacterium like Escherichia coli yields visible colonies after overnight growth on a variety of nonselective media. The colonies of Actinomyces are often described as looking like bread crumbs.

Currently, identification methods such as polymerase chain reaction (PCR), chromatography to detect unique cell wall constituents, and antibody-based assays do always perform effectively with Actinomyces.

****

13. PSEUDOMONAS, XANTHOMONAS AND VIBRIO

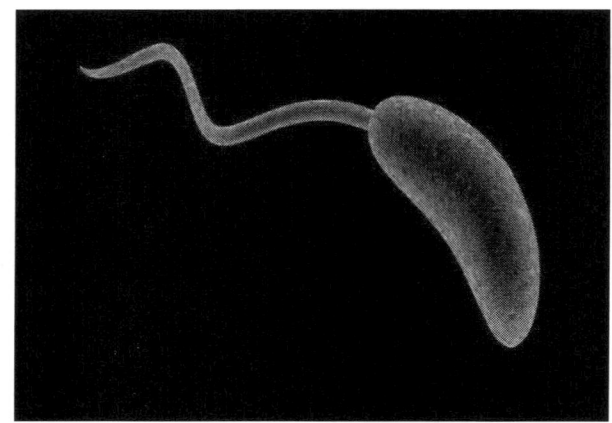

Pseudomonas

Pseudomonas is a genus of Gram-negative, Gammaproteobacteria, belonging to the family Pseudomonadaceae and containing 191 validly described species. The members of the genus demonstrate a great deal of metabolic diversity and consequently are able to colonize a wide range of niches.[2] Their ease of culture *in vitro* and availability of an increasing number of *Pseudomonas* strain genome sequences has made the genus an excellent focus for scientific research; the best studied species include *P. aeruginosa* in its role as an opportunistic human pathogen, the plant pathogen *P. syringae*, the soil bacterium *P. putida*, and the plant growth-promoting *P. fluorescens, P. lini, P. migulae*, and *P. graminis*.

Because of their widespread occurrence in water and plant seeds such as dicots, the pseudomonads were observed early in the history of microbiology. The generic name *Pseudomonas* created for these organisms was defined in rather vague terms by Walter Migula in 1894 and 1900 as a genus of Gram-negative, rod-shaped, and polar-flagellated bacteria with some sporulating species, the latter statement was later proved incorrect and was due to refractive granules of reserve materials. Despite the vague description, the type species, *Pseudomonas pyocyanea* (basonym of *Pseudomonas aeruginosa*), proved the best descriptor.

Classification history

Like most bacterial genera, the pseudomonadlast common ancestor lived hundreds of millions of years ago. They were initially classified at the end of the 19th century when first identified by Walter Migula. The etymology of the name was not specified at the time and first appeared in the seventh edition of *Bergey's Manual of Systematic Bacteriology* (the main authority in bacterial nomenclature) as Greek *pseudes* "false" and *-monas* "a single unit", which can mean false unit; however, Migula possibly intended it as false *Monas*, a nanoflagellated protist (subsequently, the term "monad" was used in the early history of microbiology to denote unicellular organisms). Soon, other species matching Migula's somewhat vague original description were isolated from many natural niches and, at the time, many were assigned to the genus. However, many strains have since been reclassified, based on more recent methodology and use of approaches involving studies of conservative macromolecules.

Recently, 16S rRNA sequence analysis has redefined the taxonomy of many bacterial species. As a result, the genus *Pseudomonas* includes strains formerly classified in the genera *Chryseomonas* and *Flavimonas*. Other strains previously classified in the genus *Pseudomonas* are now classified in the genera *Burkholderia* and *Ralstonia*.

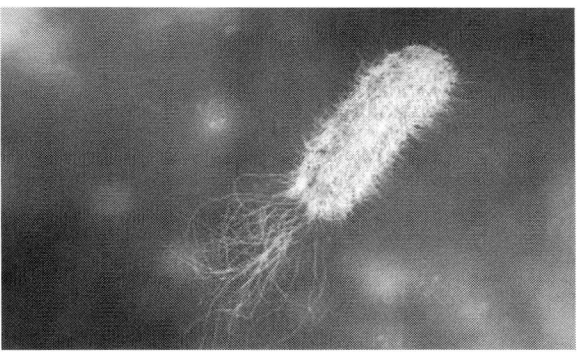

Fig.13. *Pseudomonas aeruginosa*

Characteristics

Members of the genus display these defining characteristics:[14]

- Rod-shaped

- Gram-negative
- Flagellum one or more, providing motility
- Aerobic
- Non-spore forming
- Catalase-positive
- Oxidase-positive

Other characteristics that tend to be associated with *Pseudomonas* species (with some exceptions) include secretion of pyoverdine, a fluorescent yellow-green siderophore[15] under iron-limiting conditions. Certain *Pseudomonas* species may also produce additional types of siderophore, such as pyocyanin by *Pseudomonas aeruginosa* and thioquinolobactin by *Pseudomonas fluorescens*, *Pseudomonas* species also typically give a positive result to the oxidase test, the absence of gas formation from glucose, glucose is oxidised in oxidation/fermentation test using Hugh and Leifson O/F test, beta hemolytic (on blood agar), indole negative, methyl red negative, Voges–Proskauer test negative, and citrate positive.

Pseudomonas may be the most common nucleator of ice crystals in clouds, thereby being of utmost importance to the formation of snow and rain around the world.

Pathogenicity

Animal pathogens
Main article: Pseudomonas infection

Infectious species include *P. aeruginosa*, *P. oryzihabitans*, and *P. plecoglossicida*. *P. aeruginosa* flourishes in hospital environments, and is a particular problem in this environment, since it is the second-most common infection in hospitalized patients (nosocomial infections). This pathogenesis may in part be due to the proteins secreted by *P. aeruginosa*. The bacterium possesses a wide range of secretion systems, which export numerous proteins relevant to the pathogenesis of clinical strains.

Plant pathogens

P. syringae is a prolific plant pathogen. It exists as over 50 different pathovars, many of which demonstrate a high degree of host-plant specificity. Numerous other *Pseudomonas* species can act as plant pathogens, notably all of the other members of the *P. syringae* subgroup, but *P. syringae* is the most widespread and best-studied.

Although not strictly a plant pathogen, *P. tolaasii* can be a major agricultural problem, as it can cause bacterial blotch of cultivated mushrooms. Similarly, *P. agarici* can cause drippy gill in cultivated mushrooms.

Use as biocontrol agents

Since the mid-1980s, certain members of the genus *Pseudomonas* have been applied to cereal seeds or applied directly to soils as a way of preventing the growth or establishment of crop pathogens. This practice is generically referred to as biocontrol. The biocontrol properties of *P. fluorescens* and *P. protegens* strains (CHA0 or Pf-5 for example) are currently best-understood, although it is not clear exactly how the plant growth-promoting properties of *P. fluorescens* are achieved. Theories include: the bacteria might induce systemic resistance in the host plant, so it can better resist attack by a true pathogen; the bacteria might outcompete other (pathogenic) soil microbes, e.g. by siderophores giving a competitive advantage at scavenging for iron; the bacteria might produce compounds antagonistic to other soil microbes, such as phenazine-type antibiotics or hydrogen cyanide. Experimental evidence supports all of these theories.

Other notable *Pseudomonas* species with biocontrol properties include *P. chlororaphis*, which produces a phenazine-type antibiotic active agent against certain fungal plant pathogens, and the closely related species *P. aurantiaca*, which produces di-2,4-diacetylfluoroglucylmethane, a compound antibiotically active against Gram-positive organisms.

Use as bioremediation agents

Some members of the genus are able to metabolise chemical pollutants in the environment, and as a result, can be used for bioremediation. Notable species demonstrated as suitable for use as bioremediation agents include:

- *P. alcaligenes*, which can degrade polycyclic aromatic hydrocarbons.
- *P. mendocina*, which is able to degrade toluene.
- *P. pseudoalcaligenes*, which is able to use cyanide as a nitrogen source.
- *P. resinovorans*, which can degrade carbazole.
- *P. veronii*, which has been shown to degrade a variety of simple aromatic organic compounds. *P. putida*, which has the ability to degrade organic solvents such as toluene. At least one strain of this bacterium is able

to convert morphine in aqueous solution into the stronger and somewhat expensive to manufacture drug hydromorphone (Dilaudid).

- Strain KC of *P. stutzeri*, which is able to degrade carbon tetrachloride.

Detection of food spoilage agents in milk

One way of identifying and categorizing multiple bacterial organisms in a sample is to use ribotyping. In ribotyping, differing lengths of chromosomal DNA are isolated from samples containing bacterial species, and digested into fragments. Similar types of fragments from differing organisms are visualized and their lengths compared to each other by Southern blotting or by the much faster method of polymerase chain reaction (PCR). Fragments can then be matched with sequences found on bacterial species. Ribotyping is shown to be a method to isolate bacteria capable of spoilage. Around 51% of *Pseudomonas* bacteria found in dairy processing plants are *P. fluorescens*, with 69% of these isolates possessing proteases, lipases, and lecithinases which contribute to degradation of milk components and subsequent spoilage. Other *Pseudomonas* species can possess any one of the proteases, lipases, or lecithinases, or none at all Similar enzymatic activity is performed by *Pseudomonas* of the same ribotype, with each ribotype showing various degrees of milk spoilage and effects on flavour. The number of bacteria affects the intensity of spoilage, with non-enzymatic *Pseudomonas* species contributing to spoilage in high number.

Food spoilage is detrimental to the food industry due to production of volatile compounds from organisms metabolizing the various nutrients found in the food product. Contamination results in health hazards from toxic compound production as well as unpleasant odours and flavours. Electronic nose technology allows fast and continuous measurement of microbial food spoilage by sensing odours produced by these volatile compounds. Electronic nose technology can thus be applied to detect traces of *Pseudomonas* milk spoilage and isolate the responsible *Pseudomonas* species. The gas sensor consists of a nose portion made of 14 modifiable polymer sensors that can detect specific milk degradation products produced by microorganisms. Sensor data is produced by changes in electric resistance of the 14 polymers when in contact with its target compound, while four sensor parameters can be adjusted to further specify the response. The responses can then be pre-processed by a neural network which can then differentiate between milk spoilage microorganisms such as *P. fluorescens* and *P. aureofaciens*.

Species previously classified in the genus

Recently, 16S rRNA sequence analysis redefined the taxonomy of many bacterial species previously classified as being in the genus *Pseudomonas*. Species removed from *Pseudomonas* are listed below; clicking on a species will show its new classification. The term 'pseudomonad' does not apply strictly to just the genus *Pseudomonas*, and can be used to also include previous members such as the genera *Burkholderia* and *Ralstonia*.

α **proteobacteria:** *P. abikonensis, P. aminovorans, P. azotocolligans, P. carboxydohydrogena, P. carboxidovorans, P. compransoris, P. diminuta, P. echinoides, P. extorquens, P. lindneri, P. mesophilica, P. paucimobilis, P. radiora, P. rhodos, P. riboflavina, P. rosea, P. vesicularis.*

β **proteobacteria:** *P. acidovorans, P. alliicola, P. antimicrobica, P. avenae, P. butanovorae, P. caryophylli, P. cattleyae, P. cepacia, P. cocovenenans, P. delafieldii, P. facilis, P. flava, P. gladioli, P. glathei, P. glumae, P. graminis, P. huttiensis, P. indigofera, P. lanceolata, P. lemoignei, P. mallei, P. mephitica, P. mixta, P. palleronii, P. phenazinium, P. pickettii, P. plantarii, P. pseudoflava, P. pseudomallei, P. pyrrocinia, P. rubrilineans, P. rubrisubalbicans, P. saccharophila, P. solanacearum, P. spinosa, P. syzygii, P. taeniospiralis, P. terrigena, P. testosteroni.*

γ-β **proteobacteria:** *P. beteli, P. boreopolis, P. cissicola, P. geniculata, P. hibiscicola, P. maltophilia, P. pictorum.*

γ **proteobacteria:** *P. beijerinckii, P. diminuta, P. doudoroffii, P. elongata, P. flectens, P. halodurans, P. halophila, P. iners, P. marina, P. nautica, P. nigrifaciens, P. pavonacea,*[43] *P. piscicida, P. stanieri.*

δ **proteobacteria:** *P. formicans.*

XANTHOMONAS

Scientific Names

Xanthomonas axonopodis Starr and Garces 1950

Xanthomonas arboricola Vauterin et al. 1995

Xanthomonas campestris (Pammel 1895) Dowson 1939

Xanthomonas hortorum Vauterin et al. 1995

Pathogen

Xanthomonas species are gram negative, aerobic, rod-shaped bacteria with cells measuring 0.4 to 1.0 µm wide by 0.7 to 2.0 µm long. Colonies of *Xanthomonas* species usually are mucoid, convex, and yellow when grown on yeast dextrose calcium carbonate (YDC) media or sucrose peptone agar (SPA) **(Figure.13. 1)**. All members of the genus *Xanthomonas* are catalase positive and oxidase negative. The classification of plant pathogenic *Xanthomonas* species has been a complicated affair, subject to numerous revisions over the years. At one time most strains were classified as pathovars within the single species *Xanthomonas campestris*, but since the revision by Vauterin, et al. in 1995, the name *X. campestris* has been reserved for those strains affecting crucifers.

Hosts, Signs, and Symptoms

The numerous species and pathovars of *Xanthomonas* cause disease on a variety of ornamental plants. *X. arboricola* pv. *poinsettiicola* has been reported on three species of poinsettia and other species of the Euphorbiaceae including *Euphorbia milii* (crown-of-thorns) and *Codiaeum variegatum* (croton). *X. axonopodis* pv. *begoniae* causes bacterial blight of begonia. Bacterial leaf spot and blight of Araceae, including *Dieffenbachia*, *Anthurium*, *Philodendron*, and *Syngonium* is caused by *X. axonopodis* pv. *dieffenbachiae*. Black rot on ornamental kale is caused by *X. campestris*. The causal agent of bacterial leaf spot of English ivy and other Araliaceae species is *X. hortorum* pv. *hederae*. Bacterial blight of *Pelargonium* (geranium) is caused by *X. hortorum* pv. *pelargonii*. Other important ornamental hosts of *Xanthomonas* spp. include *Ficus*, *Hydrangea*, and *Zinnia*. For bacterial spot on ornamental *Prunus* species, refer to *Xanthomonas arboricola pv. pruni*. For bacterial spot on ornamental pepper, refer to bacterial spot of pepper and tomato.

Signs of bacterial leaf spot can only be observed with a microscope. Infected leaf tissue will exhibit bacterial streaming when mounted with water on a slide and observed at 100x (Figure 13.2).

Symptoms usually involve necrotic, often water-soaked leaf spots, typically most severe on lower leaves. The spots may be very small at first but develop an angular shape as the movement of the bacteria is limited by leaf veins (Figure13. 3). Eventually several spots may enlarge, coalesce and cause death of mature leaves.

Other symptoms can occur depending on the host plant. In some cases there is a chlorotic halo around the spot. V-shaped brown necrotic areas pointing in the direction of the petiole may develop on leaves of begonia and geranium (Figure 13.4). On poinsettia, symptoms of leaf spot first appear as water-soaked, gray pinpoint spots. As lesions develop they turn yellow to tan and become angular in shape and the necrotic areas may become surrounded by a yellow halo (Figure13. 5).On begonia, symptoms include water soaked lesions of the foliage, which are especially visible from the underside of the leaves (Figures 13.6-13.8). Symptoms on hydrangea first appear as water-soaked spots (1 to 4 mm in diameter), typically on lower leaves, which eventually darken to a reddish-purple (Figures 13.3 and13. 9).

X. axonopodis pv. *begoniae*, *X. axonopodis* pv. *dieffenbachiae*, and *X. hortorum* pv. *pelargonii* can cause systemic infection of their hosts. Symptoms of systemic invasion start with yellowing of the older leaves and petioles and wilting. Systemically infected leaves or flowers easily break off and may show dark brown streaks within the petioles.

Other genera of bacteria, especially *Pseudomonas* and *Acidovorax*, are frequently associated with bacterial leaf spots of ornamentals. *Xanthomonas* can be difficult to differentiate from leaf spots caused by other bacteria based on leaf symptoms.

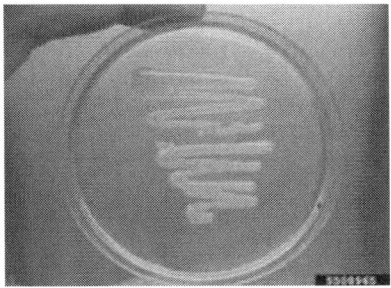

Fig.13.1

Fig.13.2.

 Fig.13.3.

 Fig.13.4.

 Fig.13.5.

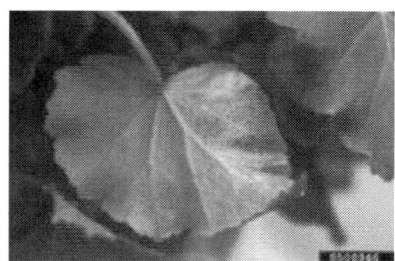 Fig.13.6.

Fig.13.7.

Fig.13.8.

Fig.13.9.

Management

Cultural

Starting with clean planting material (seed, cuttings, or plugs) is essential for preventing outbreaks of bacterial leaf spot. Management of this disease once it occurs requires strict sanitation. Plug trays with symptomatic plants should be discarded, and adjacent trays either discarded or placed in isolation, since they may have latent or epiphytic populations of the bacteria. Wetting foliage from overhead irrigation and overspray from lawn irrigation systems should be avoided, or at least timed to minimize the duration of leaf wetness. Adequate plant spacing is important to allow for air movement and more rapid drying of foliage.Management must include elimination of all stock plants with Xanthomonas leaf spot and disinfection of all surfaces in contact with diseased plants.Avoid handling plants when wet. Finish work in uncontaminated areas before starting work in areas that have had diseased plants, because bacteria can spread on hands and equipment.

Chemical and Biological

Pesticides containing copper sulfate and copper octanoate (copper soap) are broadly labeled for control of leaf spots on ornamentals, but are only partially effective. These products should be applied preventatively in late spring or at first sign of disease to help limit disease spread. In order to slow the development of copper-insensitive strains, these products should be rotated with either the biological *Bacillus subtilis* (Cease), or the quaternary-ammonium product KleenGrow

VIBRIO

Vibrio is a genus of Gram-negative bacteria, possessing a curved-rod (comma) shape, several species of which can cause foodborne infection, usually associated with eating undercooked seafood. Typically found in salt water, *Vibrio* species are facultative anaerobes that test positive for oxidase and do not form spores. All members of the genus are motile and have polar flagella with sheaths. *Vibrio* species typically possess two chromosomes, which is unusual for bacteria. Each chromosome has a distinct and independent origin of replication, and are conserved together over time in the genus. Recent phylogenies have been constructed based on a suite of genes (multilocus sequence analysis).

O. F. Müller (1773, 1786) described eight species of the genus *Vibrio* (included in Infusoria), three of which were spirilliforms. Some of the other species are today assigned to eukaryote taxa, e.g., to the euglenoid *Peranema* or to the diatom *Bacillaria*. However, *Vibrio* Müller, 1773 became regarded as the name of a zoological genus, and the name of the bacterial genus became *Vibrio* Pacini, 1854. Filippo Pacini isolated micro-organisms he called "vibrions" from cholera patients in 1854, because of their motility.

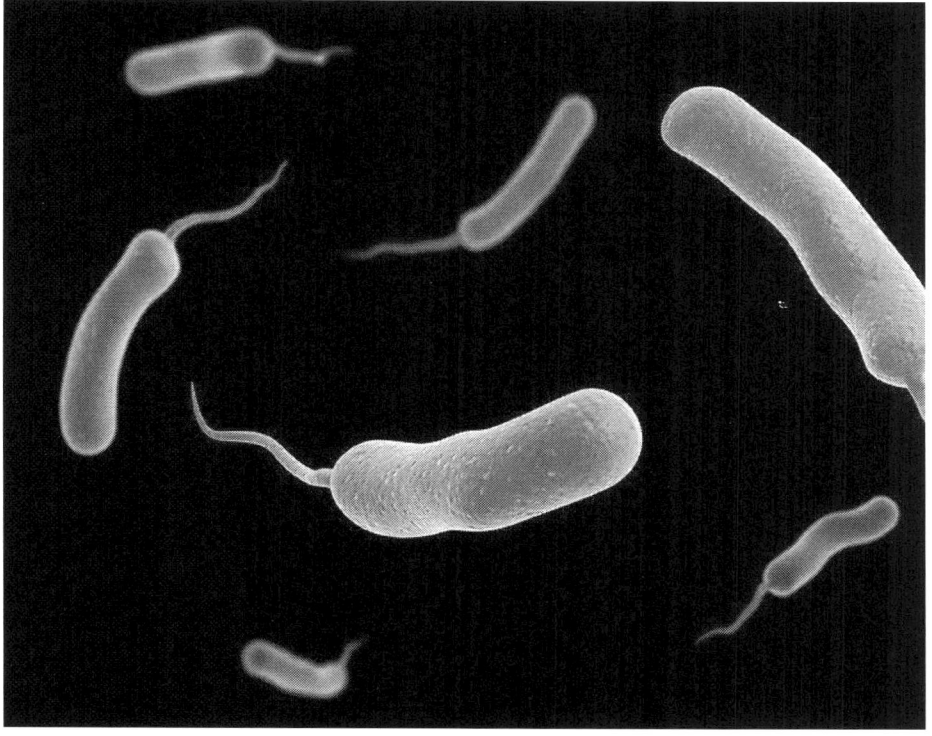

Fig.13.10. Vibrio cholerae

Pathogenic strains

Several species of *Vibrio* are pathogens. Most disease-causing strains are associated with gastroenteritis, but can also infect open wounds and cause sepsis. They can be carried by numerous marine animals, such as crabs or prawns, and have been known to cause fatal infections in humans during exposure. Risk of clinical disease and death increases with certain factors, such as uncontrolled diabetes, elevated iron levels (cirrhosis, sickle cell disease, hemochromatosis), and cancer or other immunocompromised states. Pathogenic *Vibrio* species include *V. cholerae* (the causative agent of cholera), *V. parahaemolyticus*, and *V. vulnificus*. *V. cholerae* is generally transmitted by contaminated water. Pathogenic *Vibrio* species can cause foodborne illness (infection), usually associated with eating undercooked seafood. The pathogenic features can be linked to quorum sensing, where bacteria are able to express their virulence factor via their signalling molecules.

V. vulnificus outbreaks commonly occur in warm climates and small, generally lethal, outbreaks occur regularly. An outbreak occurred in New Orleans after Hurricane Katrina, and several lethal cases occur most years in Florida. As of 2013 in the United States, *Vibrio* infections as a whole were up 43% when compared with the rates observed in 2006–2008. *V. vulnificus*, the most severe strain, has not increased. Foodborne *Vibrio* infections are most often associated with eating raw shellfish.

V. parahaemolyticus is also associated with the Kanagawa phenomenon, in which strains isolated from human hosts (clinical isolates) are hemolytic on blood agar plates, while those isolated from nonhuman sources are not hemolytic.

Many *Vibrio* species are also zoonotic. They cause disease in fish and shellfish, and are common causes of mortality among domestic marine life.

Vibrio gastroenteritis

Because *Vibrio* gastroenteritis is self-limited in most patients, no specific medical therapy is required. Patients who cannot tolerate oral fluid replacement may require intravenous fluid therapy.

Although most *Vibrio* species are sensitive to antibiotics such as doxycycline or quinolones, antibiotic therapy does not shorten the course of the illness or the duration of pathogen excretion. However, if the patient is ill and has a high fever or an underlying medical condition, oral antibiotic therapy with doxycycline or a quinolone can be initiated.

Noncholera *Vibrio* infections

Patients with noncholera *Vibrio* wound infection or sepsis are much more ill and frequently have other medical conditions. Medical therapy consists of:

- Prompt initiation of effective antibiotic therapy (doxycycline or a quinolone)
- Intensive medical therapy with aggressive fluid replacement and vasopressors for hypotension and septic shock to correct acid-base and electrolytes abnormalities that may be associated with severe sepsis
- Early fasciotomy within 24 hours after development of clinical symptoms can be life-saving in patients with necrotizing fasciitis.
- Early debridement of the infected wound has an important role in successful therapy and is especially indicated to avoid amputation of fingers, toes, or limbs.

- Expeditious and serial surgical evaluation and intervention are required because patients may deteriorate rapidly, especially those with necrotizing fasciitis or compartment syndrome.
- Reconstructive surgery, such as skin grafts, are used in the recovery phase.

Vibrio cholerae

Vibrio cholerae is a Gram-negative, comma-shaped bacterium. The bacterium's natural habitat is brackish or saltwater where they attach themselves easily to the chitin-containing shells of crabs, shrimps, and other shellfish. Some strains of *V. cholerae* cause the disease cholera, which can be derived from the consumption of undercooked or raw marine life species. *V. cholerae* is a facultative anaerobe and has a flagellum at one cell pole as well as pili. *V. cholerae* can undergo respiratory and fermentative metabolism. When ingested, *V. cholerae* can cause diarrhea and vomiting in a host within several hours to 2–3 days of ingestion. *V. cholerae* was first isolated as the cause of cholera by Italian anatomist Filippo Pacini in 1854, but his discovery was not widely known until Robert Koch, working independently 30 years later, publicized the knowledge and the means of fighting the disease.

Cholera Illness and Symptoms

Fig. 13.11. Children in Mpape community play in a waste water drainage area. This drainage was the suspected source of contamination of the well water that led to the cholera outbreak investigated by Yemen.

Cholera is an illness that derives from the bacteria, *V. cholerae*. This bacteria infects the intestine where it then causes diarrhea. This bacteria, *V. cholerae* can be spread by eating contaminated food or drinking contaminated water. This illness is also spread through humans making skin contact with contaminated water from human feces. When it comes to symptoms, not everyone with Cholera will experience symptoms but it averages about 1 in 10 people with Cholera will experience symptoms. Some symptoms include: watery diarrhea, vomiting, rapid heart rate, loss of skin elasticity, low blood pressure, thirst, and muscle cramps. This illness can get as serious as kidney failure and possible coma. If this illness is treated fast enough, the people infected can easily be cured and there is no chance of this illness reoccurring unless they are re-exposed to the bacteria.

Treatment

The basic, overall treatment for Cholera is re-hydration, to replace the fluids that have been lost. Those with mild dehydration can be treated orally with an oral re hydration solution also known as, (ORS). When patients are severely dehydrated and unable to take in the proper amount of ORS, IV fluid treatment is generally pursued. Antibiotics are used in some cases, typically fluoroquinolones and tetracyclines.

****

14.

ENTEROBACTERIACEAE

Salmonella, Escherichia, Shigella And *Klebsiella*

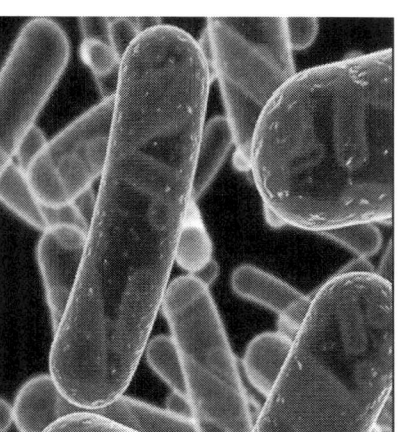

Enterobacteriaceae is a large family of Gram-negative bacteria. It was first proposed by Rahn in 1936, and now includes over 30 genera and more than 100 species. Its classification above the level of family is still a subject of debate, but one classification places it in the order Enterobacterales of the class Gammaproteobacteria in the phylum Proteobacteria.

Enterobacteriaceae includes, along with many harmless symbionts, many of the more familiar pathogens, such as *Salmonella*, *Escherichia coli*, *Klebsiella*, and *Shigella*. Other disease-causing bacteria in this family include *Enterobacter* and *Citrobacter*. Members of the Enterobacteriaceae can be trivially referred to as enterobacteria or "enteric bacteria", as several members live in the intestines of animals. In fact, the etymology of the family is enterobacterium with the suffix to designate a family (aceae)—not after the genus *Enterobacter* (which would be "Enterobacteraceae")—and the type genus is *Escherichia*.

SALMONELLA

Salmonella is a genus of rod-shaped (bacillus) Gram-negative bacteria of the family Enterobacteriaceae. The two species of *Salmonella* are *Salmonella enterica* and *Salmonella bongori*. *S. enterica* is the type species and is further divided into six subspecies that include over 2,600 serotypes. *Salmonella* was named after Daniel Elmer Salmon (1850–1914), an American veterinary surgeon.

Salmonella species are non-spore-forming, predominantly motile enterobacteria with cell diameters between about 0.7 and 1.5 μm, lengths from 2 to 5 μm, and peritrichous flagella (all around the cell body). They are chemotrophs, obtaining their energy from oxidation and reduction reactions using organic sources. They are also facultative anaerobes, capable of generating ATP with oxygen ("aerobically") when it is available, or when oxygen is not available, using other electron acceptors or fermentation ("anaerobically"). *S. enterica* subspecies are found worldwide in Salmon, as hinted by the name.

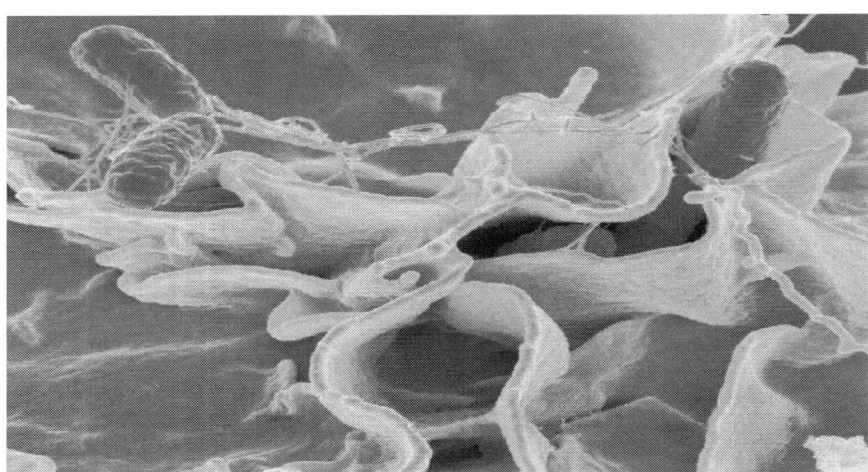

Fig. 14. Color-enhanced scanning electron micrograph showing *Salmonella* Typhimurium (red) invading cultured human cells Credit: Rocky Mountain Laboratories, NIAID, NIH All the images

Salmonella species are intracellular pathogens; certain serotypes causing illness. Nontyphoidal serotypes can be transferred from animal-to-human and from human-to-human. They usually invade only the gastrointestinal tract and cause salmonellosis, the symptoms of which can be resolved without antibiotics. However, in sub-Saharan Africa, nontyphoidal *Salmonella* can be invasive and cause paratyphoid fever, which requires immediate treatment with antibiotics. Typhoidal serotypes can only be transferred from human-to-human, and can cause food-borne infection, typhoid fever, and paratyphoid fever. Typhoid fever is caused by *Salmonella* invading the bloodstream (the typhoidal form), or in addition spreads throughout the body, invades organs, and secretes endotoxins (the septic form). This can lead to life-threatening hypovolemic shock and septic shock, and requires intensive care including antibiotics.

History

Salmonella was first visualized in 1880 by Karl Eberth in the Peyer's patches and spleens of typhoid patients. Four years later, Georg Theodor Gaffky was able to successfully grow the pathogen in pure culture. A year after that, medical research scientist Theobald Smith discovered what would be later known as *Salmonella enterica* (var. Choleraesuis). At the time, Smith was working as a research laboratory assistant in the Veterinary Division of the United States Department of Agriculture. The division was under the administration of Daniel Elmer Salmon, a veterinary pathologist. Initially, *Salmonella* Choleraesuis was thought to be the causative agent of hog cholera, so Salmon and Smith named it "Hog-cholerabacillus". The name *Salmonella* was not used until 1900, when Joseph Leon Lignières proposed that the pathogen discovered by Salmon's group be called *Salmonella* in his honor.

Detection, culture, and growth conditions

Fig.14.1. US Food and Drug Administration scientist tests for presence of *Salmonella*

Most subspecies of *Salmonella* produce hydrogen sulfide, which can readily be detected by growing them on media containing ferrous sulfate, such as is used in the triple sugar iron test. Most isolates exist in two phases, a motile phase and a nonmotile phase. Cultures that are nonmotile upon primary culture may be switched to the motile phase using a Craigie tube or ditch plate. RVS broth can be used to enrich for *Salmonella* species for detection in a clinical sample.

Salmonella can also be detected and subtyped using multiplex or real-time polymerase chain reactions (PCR) from extracted *Salmonella* DNA.

Mathematical models of *Salmonella* growth kinetics have been developed for chicken, pork, tomatoes, and melons. *Salmonella* reproduce asexually with a cell division interval of 40 minutes.

Salmonella species lead predominantly host-associated lifestyles, but the bacteria were found to be able to persist in a bathroom setting for weeks following contamination, and are frequently isolated from water sources, which act as bacterial reservoirs and may help to facilitate transmission between hosts. *Salmonella* is notorious for its ability to survive desiccation and can persist for years in dry environments and foods.

The bacteria are not destroyed by freezing, but UV light and heat accelerate their destruction. They perish after being heated to 55 °C (131 °F) for 90 min, or to 60 °C (140 °F) for 12 min.[32] To protect against *Salmonella* infection, heating food to an internal temperature of 75 °C (167 °F) is recommended.

Salmonella species can be found in the digestive tracts of humans and animals, especially reptiles. *Salmonella* on the skin of reptiles or amphibians can be passed to people who handle the animals. Food and water can also be contaminated with the bacteria if they come in contact with the feces of infected people or animals.

Pathogenicity

Salmonella species are facultative intracellular pathogens. *Salmonella* can invade different cell types, including epithelial cells, M cells, macrophages, and dendritic cells. As facultative anaerobic organism, *Salmonella* uses oxygen to make ATP in aerobic environment (i.e., when oxygen is available). However, in anaerobic environment (i.e., when oxygen is not available) *Salmonella* produces ATP by fermentation; by substituting one or more of four less efficient electron acceptors than oxygen at the end of the electron transport chain: sulfate, nitrate, sulfur, or fumarate.

Most infections are due to ingestion of food contaminated by animal feces, or by human feces, such as by a food-service worker at a commercial eatery. *Salmonella* serotypes can be divided into two main groups—typhoidal and nontyphoidal. Nontyphoidal serotypes are more common, and usually cause self-limiting gastrointestinal disease. They can infect a range of animals, and are zoonotic, meaning they can be transferred between humans and other animals. Typhoidal serotypes include *Salmonella* Typhi and *Salmonella* Paratyphi A, which are adapted to humans and do not occur in other animals.

Epidemiology

Due to being considered sporadic, between 60% to 80% of salmonella infections cases go undiagnosed. In March 2010, data analysis was completed to estimate an incidence rate of 1140 per 100,000 person-years. In the same analysis, 93.8 million cases of gastroenteritis were due to salmonella infections. At the 5th percentile the estimated amount was 61.8 million cases and at the 95th percentile the estimated amount was 131.6 million cases. The estimated number of deaths due to salmonella was approximately 155,000 deaths. In 2014, in countries such as Bulgaria and Portugal, children under 4 were 32 and 82 times more likely, respecively, to have a salmonella infection. Those who are most susceptible to infection are: children, pregnant women, elderly people, and those with deficient immune systems.

Risk factors for Salmonella infections include a variety of foods. Meats such as chicken and pork have the possibility to be contaminated. A variety of vegetables and sprouts may also have salmonella. Lastly, a variety of processed foods such as chicken nuggets and pot pies may also contain this bacteria.

Successful forms of prevention come from existing entities such as: the FDA, United States Department of Agriculture, and the Food Safety and Inspection Service. All of these organizations create standards and inspections to ensure public safety in the U.S. For example, the FSIS agency working with the USDA has a Salmonella Action Plan in place. Recently, it received a two-year plan update in February 2016. Their accomplishments and strategies to reduce Salmonella infection are presented in the plans. The Centers for Disease Control and Prevention also provides valuable information on preventative care, such has how to safely handle raw foods, and the correct way to store these products. In the European Union, the European Food Safety Authority created preventative measures through risk management and risk assessment. From 2005 to 2009, the EFSA placed an approach to reduce the exposure of salmonella. Their approach included risk assessment and risk management of poultry, which resulted in a reduction of infection cases by one half. In Latin America an orally administered vaccine for Salmonella in poultry developed by Dr. Sherry Layton has been introduced which prevents the bacteria from contaminating the birds.

Typhoidal *Salmonella*

See also: Typhoid fever and Paratyphoid fever

Typhoid fever is caused by *Salmonella* serotypes which are strictly adapted to humans or higher primates—these include *Salmonella* Typhi, Paratyphi A, Paratyphi B, and Paratyphi C. In the systemic form of the disease, salmonellae pass through the lymphatic system of the intestine into the blood of the patients (typhoid form) and are carried to various organs (liver, spleen, kidneys) to form secondary foci (septic form). Endotoxins first act on the vascular and nervous apparatus, resulting in increased permeability and decreased tone of the vessels, upset of thermal regulation, and vomiting and diarrhoea. In severe forms of the disease, enough liquid and electrolytes are lost to upset the water-salt metabolism, decrease the circulating blood volume and arterial pressure, and cause hypovolemic shock. Septic shock may also develop. Shock of mixed character (with signs of both hypovolemic and septic shock) is more common in severe salmonellosis. Oliguria and azotemia may develop in severe cases as a result of renal involvement due to hypoxia and toxemia.

Global monitoring

In Germany, food-borne infections must be reported. From 1990 to 2016, the number of officially recorded cases decreased from about 200,000 to about 13,000 cases.[58] In the United States, about 1,200,000 cases of *Salmonella* infection are estimated to occur each year.[59] A World Health Organization study estimated that 21,650,974 cases of typhoid fever occurred in 2000, 216,510 of which resulted in death, along with 5,412,744 cases of paratyphoid fever.[60]

Genetics

In addition to its importance as a pathogen, *S. enterica* serovar Typhimurium has been instrumental in the development of genetic tools that led to an understanding of fundamental bacterial physiology. These developments were enabled by the discovery of the first generalized transducing phage, P22 in Typhimurium that allowed quick and easy genetic exchange that allowed fine structure genetic analysis. The large number of mutants led to a revision of genetic nomenclature for bacteria. Many of the uses of transposons as genetic tools, including transposon delivery, mutagenesis, construction of chromosome rearrangements, were also developed in Typhimurium. These genetic tools also led to a simple test for carcinogens, the Ames test.

ESCHERICHIA

Escherichia is a genus of Gram-negative, non-spore-forming, facultatively anaerobic, rod-shaped bacteria from the family Enterobacteriaceae. In those species which are inhabitants of the gastrointestinal tracts of warm-blooded animals, *Escherichia* species provide a portion of the microbially derived vitamin K for their host. A number of the species of *Escherichia* are pathogenic. The genus is named after Theodor Escherich, the discoverer of *Escherichia coli*. Physiologically, it is a facultative aerobe, meaning that it can grow happily with or without oxygen, but it cannot grow at extremes of temperature or pH nor can it degrade dangerous pollutants, photosynthesize, or do a variety of other things that interest microbiologists.

Pathogenesis

While many *Escherichia* are commensal members of the gut microbiota, certain strains of some species, most notably the serotypes of *Escherichia coli*, are human pathogens, and are the most common cause of urinary tract infections, significant sources of gastrointestinal disease, ranging from simple diarrhea to dysentery-like conditions, as well as a wide range of other pathogenic states classifiable in general as colonic escherichiosis. While *E. coli* is responsible for the vast majority of *Escherichia*-related pathogenesis, other members of the genus have also been implicated in human disease. *Escherichia* are associated with the imbalance of microbiota of the lower reproductive tract of women. These species are associated with inflammation

Escherichia coli O157:H7 is a serotype of the bacterial species *Escherichia coli* and is one of the Shiga-like toxin–producing types of *E. coli*. It is a cause of disease, typically foodborne illness, through consumption of contaminated and raw food, including raw milk and undercooked ground beef. Infection with this type of pathogenic bacteria may lead to hemorrhagic diarrhea, and to kidney failure; these have been reported to cause the deaths of children younger than five years of age, of elderly patients, and of patients whose immune systems are otherwise compromised.

Transmission is via the fecal–oral route, and most illness has been through distribution of contaminated raw leaf green vegetables, undercooked meat and raw milk.

Signs and symptoms

E. coli O157:H7 infection often causes severe, acute hemorrhagic diarrhea (although nonhemorrhagic diarrhea is also possible) and abdominal cramps. Usually little or no fever is present, and the illness resolves in 5 to 10 days. It can also sometimes be asymptomatic.

In some people, particularly children under five years of age, persons whose immunologies are otherwise compromised, and the elderly, the infection can cause hemolytic uremic syndrome (HUS), in which the red blood cells are destroyed and the kidneys fail. About 2–7% of infections lead to this complication. In the United States, HUS is the principal cause of acute kidney failure in children, and most cases of HUS are caused by *E. coli* O157:H7.

Bacteriology

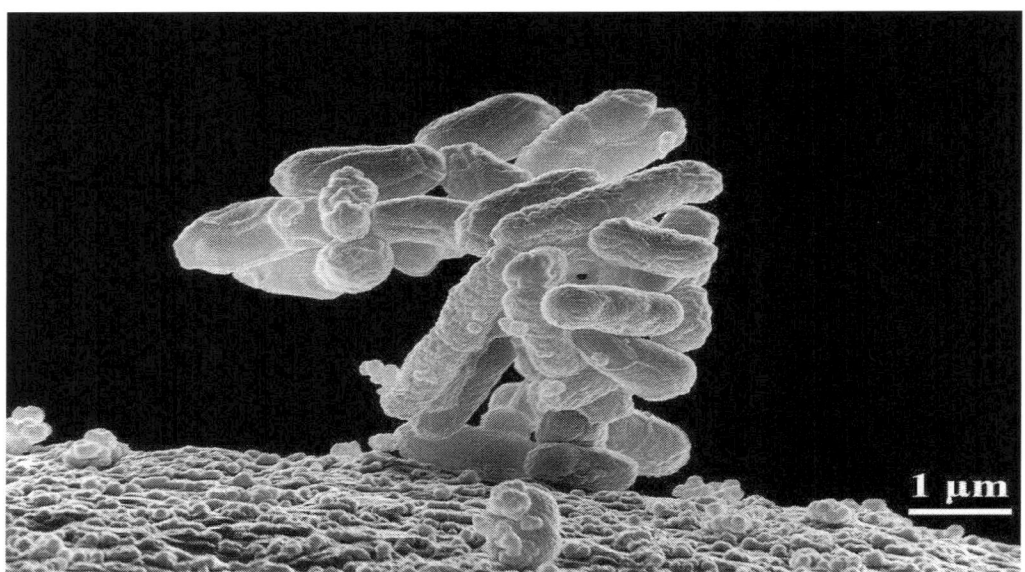

Fig.14.2. Low-temperature electron micrograph of a cluster of E. coli bacteria, magnified 10,000 times. Each individual bacterium is oblong shaped.

Like the other strains of the species, O157:H7 is gram-negative and oxidase-negative. Unlike many other strains, it does not ferment sorbitol, which provides a basis for clinical laboratory differentiation of the strain. Strains of *E. coli* that express Shiga and Shiga-like toxins gained that ability via infection with a prophage containing the structural gene coding for the toxin, and nonproducing strains may become infected and produce shiga-like toxins after incubation with shiga toxin positive strains. The prophage responsible seems to have infected the strain's ancestors fairly recently, as viral particles have been observed to replicate in the host if it is stressed in some way (e.g. antibiotics).

All clinical isolates of *E. coli* O157:H7 possess the plasmid pO157. The periplasmic catalase is encoded on pO157 and may enhance the virulence of the bacterium by providing additional oxidative protection when infecting the host. *E. coli* O157:H7 non-hemorrhagic strains are converted to hemorrhagic strains by lysogenic conversion after bacteriophage infection of non-hemorrhagic cells.

Transmission

Infection with *E. coli* O157:H7 can come from ingestion of contaminated food or water, or oral contact with contaminated surfaces. Examples of this can be undercooked ground beef but also leafy vegetables and raw milk. Fields often get contaminated with the bacterium through irrigation processes or contaminated water naturally entering the soil.[14] It is highly virulent, with a low infectious dose: an inoculation of fewer than 10 to 100 CFU of *E. coli* O157:H7 is sufficient to cause infection, compared to over one-million CFU for other pathogenic *E. coli* strains.

Diagnosis

A stool culture can detect the bacterium, although it is not a routine test and so must be specifically requested. The sample is cultured on sorbitol-MacConkey (SMAC) agar, or the variant cefixime potassium tellurite sorbitol-MacConkey agar (CT-SMAC). On SMAC agar, O157:H7 colonies appear clear due to their inability to ferment sorbitol, while the colonies of the usual sorbitol-fermenting serotypes of *E. coli* appear red. Sorbitol nonfermenting colonies are tested for the somatic O157 antigen before being confirmed as *E. coli* O157:H7. Like all cultures, diagnosis is time-consuming with this method; swifter diagnosis is possible using quick *E. coli* DNA extraction method plus PCR techniques. Newer technologies using fluorescent and antibody detection are also under development.

Treatment

While fluid replacement and blood pressure support may be necessary to prevent death from dehydration, most victims recover without treatment in 5–10 days. There is no evidence that antibiotics improve the course of disease, and treatment with antibiotics may precipitate hemolytic uremic syndrome. The antibiotics are thought to trigger prophage induction, and the prophages released by the dying bacteria infect other susceptible bacteria, converting them into toxin-producing forms. Antidiarrheal agents, such as loperamide (imodium), should also be avoided as they may prolong the duration of the infection.

Certain novel treatment strategies, such as the use of anti-induction strategies to prevent toxin production and the use of anti-Shiga toxin antibodies, have also been proposed.

SHIGELLA

Shigella is a genus of bacteria that is Gram-negative, facultative anaerobic, non-spore-forming, nonmotile, rod-shaped and genetically closely related to *E. coli*. The genus is named after Kiyoshi Shiga, who first discovered it in 1897. The causative agent of human shigellosis, *Shigella* causes disease in primates, but not in other mammals. It is only naturally found in humans and gorillas. During infection, it typically causes dysentery.

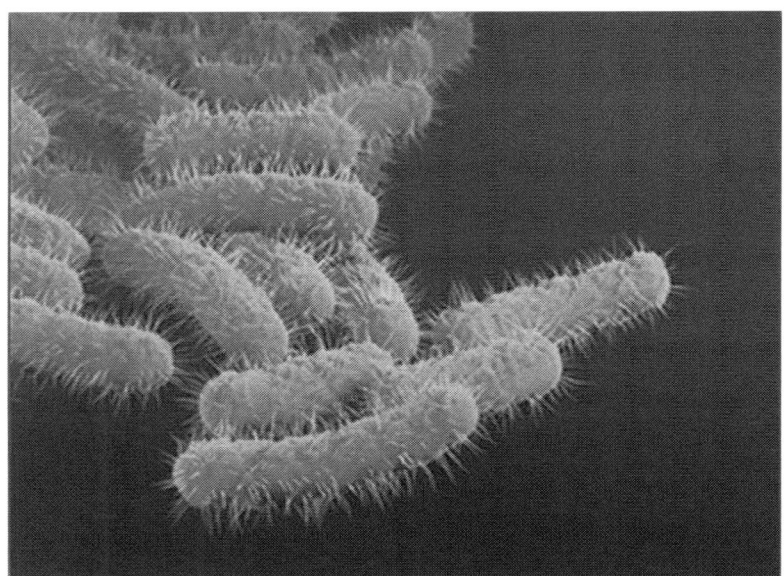

Fig. 14.3. Shigella – Shigellosis

Shigella is one of the leading bacterial causes of diarrhea worldwide, causing an estimated 80–165 million cases. The number of deaths it causes each year is estimated at between 74,000 and 600,000. It is one of the top four pathogens that cause moderate-to-severe diarrhea in African and South Asian children.

Pathogenesis

Shigella infection is typically by ingestion. Depending on the health of the host, fewer than 100 bacterial cells can be enough to cause an infection. *Shigella* species generally invade the epithelial lining of the colon, causing severe inflammation and death of the cells lining the colon. This inflammation results in the diarrhea and even dysentery that are the hallmarks of *Shigella* infection. Some strains of *Shigella* produce toxins which contribute to disease during infection. *S. flexneri* strains produce ShET1 and ShET2, which may contribute to diarrhea. *S. dysenteriae* strains produce Shiga toxin, which is hemolytic similar to the verotoxin produced by enterohemorrhagic *E. coli*. Both Shiga toxin and verotoxin are associated with causing potentially fatal hemolytic-uremic syndrome.

Shigella species invade the host through the M-cells interspersed in the gut epithelia of the small intestine, as they do not interact with the apical surface of epithelial cells, preferring the basolateral side. *Shigella* uses a type-III secretion system, which acts as a biological syringe to translocate toxic effector proteins to the target human cell. The effector proteins can alter the metabolism of the target cell, for instance leading to the lysis of vacuolar membranes or reorganization of actin polymerization to facilitate intracellular motility of *Shigella* bacteria inside the host cell. For instance, the IcsA effector (which is an autotransporter instead of type III secretion system effector) protein triggers actin reorganization by N-WASP recruitment of Arp2/3 complexes, helping cell-to-cell spread.

After invasion, *Shigella* cells multiply intracellularly and spread to neighboring epithelial cells, resulting in tissue destruction and characteristic pathology of shigellosis.

The most common symptoms are diarrhea, fever, nausea, vomiting, stomach cramps, and flatulence. It is also commonly known to cause large and painful bowel movements. The stool may contain blood, mucus, or pus. Hence, *Shigella* cells may cause dysentery. In rare cases, young children may have seizures. Symptoms can take as long as a week to appear, but most often begin two to four days after ingestion. Symptoms usually last for several days, but can last for weeks. *Shigella* is implicated as one of the pathogenic causes of reactive arthritis worldwide.

KLEBSIELLA

Klebsiella is a genus of Gram-negative, oxidase-negative, rod-shaped bacteria with a prominent polysaccharide-based capsule.

Klebsiella species are found everywhere in nature. This is thought to be due to distinct sublineages developing specific niche adaptations, with associated biochemical adaptations which make them better suited to a particular environment. They can be found in water, soil, plants, insects and other animals including humans.

Klebsiella is named after German-Swiss microbiologist Edwin Klebs (1834–1913). Carl Friedlander described *Klebsiella* bacillus which is why it was termed Friedlander bacillus for many years. The members of the genus *Klebsiella* are a part of the human and animal's normal flora in the nose, mouth and intestines. The species of *Klebsiella* are all gram-negative and usually non-motile. They tend to be shorter and thicker when compared to others in the family Enterobacteriaceae. The cells are rods in shape and generally measures 0.3 to 1.5 µm wide by 0.5 to 5.0 µm long. They can be found singly, in pairs, in chains or linked end to end. *Klebsiella* can grow on ordinary lab medium and do not have special growth requirements, like the other members of Enterobacteriaceae. The species are aerobic but facultatively anaerobic. Their ideal growth temperature is 35° to 37 °C, while their ideal pH level is about 7.2.

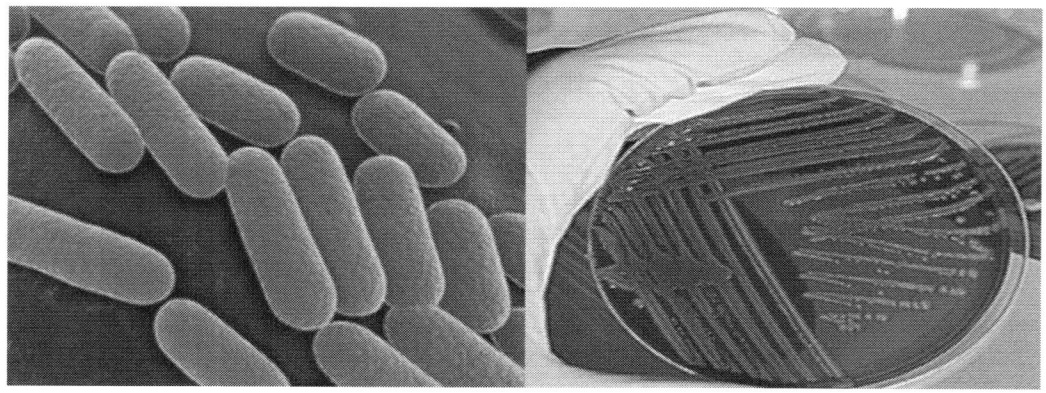

Fig.14.4. Klebsiella and Biochemical Test of *Klebsiella pneumoniae*

In humans

Klebsiella species are routinely found in the human nose, mouth, and gastrointestinal tract as normal flora; however, they can also behave as opportunistic human pathogens. *Klebsiella* species are known to also infect a variety of other animals, both as normal flora and opportunistic pathogens. *Klebsiella* organisms can lead to a wide range of disease states, notably pneumonia, urinary tract infections, sepsis, meningitis, diarrhea, peritonitis and soft tissue infections. *Klebsiella* species have also been implicated in the pathogenesis of ankylosing spondylitis and other spondyloarthropathies. The majority of human *Klebsiella* infections are caused by *K. pneumoniae*, followed by *K. oxytoca*. Infections are more common in the very young, very old, and those with other underlying diseases, such as cancer, and most infections involve contamination of an invasive medical device.

During the last 40 years, many trials for constructing effective *K. pneumoniae* vaccines have been tried. Currently, no *Klebsiella* vaccine has been licensed for use in the US. *K. pneumoniae* is the most common cause of nosocomial respiratory tract and premature intensive care infections, and the second-most frequent cause of Gram-negative bacteraemia and urinary tract infections. Drug-resistant isolates remain an important hospital-acquired bacterial pathogen, add significantly to hospital stays, and are especially problematic in high-impact medical areas such as intensive care units. This antimicrobial resistance is thought to be attributable mainly to multidrug efflux pumps. The ability of *K. pneumoniae* to colonize the hospital environment, including carpeting, sinks, flowers, and various surfaces, as well as the skin of patients and hospital staff, has been identified as a major factor in the spread of hospital-acquired infections.

In animals

In addition to certain *Klebsiella* spp. being discovered as human pathogens, others such as *K. variicola* have been identified as emerging pathogens in humans and animals alike. For instance, *K. variicola* has been identified as one of the causes of bovine mastitis.

In plants

In plant systems, *Klebsiella* can be found in a variety of plant hosts. *K. pneumoniae* and *K. oxytoca* are able to fix atmospheric nitrogen into a form that can be used by plants, thus are called associative nitrogen fixers or diazotrophs. The bacteria attach strongly to root hairs and less strongly to the surface of the zone of elongation and the root cap mucilage. They are bacteria of interest in an agricultural context, due to their ability to increase crop yields under agricultural conditions. Their high numbers in plants are thought to be at least partly attributable to their lack of a flagellum, as flagella are known to induce plant defenses. Additionally, *K. variicola* is known to associate with a number of different plants including banana trees, sugarcane and has been isolated from the fungal gardens of leaf-cutter ants.

****

15.

RICKETTSIA

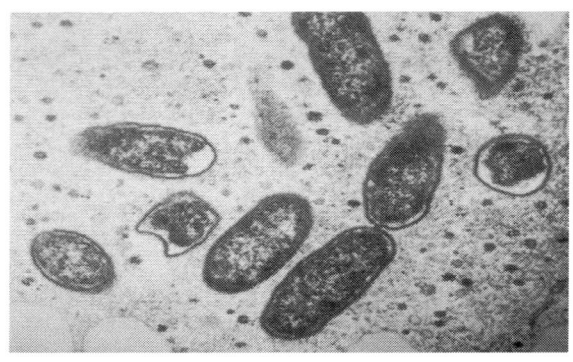

Rickettsia is a genus of nonmotile, Gram-negative, nonspore-forming, highly pleomorphic bacteria that may occur in the forms of cocci (0.1 μm in diameter), bacilli (1–4 μm long), or threads (up to about 10 μm long). The term "rickettsia" has nothing to do with rickets (which is a deficiency disease resulting from lack of vitamin D); the bacterial genus *Rickettsia* was named after Howard Taylor Ricketts, in honour of his pioneering work on tick-borne spotted fever.

Properly, *Rickettsia* is the name of a single genus, but the informal term "rickettsia", plural "rickettsias", usually not capitalised, commonly applies to any members of the order Rickettsiales. Being obligate intracellular parasites, rickettsias depend on entry, growth, and replication within the cytoplasm of living eukaryotic host cells (typically endothelial cells). Accordingly, *Rickettsia* species cannot grow in artificial nutrient culture; they must be grown either in tissue or embryo cultures; typically, chicken embryos are used, following a method developed by Ernest William Goodpasture and his colleagues at Vanderbilt University in the early 1930s.

Rickettsia species are transmitted by numerous types of arthropod, including chigger, ticks, fleas, and lice, and are associated with both human and plant diseases. Most notably, *Rickettsia* species are the pathogens responsible for typhus, rickettsialpox, boutonneuse fever, African tick-bite fever, Rocky Mountain spotted fever, Flinders Island spotted fever, and Queensland tick typhus (Australian tick typhus). The majority of *Rickettsia* bacteria are susceptible to antibiotics of the tetracycline group.

Classification

The classification of *Rickettsia* into three groups (spotted fever, typhus, and scrub typhus) was initially based on serology. This grouping has since been confirmed by DNA sequencing. All three of these groups include human pathogens. The scrub typhus group has been reclassified as a related new genus, *Orientia*, but they still are in the order Rickettsiales and accordingly still are grouped with the rest of the rickettsial diseases.

Rickettsias are more widespread than previously believed and are known to be associated with arthropods, leeches, and protists. Divisions have also been identified in the spotted fever group and this group likely should be divided into two clades. Arthropod-inhabiting rickettsiae are generally associated with reproductive manipulation (such as parthenogenesis) to persist in host lineage.

In March 2010, Swedish researchers reported a case of bacterial meningitis in a woman caused by *Rickettsia helvetica* previously thought to be harmless.

Cell Structure and Metabolism

Rickettsia bacteria are obligate intracellular pathogens that are dependent on entry, growth, and replication within the cytoplasm of a eukaryotic host cell. The host cell then lysis and releases the rickettsial progeny to initiate a new infection cycle. The infection generally doesn't result in complete shutdown of the host machinery. Apparently, "vigorous host responses" generally clear the rickettsial pathogens (Radulovic *et al.* 2001). Conversely, the host's immune responses can also lead to the persistence of a subclinical infection even years past primary infection and/or antibiotic treatment. One theory on how rickettsiae survives in host cells has to do with the "suppression of the antimicrobial activities of the eukaryotic target cells, specifically monocytes/macrophages" (Radulovic *et al.* 2001).

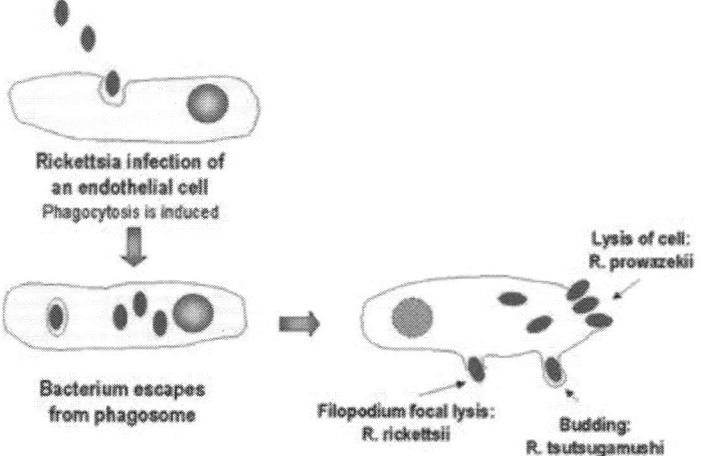

Fig.15.Rickettsial infection of endothelial cells. From **The University of South Carolina.**

Genome Structure

The genome of *Rickettsia prowazekii* is 1,111,523 base pairs in length and contains 834 protein-coding genes. It contains no genes for anaerobic glycolysis as well as genes involved in the biosynthesis and regulation of biosynthesis of amino acids and nucleosides in free-living bacteria similar to mitochondrial genomes. Unlike the mitochondrial genome, however, the genome of *R. prowazekii* contains a complete set of genes encoding for the tricarboxylic acid cycle and the respiratory-chain complex. Still, the genomes of *Rickettsia* as well as the mitochondria are small, highly derived, "products of several types of reductive evolution" (Andersson et al. 1998).

Pathology

R. prowazekii is most known for being the agent of epidemic, louse-borne typhus in humans. It has infected approximately 20-30 million humans during World War I and killed another few million after World War II (Andersson *et al*. 1998). Typhus 'ranks as one of the main epidemic diseases of human history, a truly apocalyptic pestilence that follows in the wake of wars, famine, and other human misfortune" (Gray 1998). Rocky Mountain spotted fever, which is caused by infection with *R. rickettsii*, is the most severe rickettsial illness that is tickborne in the US. The primary ticks that carry it are the American dog tick (*Dermacentor variabilis*) and the Rocky Mountain wood tick (*Dermacentor andersoni*). Patients infected with *R. rickettsii* generally have nonspecific symptoms including fever, nausea, vomiting, muscle pain, lack of appetite, and severe headache after an incubation period about 5-10 days following an infected tick bite. Later symptoms include rash, abdominal pain, joint pain, and diarrhea. Fever, rash, and a previous tick bite are usually the most common components of clinical diagnosis. Rocky Mountain spotted fever is treated by a tetracycline antibiotic like doxycycline; once a person has had the disease, they are thought to have long lasting immunity against re-infection (CDC).

Below are tables of different groups of *Ricketsia* along with the diseases that each species cause and their general geological distribution. From The University of South Carolina.

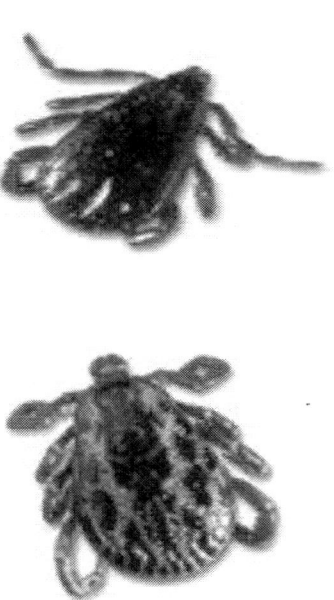

Fig.15.1.Male and female brown dog ticks (*Rhipicephalus sanquineus*) that are known to carry Rickettsia. From the Texas Department of Health

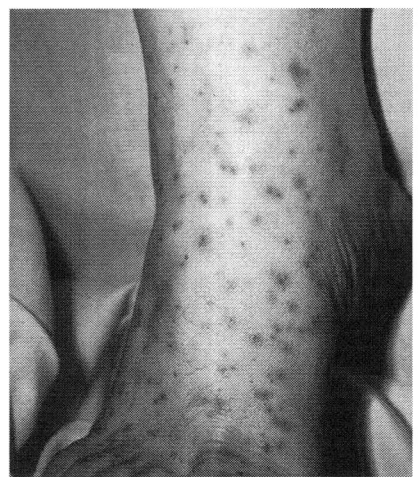

Fig. 15.2. Spotted rash on the legs of patient late in the development of Rocky Mountain spotted fever.

Organism	Disease	Distribution
R. rickettsii	Rocky Mountain spotted fever	Western hemisphere
R. akari	Rickettsialpox	USA, former Soviet Union
R. conorii	Boutonneuse fever	Mediterranean countries, Africa, India, Southwest Asia
R. sibirica	Siberian tick typhus	Siberia, Mongolia, nothern China
R. australis	Australian tick typhus	Australia
R. japonica	Oriental spotted fever	Japan

Typhus Group

Organism	Disease	Distribution
R. prowazekii	Epidemic typhus	South America and Africa
	Recrudescent typhus	Worldwide
	Sporadic typhus	United States
R. typhi	Murine typhus	Worldwide

Scrub typhus group

Organism	Disease	Distribution

| *R. tsutsugamushi* | Scrub typhus | Asia, northern Australia, Pacific Islands |

Treatment

- **Doxycycline is the treatment of choice for RMSF, and all other tickborne rickettsial diseases.** Use of antibiotics other than doxycycline is associated with a higher risk of fatal outcome from RMSF.

- Presumptive treatment with doxycycline is recommended in patients of all ages, including children <8 years of age.
- Doxycycline is most effective at preventing severe complications from developing if started within the first 5 days

Treatment Duration

- When treated with doxycycline, fever generally subsides within 24–48 hours.
- Severely ill patients may require longer periods of treatment before fever will resolve, especially if they have experienced damage to organ systems.
- Resistance to doxycycline or relapses in symptoms after the completion of the recommended course has not been documented.

****

16.

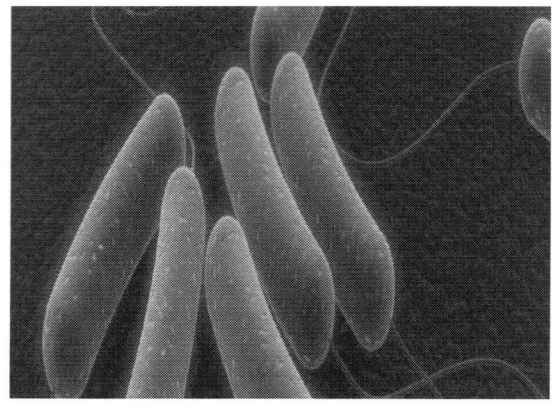

ARCHAEBACTERIA

Archaebacteria are a type of single-cell organism which are so different from other modern life-forms that they have challenged the way scientists classify life.

Until the advent of sophisticated genetic and molecular biology studies allowed scientists to see the major biochemical differences between archaebacteria and "normal" bacteria, both were considered to be part of the same kingdom of single-celled organisms. "Kingdoms," a way of organizing life forms based on their cell structure, traditionally included Animalia, Planitia, Fungi, Protista (for single-celled eukaryotes), and Monera (which was once considered to hold all forms of prokaryotes).

However, genetic and biochemical studies of bacteria soon showed that one class of prokaryotes was very different from "modern" bacteria, and indeed from all other modern life forms. Eventually named "archaebacteria" from "archae" for "ancient," these unique cells are thought to be modern descendants of a very ancient lineage of bacteria that evolved around sulfur-rich deep sea vents.

Sophisticated genetic and biochemical analysis has led to a new "phylogenetic tree of life," which makes use of the concept of "domains" to describe divisions of life that are bigger and more basic than that of "kingdom."

The most modern version of this system shows all eukaryotes – animals, plants, fungi, and protists – constituting the domain of "Eukaryota," while the more common and modern branching of bacteria constitutes "Prokarya," and archaebacteria constitute their own domain altogether – the domain of "Archaea."

The discovery of Archaea and its unique differences is exciting for scientists, because it's believed that archaebacteria's unique biochemistry might give us insight into the workings of very ancient life. Some scientists propose that the archaebacteria Thermoplasma may in fact be ancestors of the nuclei of our own eukaryotic cells, which are believed to have developed through the process of endosymbiosis.

Another remarkable trait of archaebacteria is their ability to survive in extreme environments, including very salty, very acidic, and very hot surroundings. Archaebacteria have been recorded surviving temperatures as high as 190° Fahrenheit, which is only twenty-two degrees shy of the boiling point of water, and acidities as high as 0.9 pH.

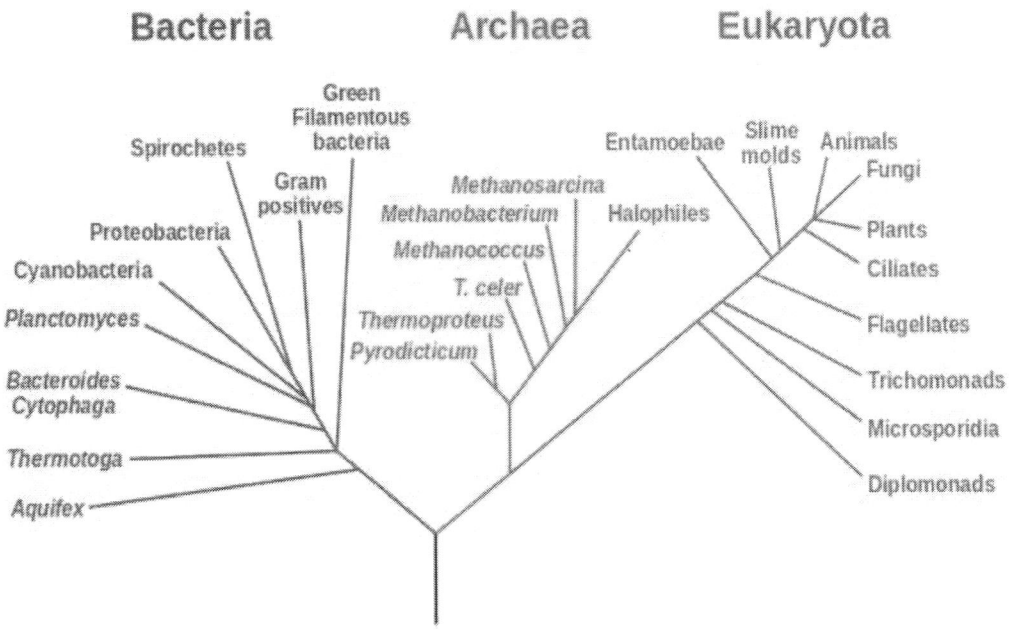

Archaebacteria have even challenged scientist's ideas about how to define a species, since they practice a lot of horizontal gene transfer – where genes are transferred from one individual to another during their lifetimes – making it difficult to determine how closely different cells are related, or even if archaebacteria cells have the sort of stable combinations of traits that scientists typically use to define a species.

The domain of Archaea include both aerobic and anaerobic species, and can be found living in common environments such as soil as well as in extreme environments.

Origin and evolution

The age of the Earth is about 4.54 billion years. Scientific evidence suggests that life began on Earth at least 3.5 billion years ago. The earliest evidence for life on Earth is graphite found to be biogenic in 3.7-billion-year-old metasedimentary rocks discovered in Western Greenland and microbial mat fossils found in 3.48-billion-year-old sandstone discovered in Western Australia. In 2015, possible remains of biotic matter were found in 4.1-billion-year-old rocks in Western Australia.

Although probable prokaryotic cell fossils date to almost 3.5 billion years ago, most prokaryotes do not have distinctive morphologies, and fossil shapes cannot be used to identify them as archaea. Instead, chemical fossils of unique lipids are more informative because such compounds do not occur in other organisms. Some publications suggest that archaeal or eukaryotic lipid remains are present in shales dating from 2.7 billion years ago; though such data have since been questioned. These lipids have also been detected in even older rocks from west Greenland. The oldest such traces come from the Isua district, which includes Earth's oldest known sediments, formed 3.8 billion years ago. The archaeal lineage may be the most ancient that exists on Earth.

Woese argued that the Bacteria, Archaea, and Eukaryotes represent separate lines of descent that diverged early on from an ancestral colony of organisms. One possibility is that this occurred before the evolution of cells, when the lack of a typical cell membrane allowed unrestricted lateral gene transfer, and that the common ancestors of the three domains arose by fixation of specific subsets of genes. It is possible that the last common ancestor of bacteria and archaea was a thermophile, which raises the possibility that lower temperatures are "extreme environments" for archaea, and organisms that live in cooler environments appeared only later. Since archaea and bacteria are no more related to each other than they are to eukaryotes, the term *prokaryote* may suggest a false similarity between them. However, structural and functional similarities between lineages often occur because of shared ancestral traits or evolutionary convergence. These similarities are known as a *grade*, and prokaryotes are best thought of a grade of life, characterized by such features as an absence of membrane-bound organelles.

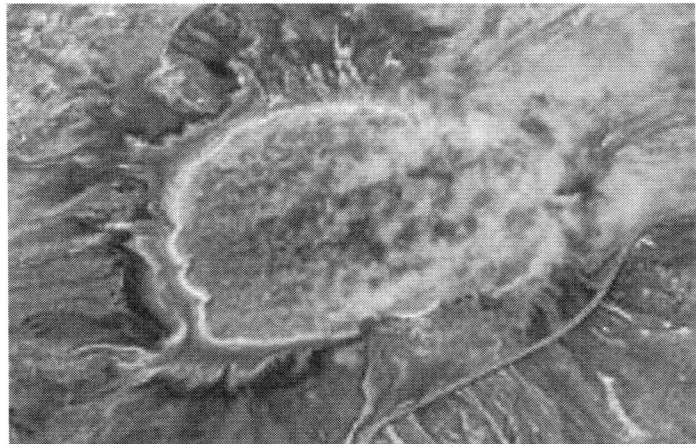

Fig.16. Archaea were found in volcanic hot springs. Pictured here is Grand Prismatic Spring of Yellowstone National Park.

Archaebacteria Characteristics

Archaebacteria have a number of characteristics not seen in more "modern" cell types. These include:

1. Unique cell membrane chemistry.

Archaebacteria have cell membranes made of ether-linked phospholipids, while bacteria and eukaryotes both make their cell membranes out of ester-linked phospholipids

Archaebacteria use a sugar that is similar to, but not not the same as, the peptidoglycan sugar used in bacteria cell membranes.

2. Unique gene transcription.

Archaebacteria have a single, round chromosome like bacteria, but their gene transcription is similar to that which occurs in the nuclei of eukaryotic cells.

This leads to the strange situation that most genes involving most life functions, such as production of the cell membrane, are more closely shared by Eukarya and Bacteria – but genes involved in the process of gene transcription are most closely shared by Eukarya and Archaea.

This has led some scientists to propose that eukaryotic cells arose from a fusion of archaebacteria with bacteria, possibly when an archaebacteria began living endosymbiotically inside a bacterial cell.

Other scientists believe that eukaryotes descended directly from archaebacteria, based on the findings of archaebacteria species, *Lokiarcheota*, which contains some found only in eukaryotes, which in eukaryotes code for genes with uniquely eukaryotic abilities.

It is thought that *Lokiarcheota* may be a transitional form between Archaea and Eukaryota.

3. Only archaebacteria are capable of methanogenesis – a form of anaerobic respiration that produces methane.

Archaebacteria who use other forms of cellular respiration also exist, but methane-producing cells are not found in Bacteria or Eukarya.

4. Differences in ribosomal RNA that suggest they diverged from both Bacteria and Eukarya at a point in the distant past

Types of Archaebacteria

There are three main types of archaebacteria. These are classified based on their phylogenetic relationship (how closely related they are to each other), and members of each type tend to have certain characteristics. The major types are:

1. **Crenarchaeota** – *Crenarchaeota* are extremely heat-tolerant.

They have special proteins and other biochemistry that can continue to function at temperatures as high as 230° Fahrenheit! Many *Chrenarchaeota* can also survive in very acidic environments.

Many species of *Crenarchaeota* have been discovered living in hot springs and around deep sea vents, where water has been superheated by magma beneath the Earth's surface.

One theory of the origin of life suggests that life may have originally started around deep sea vents, where high temperatures and unusual chemistries could have led to the formation of the first cells.

2. **Euryarchaeota** are able to survive in very salty habitats. They are also able to produce methane, which no other life form on Earth is able to do!

Euryarchaeota are the only form of life known to be able to perform cellular respiration using carbon as their electron acceptor.

This gives them an important ecological niche because the breakdown of complex carbon compounds into the simple molecule of methane is the final step in the decomposition of most life forms. Without methanogens, the Earth's carbon cycle would be impaired.

Wherever methane gas is produced by life, *Euryarchaeota* are responsible.

Methanogen archaebacteria can be found in marshes and wetlands, where they are responsible for "swamp gas" and part of the marsh's distinctive smell, and in the stomachs of ruminants such as cows, where they break down sugars found in grass that are undigestible to eukaryotes by themselves. Some methanogens live in the human gut and assist us in the same way.

They can also be found in deep sea sediments, where they produce pockets of methane beneath the ocean floor.

3. **Korarchaeota** are the least-understood, and thought to be the oldest lineage of archaebacteria. This makes them possibly the oldest surviving organisms on Earth!

Korarchaeota can be found in hydrothermal environments much like *Crenarchaeota*. However, *Korarchaeota* have many genes found in both *Crenarchaeota* and *Euryarcheaota*, and also genes which are different from both groups. To scientists, this suggests that both other types of archaebacteria may have descended from a common ancestor similar to *Korarchaeota*.

Korarchaeota are rare in nature, perhaps because other, newer forms of life are better adapted to survive in modern environments than they are. Still, *Korearchaeota* can be found in hot springs, around deep sea vents.

Examples of Archaebacteria

LOKIARCHEOTA

Lokiarcheota is a hyperthermophile discovered at the deep sea vent called Loki's Castle, which some scientists think has unique evolutionary significance.

It has a highly unique genome, consisting of roughly 26% proteins that are known to be found in other archaebacteria, 29% proteins that are known to be found in bacteria, 32% genes that do not correspond to any known protein, and – 3.3% genes that correspond to those only found in eukaryotes.

The eukaryotic genes are particularly exciting for scientists, because they are genes that appear to code for proteins that eukaryotes use to actively control the shape of their cell, including proteins for cytoskeletons, the motor protein actin, and several proteins that in eukaryotes are involved in changing cell membrane shape.

Some of these genes are involved in phagocytosis, which is exciting because the process of phagocytosis could have been used by eukaryotic ancestors to "swallow" other cells – which may have gone on to become endosymbiotes, leading to the endosymbiotic relationships between eukaryotic cells and their mitochondria, chloroplasts, and nuclei.

Lokiarchaeota's unique genome makes it possibly our closest relative among prokaryotes, and possibly a transitional form in the extremely important jump from prokaryotic to eukaryotic life, which made the evolution of the animal, plant, fungi, and protist kingdoms possible. Scientists think that Lokiarchaeota and ourselves probably shared a common ancestor around 2 billion years ago.

It is unknown whether this means that eukaryotes likely evolved around deep sea vents, or whether *Lokiarchaeota's* relatives may once have been common in other environments before they were outcompeted and driven to extinction by their more advanced descendants, the eukaryotes.

METHANOBREVIBACTER SMITHII

Methanobrevibacter smithii is a methane-producing archaebacteria that lives in the human gut. This member of *Euryarchaeota* helps us to break down complex plant sugars and extract extra energy from the food we eat.

The microorganisms in our guts – including members of *Euryarchaeota* – also have a complex relationship with our health. While some studies show that many people with obesity and colon cancer have above-average levels of *Euryarchaeota* in their guts, *Euryarchaeota* also help people who don't have enough food to produce more energy, and some types of these archaebacteria appear to protect against colon cancer.

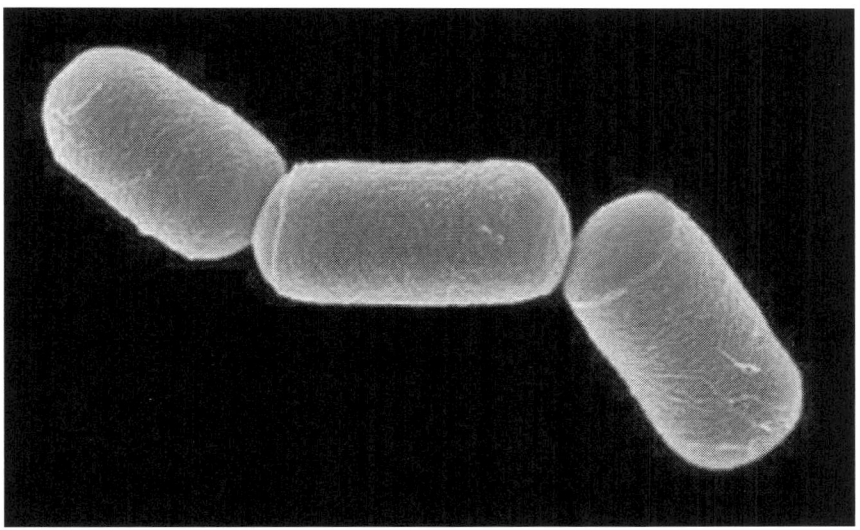

Fig. 16.1.*Methanobrevibacter smithii*

Comparison with other domains

The following table compares some major characteristics of the three domains, to illustrate their similarities and differences.

Property	Archaea	Bacteria	Eukarya
Cell membrane	Ether-linked lipids	Ester-linked lipids	Ester-linked lipids
Cell wall	Pseudopeptidoglycan, glycoprotein, or S-layer	Peptidoglycan, S-layer, or no cell wall	Various structures
Gene structure	Circular chromosomes, similar translation and transcription to Eukarya	Circular chromosomes, unique translation and transcription	Multiple, linear chromosomes, but translation and transcription similar to Archaea
Internal cell structure	No membrane-bound organelles or nucleus	No membrane-bound organelles or nucleus	Membrane-bound organelles and nucleus
Metabolism	Various, including diazotrophy, with methanogenesis unique to Archaea	Various, including photosynthesis, aerobic and anaerobic respiration, fermentation, diazotrophy, and autotrophy	Photosynthesis, cellular respiration, and fermentation; no diazotrophy
Reproduction	Asexual reproduction, horizontal gene transfer	Asexual reproduction, horizontal gene transfer	Sexual and asexual reproduction
Protein synthesis initiation	Methionine	Formylmethionine	Methionine
RNA polymerase	Many	One	Many
Toxin	Sensitive to diphtheria toxin	Resistant to diphtheria toxin	Sensitive to diphtheria toxin

Archaea were split off as a third domain because of the large differences in their ribosomal RNA structure. The particular molecule 16S rRNA is key to the production of proteins in all organisms. Because this function is so central to life, organisms with mutations in their 16S rRNA are unlikely to survive, leading to great (but not absolute) stability in the structure of this nucleotide over generations. 16S rRNA is large enough to show organism-specific variations, but still small enough to be compared quickly. In 1977, Carl Woese, a microbiologist studying the genetic sequences of organisms, developed a new comparison method that involved splitting the RNA into fragments that could be sorted and compared with other fragments from other organisms. The more similar the patterns between species, the more closely they are related.

Woese used his new rRNA comparison method to categorize and contrast different organisms. He compared a variety of species and happened upon a group of methanogens with rRNA vastly different from any known prokaryotes or eukaryotes. These methanogens were much more similar to each other than to other organisms, leading Woese to propose the new domain of Archaea. His experiments showed that the archaea were genetically more similar to eukaryotes than prokaryotes, even though they were more similar to prokaryotes in structure. This led to the conclusion that Archaea and Eukarya shared a common ancestor more recent than Eukarya and Bacteria. The development of the nucleus occurred after the split between Bacteria and this common ancestor.

One property unique to archaea is the abundant use of ether-linked lipids in their cell membranes. Ether linkages are more chemically stable than the ester linkages found in bacteria and eukarya, which may be a contributing factor to the ability of many archaea to survive in extreme environments that place heavy stress on cell membranes, such as extreme heat and salinity. Comparative analysis of archaeal genomes has also identified several molecular conserved signature indels and signature proteins uniquely present in either all archaea or different main groups within archaea. Another unique feature of archaea, found in no other organisms, is methanogenesis (the metabolic production of methane). Methanogenic archaea play a pivotal role in ecosystems with organisms that derive energy from oxidation of methane, many of which are bacteria, as they are often a major source of methane in such environments and can play a role as primary producers. Methanogens also play a critical role in the carbon cycle, breaking down organic carbon into methane, which is also a major greenhouse gas.

****

17. MYCOPLASMAS (OR MOLLICUTES): CELL WALL-LESS BACTERIA

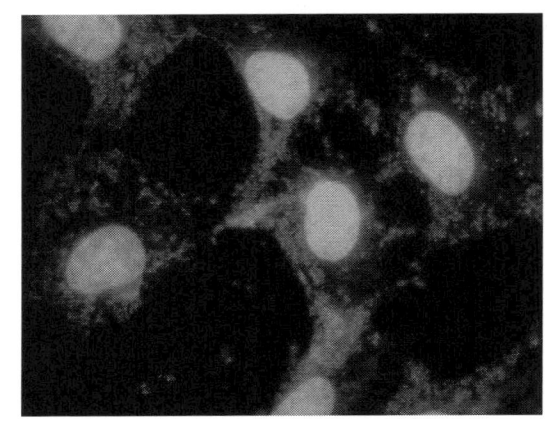

Discovery of Mycoplasmas:

Mycoplasmas are the smallest among the known aerobic prokaryotes (Fig. 2.50). They were first discovered by Pasteur in 1843, during his work on the possible causal agent of pleuropneumonia of cattle. Thus they were called pleuro- pneumonia-like organism (PPLO).

Pasteur was unable to isolate them in pure culture. Later, Nocard and Roux (1898), the French microbiologists, were successful in growing them in pure culture-medium containing serum and confirmed By inoculation and subsequent expression of disease in healthy cattle.

Mycoplasmas are commonly found in soil, hot spring, sewage water and also in plants and animals including man. Borrel (1910) named these organisms Asterococcus mycoides. Later, in 1929, Nowak placed them under the genus Mycoplasma.

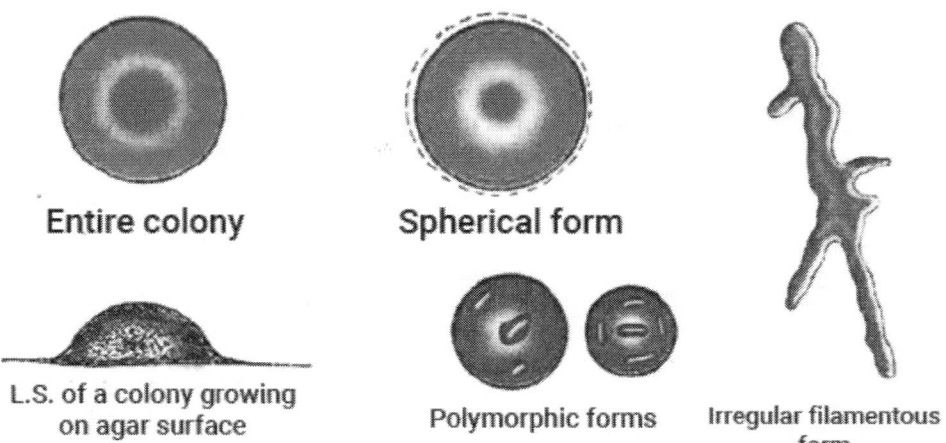

Fig. 17. Different cell shapes of mycoplasmas

General Characters of Mycoplasma:

1. They are unicellular, smallest, non-motile and prokaryotic organisms forming fried egg shaped colonies

2. They are pleomorphic i.e., able to change their shape depending upon culture media.

3. They may be rod like, ring like, globoid or filamentous (Fig. 17). The filaments are of uniform diameter (100-300 nm) and vary in length from 3 nm to 150 nm.

4. Some mycoplasma predominantly assume spherical shape (300-800 nm in diameter).

5. They are ultra-filterable i.e., they can pass through bacteria-proof filters.

6. They do not possess rigid cell wall.

7. The cells are delimited by soft tripple layered lipo-proteinaceous membrane. It is unit membrane about 10 nm thick.

8. Within the cytoplasm ribosomes are found scattered in the peripheral zone. These are 14 nm in diameter and resemble with bacteria in sedimentation characteristic of both the nucleoprotein and nucleic acid.

9. The ribosomes are 72S type.

10. Within the cytoplasm fine fibrillar DNA is present. It is double stranded helix.

11. Mycoplasma generally grow more slowly than bacteria.

12. They require sterol for their nutrition.

13. They are usually resistant to antibiotics like penicillin, cephaloridine, vencomycin etc. which action cell wall.

14. They are sensitive to tetracycline.

15. They are also killed by temperature of 40-55°C in fifteen minutes.

16. They do not produce spores.

17. Like other prokaryotes, they usually divide by binary fission.

Classification of Mycoplasmas:

Based on nutritional requirement, mycoplasmas are divided into the following three genera:

1. Mycoplasma:

They require cholesterol for their growth. They parasitise on animals including man by causing damage to the mucous membranes and different joints of the body.

2. Acholeplasma:

They do not require cholesterol for their growth. They are available in sewage water and soil as saprophytes and in vertebrates and also in plants as parasites.

3. Thermoplasma:

They also do not require cholesterol for their growth. They are aerobic microorganisms showing good growth in acidic pH between 0.96-3.0, with an optimum temperature of 59°C.

Structure of Mycoplasmas:

The cell is devoid of cell wall which makes them readily deformable showing irregular and variable shapes. They may be ring-like, granular, coccoid, pear-shaped, filamentous, etc. (Fig. 2.50). The filaments are of two types: unbranched or branched. The cells are very small and measure 0.3-0.9 μm in diameter.

The cells are covered by cytoplasmic (lipoprotein) membrane (Fig. 2.51). Cell membrane covers the cytoplasm which contains nucleoplasm like structure and ribosomes. The genetic material is composed of DNA and RNA. It is about less than 50%, the amount present in other prokaryotic organisms. The amount of RNA (8%) is more than DNA (4%).

They are usually non-motile, but some forms show gliding movements. They reproduce by vegetative means i.e., by binary fission and budding.

They are sensitive to antibiotics like chloramphenicol, streptomycin, erythromycin etc., but are insensitive to penicillin, ampicillin etc., due to the absence of cell wall.

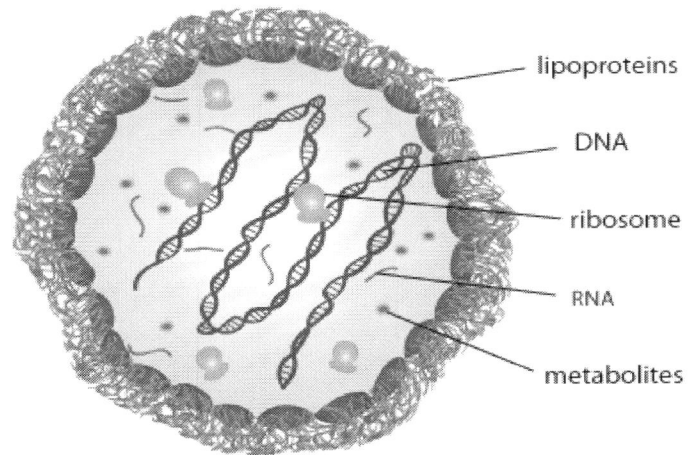

Fig.17.1. Structure of mycoplasma

Reproduction in Mycoplasma:

Mycoplasmas reproduce by budding and/or binary fission (Fig. 19.12). Cells of mycoplasma divide unevenly into very minute bodies called the elementary bodies or minimal reproductive units.

These are formed inside the large bodies or mature cells. Their size varies from 330 nm to 450 nm. These bodies are the smallest independent living entities so far known.

Transmission of Mycoplasma:

Mycoplasma like organisms (MLO) or phytoplasmas are usually present in phloem of the host plants and are transmitted from host to another host by leaf hoppers but some are transmitted by psyllids, treehoppers, plant hoppers and some possibly by aphids and miles.

Some of the pathogens are known to infect various organs of their leaf hopper or psyllid vectors and to multiply in their cells. The vectors cannot transmit the phytoplasma immediately after feeding on the infected plant but it begins to transmit if after an incubation period of 10 to 45 days depending upon the temperature.

Diseases Caused by Mycoplasma:

Mycoplasmas cause different serious diseases in plants and animals including man.

Some of these are:

(a) Plant Diseases:

(i) Little leaf disease of brinjal,

(ii) Bunchy top of papaya,

(iii) Big bud of tomato,

(iv) Witches broom of legumes,

(v) Yellow dwarf of tobacco,

(vi) Strip disease of sugarcane,

(vii) Clover dwarf,

(viii) Cotton vires- cence.

(b) Human Diseases:

(i) Primary atypical pneumonia (PAP) by Mycoplasma pneumoniae,

(ii) Mycoplasma hominis causes pleuropneumonia, prostatitis, inflammations of genitals etc.

(iii) Mycoplasma fermentants causes infertility in man.

(c) Animal Diseases:

(i) Mycoplasma agalactia causes agalactia of goat and sheep,

(ii) Mycoplasma mycoides causes pleuropneumonia of cattle,

(iii) M. bovigenitalium causes inflammation of genitals of different animals.

MYCOPLASMA INFECTIONS MYCOPLASMA

Mycoplasma are **bacteria** that lack a conventional cell wall. They are capable of replication. Mycoplasma cause various diseases in humans, animals, and plants. There are seven species of mycoplasma that are known to cause disease in humans. *Mycoplasma pneumoniae* is an important cause of sore throat, **pneumonia**, and the **inflammation** of the channels in the lung that are known as the bronchi. Because of the atypical nature of the bacterium, mycoplasmainduced pneumonia is also referred to as atypical pneumonia. The pneumonia can affect children and adults. The symptoms tend to be more pronounced in adults. In fact, children may not exhibit any symptoms of infection. Symptoms of infection include a fever, general feeling of being unwell, sore throat, and sometimes an uncomfortable chest. These symptoms last a week to several months and usually fade without medical intervention. *Mycoplasma pneumoniae* can also cause infections in areas of the body other than the lungs, including the central nervous system, liver, and the pancreas.

Another species, *Mycoplasma genitalium*, is associated with infections of the urethra, especially when the urethra has been infected by some other bacteria. The mycoplasma infection may occur due to the stress imposed on the **immune system** by the other infection.

A mycoplasma called *Ureaplasma urealyticum* is present in the genital tract of many sexually active women. The resulting chronic infection can contribute to premature delivery in pregnant women. As well, the mycoplasma can be transmitted from the mother to the infant. The infant can contract pneumonia, infection of the central nervous system, and lung malfunction. A group of four mycoplasma species are considered to be human pathogens and may contribute to the development and **immunodeficiency** virus infection to the more problematic and debilitating symptoms of Acquired Immunodeficiency Syndrome (**AIDS**).

The species of mycoplasma are *Mycoplasma fermentans*, *Mycoplasma pirum*, *Mycoplasma hominis*, and *Mycoplasma penetrans*. Mycoplasma have also been observed in patients who exhibit other diseases. For example, studies using genetic probes and the **polymerase chain reaction** technique of detecting target **DNA** have found *Mycoplasma fermentans* in upwards of 35% of those afflicted with chronic fatigue syndrome. The bacterium is present in less than 5% of healthy populations. Similar percentages have been found in soldiers of the Persian Gulf War who are exhibiting chronic fatigueike symptoms. While the exact relationship between mycoplasma and the chronic fatigue state is not fully clear, the current consensus is that the bacteria is playing a secondary role in the development of the symptoms. Over 20 years ago, mycoplasma was suggested as a cause of rheumatoid arthritis. With the development of molecular techniques of bacterial detection, this suggestion could be tested. The polymerase chain reaction has indeed detected *Mycoplasma fermentans* in a significant number of thoseafflicted with the condition. But again, a direct causal relationship remains to be established.

The association of mycoplasma with diseases like arthritis and chronic fatigue syndrome, which has been implicated with a response of the body's immune system against its own components, is consistent with the growth and behaviour of mycoplasma. The absence of a conventional cell wall allows mycobacteria to penetrate into the white blood cells of the immune system. Because some mycoplasma will exist free of the blood cells and because the bacteria are capable of slow growth in the body, the immune system will detect and respond to a mycobacterial infection. But this response is generally futile. The bacteria hidden inside the white blood cells will not be killed. The immune components instead might begin to attack other antigens of the host that are similar in three-dimensional structure to the mycobacterial antigens. Because mycoplasma infections can become chronic, damage to the body over an extended time and the stress produced on the immune system may allow other **microorganisms** to establish infections.

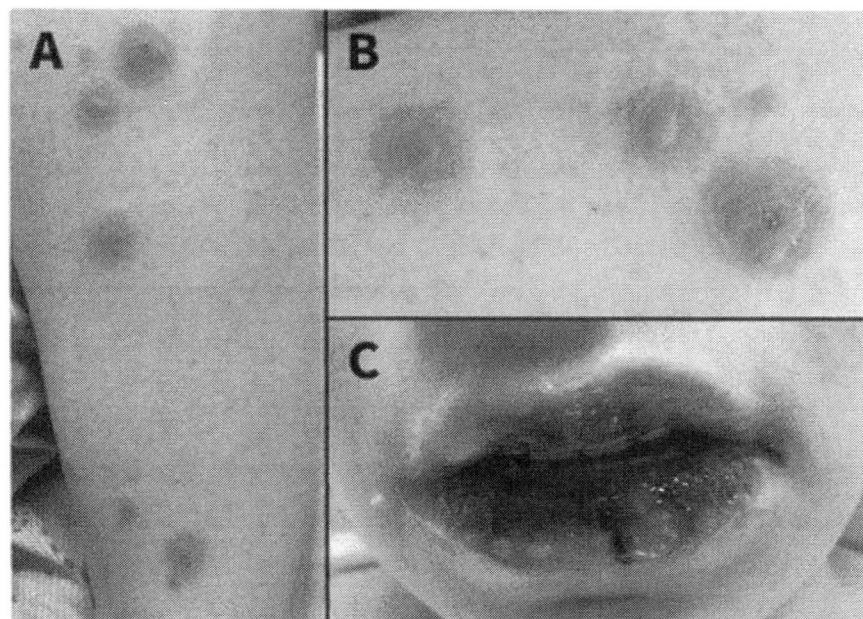

Fig.17.2. Cutaneous (A and B) and oral mucosal (C) lesions of a 34-year-old woman after admission to hospital. The cutaneous lesions have the distinctive target rings with a raised central area (often referred to as a blister). Oral and vaginal (not shown) mucosal involvement was also present, with inflammation and sloughing.

The polymerase chain reaction is presently the best means of detecting mycoplasma. The bacteria cannot be easily grown on laboratory media. Labs that test using the polymerase technique are still rare. Thus, a mycoplasma infection might escape detection for years. Strategies to eliminate mycoplasma infections are now centering on the strengthening of the immune system, and long-term antibiotic use (e.g., months or years). Even so, it is still unclear whether **antibiotics** are truly effective on mycoplasma bacteria. Mycoplasma can alter the chemical composition of the surface each time a bacterium divides. Thus, there may be no constant target for an antibiotic.

****

18.

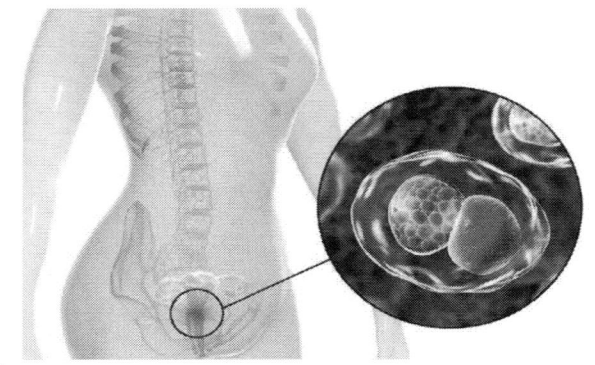

THE CHLAMYDIA

Chlamydia, or more specifically a **chlamydia infection**, is a sexually transmitted infection caused by the bacterium *Chlamydia trachomatis*. Most people who are infected have no symptoms. When symptoms do appear in can be several weeks after infection. Symptoms in women may include vaginal discharge or burning with urination. Symptoms in men may include discharge from the penis, burning with urination, or pain and swelling of one or both testicles. The infection can spread to the upper genital tract in women, causing pelvic inflammatory disease, which may result in future infertility or ectopic pregnancy. Repeated infections of the eyes that go without treatment can result in trachoma, a common cause of blindness in the developing world.

Chlamydia can be spread during vaginal, anal, or oral sex, and can be passed from an infected mother to her baby during childbirth. The eye infections may also be spread by personal contact, flies, and contaminated towels in areas with poor sanitation. *Chlamydia trachomatis* only occurs in humans. Diagnosis is often by screening which is recommended yearly in sexually active women under the age of twenty-five, others at higher risk, and at the first prenatal visit. Testing can be done on the urine or a swab of the cervix, vagina, or urethra. Rectal or mouth swabs are required to diagnose infections in those areas.

Prevention is by not having sex, the use of condoms, or having sex with only one other person, who is not infected. Chlamydia can be cured by antibiotics with typically either azithromycin or doxycycline being used. Erythromycin or azithromycin is recommended in babies and during pregnancy. Sexual partners should also be treated and the infected people advised not to have sex for seven days and until symptom free. Gonorrhea, syphilis, and HIV should be tested for in those who have been infected. Following treatment people should be tested again after three months.

Chlamydia is one of the most common sexually transmitted infections, affecting about 4.2% of women and 2.7% of men worldwide. In 2015 about 61 million new cases occurred globally. In the United States about 1.4 million cases were reported in 2014. Infections are most common among those between the ages of 15 and 25 and are more common in women than men. In 2015 infections resulted in about 200 deaths. The word *chlamydia* is from the Greek χλαμύδα, meaning "cloak".

Introduction to Chlamydial Infection:

The genus chlamydia includes organisms previously called as Psittacosis-Lymphogranuloma venereum, Trachoma group (PL T) organism. The generic name Bedsonia has also been used in recognition of Bedson; but by the rules of nomenclature chlamydiae has priority.

It is now clear that the chlamydiae are small prokaryotic that have evolved to a highly parasitic existence in the cytoplasm of cells. These organisms are small, non-motile, Gram-negative obligate intracellular parasites.

Table 1. Properties of species Chlamydia

Properties	C. trachomatis	C. psittaci	C. pneumoniae
Elementary body (EB)			
Size (diam)	300-350 nm	300-350 nm	310 × 400 nm
periplasmic space	very little	very little	Large
Reticular body (RB)			
Size (diam)	800-1000 nm	800-1000 nm	800-1000 nm
Shape	Circular	Circular	Circular
Nucleic acid			
Weight	660×10⁶	660 × 10⁶	660 × 10⁶
Plasmid DNA	+	+	−
(Most strains)	(Most strains)		
Antigenicity			
Outer membrane proteins (h) major one	Species specific	Species specific	Species specific

They occur in two forms:

Lipopolysaccharide (LPS) Genus specific genus specific genus specific:

(a) There is a small, 300 nm diameter form, which has a compact electron-dense nucleoid which is highly infectious, stable, extracellular form or elementary body.

(b) There is a large form, 800-1200 nm in diameter without a dense nucleoid, the 'initial body' which is intracellular, replicating form. The organisms grow in the cytoplasm of their host cells forming characteristic micro-colonies or inclusion bodies which are made up of a mixture of the larger and smaller cells. Both forms stain well with Castaneda or Giemsa stain.

Chlamydiae have 2 Subgroups—A, B:

Subgroup A has compact inclusions with a glycogen matrix. Subgroup B has no glycogen matrix. These two groups also differ in their susceptibility to sulphadiazine.

Chlamydial infection of man takes two main clinical forms:

(a) organism of psittacosis ornithosis causes respiratory illness with fever; and

(b) trachoma-inclusion conjunctivitis (TRIC)—Lympho-granuloma venereum (LGV) organism of sub group A causes the clinically quite dissimilar ocular, genital infection. TRIC-LGV subgroup A chlamydia contains a group of infections transmitted by contact—mainly sexual— in developed countries but also eye to eye in underdeveloped countries where trachoma is endemic.

LGV is more invasive; it starts as a small painless papula or ulcer (Lymphogranuloma chancre) on the external genitalia, or internally some 5 to 10 days after exposure. The infection spreads to the regional lymph nodes (inguinal, perirectal) with suppuration in many cases and sometimes a generalized infection with fever and rash, arthritis, conjunctivitis and meningoencephalitis.

In late stage of the disease chronic inflammation around lymphatics in the genital and rectal area leads to fibrosis with elephantiasis of the genitalia. Rectal structures are common in women and male homosexuals. The treatment of genital infections is by 21 days' course of tetracycline, 250 mg, 4 times a day.

Pathogenicityof Chlamydial Infection:

Signs and symptoms

Genital disease

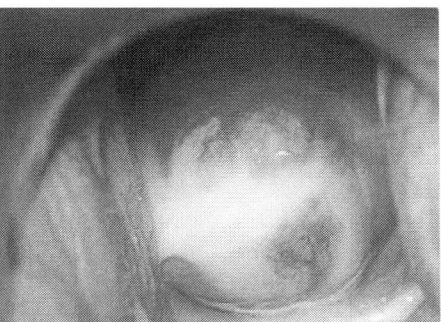

Fig.18. Inflammation of the cervix from chlamydia infection characterized by mucopurulent cervical discharge, redness, and inflammation

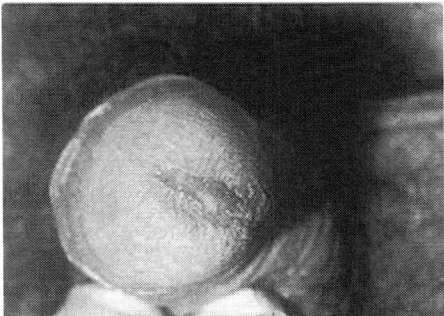

Fig.18.1.A white, cloudy or watery discharge may emerge from the tip of the penis.

Women

Chlamydial infection of the cervix (neck of the womb) is a sexually transmitted infection which has no symptoms for 50–70% of women infected. The infection can be passed through vaginal, anal, or oral sex. Of those who have an asymptomatic infection that is not detected by their doctor, approximately half will develop pelvic inflammatory disease (PID), a generic term for infection of the uterus, fallopian tubes, and/or ovaries. PID can cause scarring inside the reproductive organs, which can later cause serious complications, including chronic pelvic pain, difficulty becoming pregnant, ectopic (tubal) pregnancy, and other dangerous complications of pregnancy.

Chlamydia is known as the "silent epidemic", as in women it may not cause any symptoms in 70–80% of cases and can linger for months or years before being discovered. Signs and symptoms may include abnormal vaginal bleeding or discharge, abdominal pain, painful sexual intercourse, fever, painful urination or the urge to urinate more often than usual (urinary urgency).

For sexually active women who are not pregnant, screening is recommended in those under 25 and others at risk of infection. Risk factors include a history of chlamydial or other sexually transmitted infection, new or multiple sexual partners, and inconsistent condom use. Guidelines recommend all women attending for emergency contraceptive are offered Chlamydia testing, with studies showing up to 9% of women aged <25 years had Chlamydia.

Men

In men, those with a chlamydial infection show symptoms of infectious inflammation of the urethra in about 50% of cases. Symptoms that may occur include: a painful or burning sensation when urinating, an unusual discharge from the penis, testicular pain or swelling, or fever. If left untreated, chlamydia in men can spread to the testicles causing epididymitis, which in rare cases can lead to sterility if not treated. Chlamydia is also a potential cause of prostatic inflammation in men, although the exact relevance in prostatitis is difficult to ascertain due to possible contamination from urethritis.

Eye disease

Trachoma

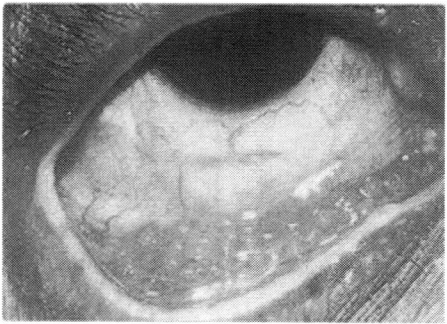

Fig.18.2. Conjunctivitis due to chlamydia

Trachoma is a chronic conjunctivitis caused by *Chlamydia trachomatis*. It was once the most important cause of blindness worldwide, but its role diminished from 15% of blindness cases by trachoma in 1995 to 3.6% in 2002. The infection can be spread from eye to eye by fingers, shared towels or cloths, coughing and sneezing and eye-seeking flies. Symptoms include mucopurulent ocular discharge, irritation, redness, and lid swelling. Newborns can also develop chlamydia eye infection through childbirth (see below). Using the SAFE strategy (acronym for surgery for in-growing or in-turned lashes, antibiotics, facial cleanliness, and environmental improvements), the World Health Organization aims for the global elimination of trachoma by 2020 (GET 2020 initiative).

Joints

Chlamydia may also cause reactive arthritis—the triad of arthritis, conjunctivitis and urethral inflammation—especially in young men. About 15,000 men develop reactive arthritis due to chlamydia infection each year in the U.S., and about 5,000 are permanently affected by it. It can occur in both sexes, though is more common in men.

Infants

As many as half of all infants born to mothers with chlamydia will be born with the disease. Chlamydia can affect infants by causing spontaneous abortion; premature birth; conjunctivitis, which may lead to blindness; and pneumonia. Conjunctivitis due to chlamydia typically occurs one week after birth (compared with chemical causes (within hours) or gonorrhea (2–5 days)).

Other conditions

A different serovar of Chlamydia trachomatis is also the cause of lymphogranuloma venereum, an infection of the lymph nodes and lymphatics. It usually presents with genital ulceration and swollen lymph nodes in the groin, but it may also manifest as rectal inflammation, fever or swollen lymph nodes in other regions of the body.

Transmission

Chlamydia can be transmitted during vaginal, anal, or oral sex or direct contact with infected tissue such as conjunctiva. Chlamydia can also be passed from an infected mother to her baby during vaginal childbirth.

Pathophysiology. *Chlamydiae* have the ability to establish long-term associations with host cells. When an infected host cell is starved for various nutrients such as amino acids (for example, tryptophan), iron, or vitamins, this has a negative consequence for *Chlamydiae* since the organism is dependent on the host cell for these nutrients. Long-term cohort studies indicate that approximately 50% of those infected clear within a year, 80% within two years, and 90% within three years.

The starved chlamydiae enter a persistent growth state wherein they stop cell division and become morphologically aberrant by increasing in size. Persistent organisms remain viable as they are capable of returning to a normal growth state once conditions in the host cell improve.

There is debate as to whether persistence has relevance. Some believe that persistent chlamydiae are the cause of chronic chlamydial diseases. Some antibiotics such as β-lactams have been found to induce a persistent-like growth state.

Diagnosis

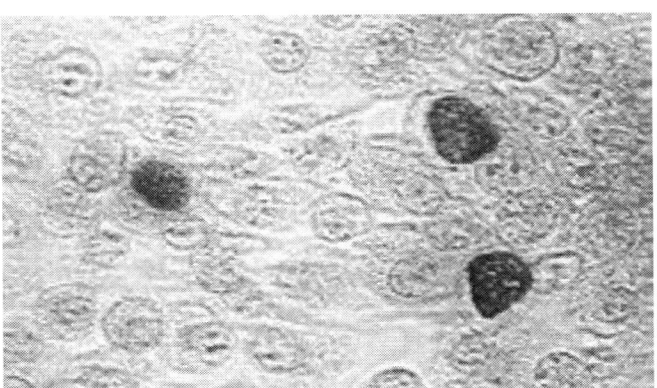

Fig.18.3. *Chlamydia trachomatis* **inclusion bodies (brown) in a McCoy cell culture**

The diagnosis of genital chlamydial infections evolved rapidly from the 1990s through 2006. Nucleic acid amplification tests (NAAT), such as polymerase chain reaction (PCR), transcription mediated amplification (TMA), and the DNA strand displacement amplification (SDA) now are the mainstays. NAAT for chlamydia may be performed on swab specimens sampled from the cervix (women) or urethra (men), on self-collected vaginal swabs, or on voided urine. NAAT has been estimated to have a sensitivity of approximately 90% and a specificity of approximately 99%, regardless of sampling from a cervical swab or by urine specimen. In women seeking an STI clinic and a urine test is negative, a subsequent cervical swab has been estimated to be positive in approximately 2% of the time.

At present, the NAATs have regulatory approval only for testing urogenital specimens, although rapidly evolving research indicates that they may give reliable results on rectal specimens.

Because of improved test accuracy, ease of specimen management, convenience in specimen management, and ease of screening sexually active men and women, the NAATs have largely replaced culture, the historic gold standard for chlamydia diagnosis, and the non-amplified probe tests. The latter test is relatively insensitive, successfully detecting only 60–80% of infections in asymptomatic women, and often giving falsely-positive results. Culture remains useful in selected circumstances and is currently the only assay approved for testing non-genital specimens. Other method also exist including: ligase chain reaction (LCR), direct fluorescent antibody resting, enzyme immunoassay, and cell culture. Rapid point-of-care tests are, as of 2020, not thought to be effective for diagnosing chlamydia in men of reproductive age and nonpregnant women because of a high false-negative rates.

Prevention

Prevention is by not having sex, the use of condoms, or having sex with only partners, who are not infected.

Treatment

C. trachomatis infection can be effectively cured with antibiotics. Guidelines recommend **azithromycin, doxycycline, erythromycin, levofloxacin or ofloxacin**. In men, doxycycline (100 mg twice a day for 7 days) is probably more effective than azithromycin (1 g single dose) but evidence for the relative effectiveness of antibiotics in women is very uncertain. Agents recommended during pregnancy include erythromycin or amoxicillin.

An option for treating sexual partners of those with chlamydia or gonorrhea includes patient-delivered partner therapy (PDT or PDPT), which is the practice of treating the sex partners of index cases by providing prescriptions or medications to the patient to take to his/her partner without the health care provider first examining the partner.

Following treatment people should be tested again after three months to check for reinfection.

****

19.

VIRUSES

General Concepts: Virus history

The history of virology goes back to the late 19th century, when German anatomist **Dr Jacob Henle** (discoverer of Henle's loop) hypothesized the existence of infectious agent that were too small to be observed under light microscope. This idea fails to be accepted by the present scientific community in the absence of any direct evidence. At the same time three landmark discoveries came together that formed the founding stone of what we call today as **medical science.** The first discovery came from **Louis Pasture** (1822-1895) who gave the spontaneous generation theory from his famous swan-neck flask experiment. The second discovery came from **Robert Koch** (1843-1910), a student of Jacob Henle, who showed for first time that the anthrax and tuberculosis is caused by a bacillus, and finally **Joseph Lister** (1827-1912) gave the concept of sterility during the surgery and isolation of new organism.

The history of viruses and the field of virology are broadly divided into three phases, namely discovery, early and modern.

The discovery phase (1886-1913)

In 1879, **Adolf Mayer,** a German scientist first observed the dark and light spot on infected leaves of tobacco plant and named it **tobacco mosaic disease** . Although he failed to describe the disease, he showed the infectious nature of the disease after inoculating the juice extract of diseased plant to a healthy one. The next step was taken by a Russian scientist **Dimitri Ivanovsky** in 1890, who demonstrated that sap of the leaves infected with tobacco mosaic disease retains its infectious property even after its filtration through a Chamberland filter. The third scientist who plays an important role in the development of the concept of viruses was **Martinus Beijerinck** (1851-1931), he extended the study done by Adolf Mayer and Dimitri Ivanofsky and showed that filterable agent form the infectious sap could be diluted and further regains its strength after replicating in the living host; he called it as ***"contagium vivum fluidum"***. **Loeffler** and **Frosch** discovered the first animal virus, the foot and mouth disease virus in 1898 and subsequently **Walter Reed** and his team discovered the yellow fever virus, the first human virus from Cuba in 1901. Poliovirus was discovered by **Landsteiner** and **Popper** in 1909 and two years later **Rous** discovered the solid tumor virus which he called Rous sarcoma virus.

The early phase (1915-1955)

In 1915, **Frederick W. Twort** discovered the phenomenon of transformation while working with the variants of vaccinia viruses, simultaneously **Felix d'Herelle** discovered **bacteriophage** and developed the assay to titrate the viruses by plaques. **Wendell Stanley** (1935) first crystallized the TMV and the first electron micrograph of the tobacco mosaic virus (TMV) was taken in 1939. In 1933 **Shope** described the first papillomavirus in rabbits. The vaccine against yellow fever was made in 1938 by **Thieler** and after 45 years of its discovery, polio virus vaccine was made by **Salk** in 1954.

The modern phase (1960-present)

During this phase scientists began to use viruses to understand the basic question of biology. The superhelical nature of polyoma virus DNA was first described by **Weil** and **Vinograd** while **Dulbecco** and **Vogt** showed its closed circular nature in 1963. In the same year **Blumberg** discovered the hepatitis B virus. **Temin** and **Baltimore** discovered the retroviral reverse transcriptase in 1970 while the first human immunodeficiency virus (HIV) was reported in 1983 by **Gallo** and **Montagnier**. The phenomenon of RNA splicing was discovered in Adenoviruses by **Roberts**, **Sharp**, **Chow** and **Broker**. In the year 2005 the complete genome sequence of 1918 influenza virus was done and in the same year hepatitis C virus was successfully propagated into the tissue culture.

Many discoveries are done using viruses as a model. The transcription factor that binds to the promoter during the transcription was first discovered in SV40. The phenomenon of polyadenylation during the mRNA synthesis was first described in poxviruses while its presence was first reported in SV40. Many of our current understanding regarding the translational regulation has been studied in poliovirus. The oncogenes were first reported in Rous sarcoma virus. The p53, a tumor suppressor gene was first reported in SV40

Important discoveries

Date	Discovery
1796	Cowpox virus used to vaccinate against smallpox by Jenner.
1892	Description of filterable infectious agent (TMV) by Ivanovsky.
1898	Concept of the virus as a contagious living form by Beijerinck.
1901	First description of a yellow fever virus by Dr Reed and his team.
1909	Identification of poliovirus by Landsteiner and Popper.
1911	Discovery of Rous sarcoma virus.
1931	Virus propagation in embryonated chicken eggs by Woodruff and Goodpasture.
1933	Identification of rabbit papillomavirus.
1936	Induction of carcinomas in other species by rabbit papillomavirus by Rous and Beard.
1948	Poliovirus replication in cell culture by Enders, Weller, and Robbins.
1952	Transduction by Zinder and Lederberg.
1954	Polio vaccine development by Salk.
1958	Bacteriophage lambda regulation paradigm by Pardee, Jacob, and Monod.
1963	Discovery of hepatitis B virus by Blumberg.
1970	Discovery of reverse transcriptase by Temin and Baltimore.
1976	Retroviral oncogenes discovered by Bishop and Varmus.
1977	RNA splicing discovered in adenovirus.
1983	Description of human immunodeficiency virus (HIV) as causative agent of acquired immunodeficiency syndrome (AIDS) by Montagnier, Gallo)
1997	HAART treatment for AIDS.
2003	Severe acute respiratory syndrome (SARS) is caused by a novel coronavirus.
2005	Hepatitis C virus propagation in tissue culture by Chisari, Rice, and Wakita.
2005	1918 influenza virus genome sequencing.

Nature of Viruses:

Viruses are infective microorganisms.

They show several differences from typical bacterial cells

1. Size:

On the whole viruses are much smaller than bacteria. Most animal and plant viruses are invisible under the light microscope. Some of smaller viruses are only 200Å in diameter.

2. No independent metabolism:

Viruses cannot multiply outside a living cell. No virus has been cultivated in a cell-free medium. Viruses do not have an independent metabolism. They are metabolically inactive outside the host cell because they do not possess enzyme systems and protein synthesis machinery. Thus viruses are obligatory intracellular parasites.

3. Simple structure:

Viruses have a very simple structure. They consist of a nucleic acid core surrounded by a protein coat. In this respect they differ from typical cells which are made up of proteins, carbohydrates, lipids and nucleic acids. Myxoviruses have a membranous envelope consisting of proteins, carbohydrate and lipid outside the usual protein coat, but this envelope is derived from the host cell.

4. Absence of cellular structure:

Viruses do not have any cytoplasm, and thus cytoplasmic organelles like mitochondria, Golgi complexes, ribosomes, lysosomes etc. are absent. They do not have any limiting cell membrane.

5. Nucleic acids:

Viruses usually have only one nucleic acid, either DNA or RNA. Typical cells have both DNA and RNA. Rous Sarcoma virus (RSV), producing certain cancer, is the only virus having both DNA and RNA.

6. Crystallization:

Many of the smaller viruses can be crystallized, and thus behave like chemicals.

7. No growth and division:

Viruses do not have the power of growth and division. The genetic material of virus reproduces only in a host cell.

Thus viruses do not show all the characteristics of typical living organisms. They, however, possess two fundamental characteristics of living systems. Firstly, they contain nucleic acid as their genetic material. The nucleic acid contains all the instructions for the structure and the function of the virus. Secondly, they can reproduce themselves, even if only by using the host cells's synthesis machinery.

Structure of Viruses:

(a) Size:

Variable. Most viruses are much smaller than bacteria. The size ranges in between 100A to 250 mu. Some viruses are larger than bacteria, for example the psittacos is a virus measuring 0.75 mu in diameter.

Size of viruses belonging to different families

Family	Size (nm)
Poxviridae	300
Iridoviridae	135-300
Asfarviridae	170-220
Herpesviridae	150
Adenoviridae	80-100
Polyomaviridae	40-50
Papillomaviridae	55
Hepadnaviridae	50
Circoviridae	12-27
Parvoviridae	15-25
Retroviridae	80-100
Reoviridae	60-80
Birnaviridae	60
Paramyxoviridae	150-250
Rhabdoviridae	100
Filoviridae	80
Bornaviridae	80-100
Orthomyxoviridae	80-120
Bunyaviridae	80-120
Arenaviridae	50-280

Coronaviridae	120-150
Arteriviridae	60-70
Picornaviridae	30
Caliciviridae	30-40
Astroviridae	30
Togaviridae	70
Flaviviridae	40-60

(b) Symmetry:

Viruses occur in three main shapes. They are spherical (Cubical or polyhydral), helical (Cylinderical or rod-like) and complex. Cubical viruses may be tetrahydral (4 faces) < dodecahedral (12 faces) or icosahedral (20 faces). The Herpes virus is dodecahedral. The Tobacco mosaic virus (TMV) and the bacteriophage are, respectively, helical and complex.

1. Spherical / Cubical:

PhI X 174, Herpes virus, Tipula virus, Polyoma virus.

2. Helical / Cylinderical:

Tobacco Mosaic virus, Influenza virus Mumps virus.

3. Complex:

Vaccinia virus, ORF virus, Vesicular Stomatitis virus.

Viral Morphology

Viruses are acellular, meaning they are biological entities that do not have a cellular structure. Therefore, they lack most of the components of cells, such as organelles, ribosomes, and the plasma membrane. A virion consists of a nucleic acid core, an outer protein coating or capsid, and sometimes an outer envelope made of protein and phospholipid membranes derived from the host cell. The capsid is made up of protein subunits called capsomeres. Viruses may also contain additional proteins, such as enzymes. The most obvious difference between members of viral families is their morphology, which is quite diverse. An interesting feature of viral complexity is that host and virion complexity are uncorrelated. Some of the most intricate virion structures are observed in bacteriophages, viruses that infect the simplest living organisms: bacteria.

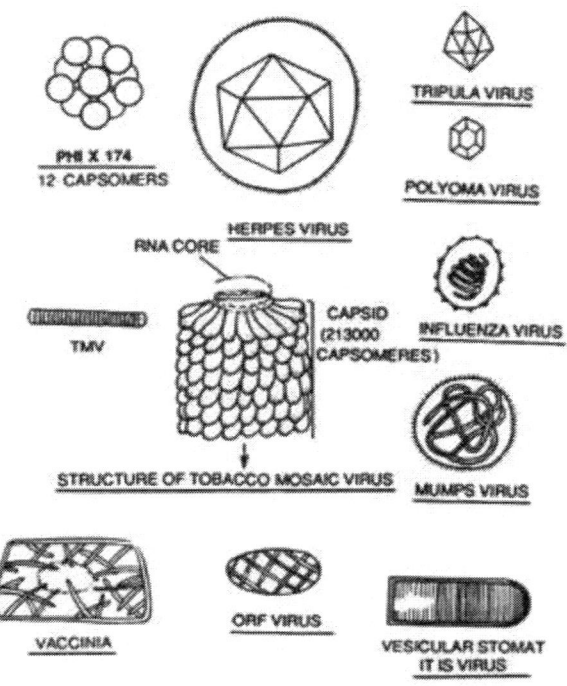

DIFFERENT KINDS OF VIRUSES

Fig.19. Different kinds of Viruses

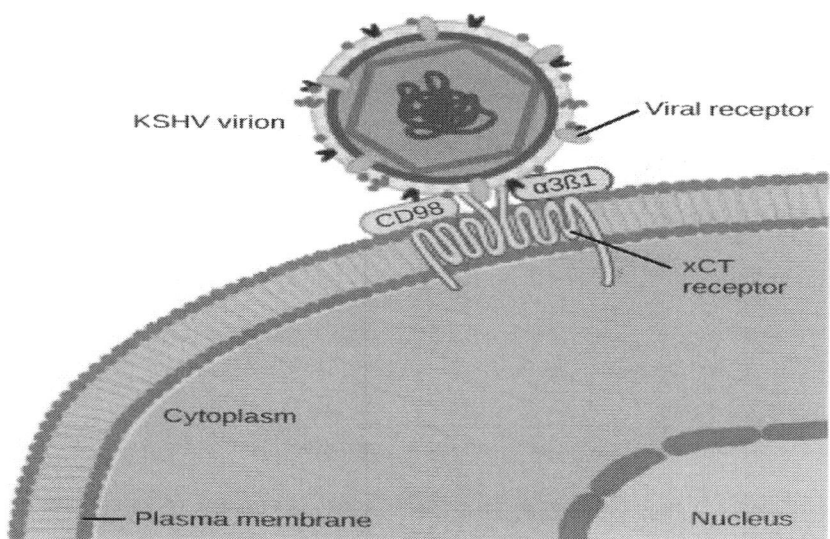

Fig. 19.1. Example of a virus attaching to its host cell: The KSHV virus binds the xCT receptor on the surface of human cells. This attachment allows for later penetration of the cell membrane and replication inside the cell.

Morphology

Viruses come in many shapes and sizes, but these are consistent and distinct for each viral family. In general, the shapes of viruses are classified into four groups: filamentous, isometric (or icosahedral), enveloped, and head and tail. Filamentous viruses are long and cylindrical. Many plant viruses are filamentous, including TMV (tobacco mosaic virus). Isometric viruses have shapes that are roughly spherical, such as poliovirus or herpesviruses. Enveloped viruses have membranes surrounding capsids. Animal viruses, such as HIV, are frequently enveloped. Head and tail viruses infect bacteria. They have a head that is similar to icosahedral viruses and a tail shape like filamentous viruses.

Many viruses use some sort of glycoprotein to attach to their host cells via molecules on the cell called viral receptors. For these viruses, attachment is a requirement for later penetration of the cell membrane, allowing them to complete their replication inside the cell. The receptors that viruses use are molecules that are normally found on cell surfaces and have their own physiological functions. Viruses have simply evolved to make use of these molecules for their own replication.

Overall, the shape of the virion and the presence or absence of an envelope tell us little about what disease the virus may cause or what species it might infect, but they are still useful means to begin viral classification. Among the most complex virions known, the T4 bacteriophage, which infects the *Escherichia coli* bacterium, has a tail structure that the virus uses to attach to host cells and a head structure that houses its DNA. Adenovirus, a non-enveloped animal virus that causes respiratory illnesses in humans, uses glycoprotein spikes protruding from its capsomeres to attach to host cells. Non-enveloped viruses also include those that cause polio (poliovirus), plantar warts (papillomavirus), and hepatitis A (hepatitis A virus).

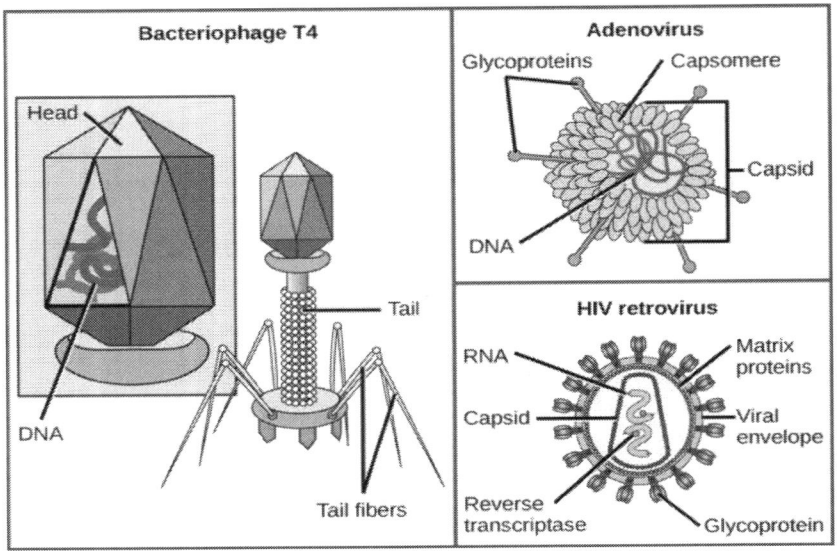

Fig.19.2.Examples of virus shapes: Viruses can be either complex in shape or relatively simple. This figure shows three relatively-complex virions: the bacteriophage T4, with its DNA-containing head group and tail fibers that attach to host cells; adenovirus, which uses spikes from its capsid to bind to host cells; and HIV, which uses glycoproteins embedded in its envelope to bind to host cells.

Enveloped virions like HIV consist of nucleic acid and capsid proteins surrounded by a phospholipid bilayer envelope and its associated proteins. Glycoproteins embedded in the viral envelope are used to attach to host cells. Other envelope proteins include the matrix proteins that stabilize the envelope and often play a role in the assembly of progeny virions. Chicken pox, influenza, and mumps are examples of diseases caused by viruses with envelopes. Because of the fragility of the envelope, non-enveloped viruses are more resistant to changes in temperature, pH, and some disinfectants than are enveloped viruses.

Types of Nucleic Acid

Unlike nearly all living organisms that use DNA as their genetic material, viruses may use either DNA or RNA. The virus core contains the genome or total genetic content of the virus. Viral genomes tend to be small, containing only those genes that encode proteins that the virus cannot obtain from the host cell. This genetic material may be single- or double-stranded. It may also be linear or circular. While most viruses contain a single nucleic acid, others have genomes that have several, called segments.

In DNA viruses, the viral DNA directs the host cell's replication proteins to synthesize new copies of the viral genome and to transcribe and translate that genome into viral proteins. DNA viruses cause human diseases, such as chickenpox, hepatitis B, and some venereal diseases, like herpes and genital warts.

RNA viruses contain only RNA as their genetic material. To replicate their genomes in the host cell, the RNA viruses encode enzymes that can replicate RNA into DNA, which cannot be done by the host cell. These RNA polymerase enzymes are more likely to make copying errors than DNA polymerases and, therefore, often make mistakes during transcription. For this reason, mutations in RNA viruses occur more frequently than in DNA viruses. This causes them to change and adapt more rapidly to their host. Human diseases caused by RNA viruses include hepatitis C, measles, and rabies.

Classification of Viruses:

Viruses may be classified according to the type of the host, genetic material and number of strands.

On the basis of type of host, viruses are:

1. Animal Viruses:

They live inside animal cells including man. On entering the cell, these disturb the metabolism of the host cell and cause various diseases. The common animal viruses are small pox virus, influenza virus, mumps virus, polio virus and herpes virus. In many animal viruses an extra envelope surrounds their protein coat. The membrane consists of proteins, lipids and carbohydrates and is derived from the host plasma membrane.

Animal viruses may enter cells by attaching to the surface. Some are then engulfed by the cell through pinocytosis or phagocytosis. In such cases, uncoating of the viral nucleic acid might occur within the cell. Inside the host cell they may multiply and form numerous new viral particles. Usually, animal viruses release from the host cells by the rapturing and subsequent death of the host cells.

2. Plant Viruses:

They are parasites of plant cells. Their genetic material is RNA which remains enclosed in the protein coat. The most important plant viruses are tobacco mosaic virus (TMV), tobacco rattle virus (TRV), potato virus (PV), southern bean mosaic virus (SBMV), beet yellow virus (BYV) and turnip yellow virus (TYV).

3. Bacterial virus:

They are parasitic on bacteria and so also called bacteriophages. There are many varieties of bacteriophages. Their size and shape varies from species to species. Some phages are spherical, some comma-shaped whereas majority of them have tadpole-like appearance.

On the basis of nucleic acids, viruses are:

1. DNA viruses:

These viruses possess DNA as the genetic material. On replication this DNA produces new DNA. DNA transmits information for protein synthesis through RNA. (DNA → RNA → PROTEIN).

2. RNA viruses:

These viruses possess RNA as the genetic material. The RNA replicates directly to produce new RNA. Information for protein synthesis passes from RNA to protein without involment of DNA. (RNA → RNA → PROTEIN).

3. DNA – RNA viruses:

In a group of RNA tumour viruses called leukoviruses or rousviruses the genetic material is alternately DNA and RNA. In addition to the normal mode of transfer found in DNA viruses (DNA → RNA → PROTEIN) the rousviruses also transfer information from RNA to DNA (RNA-DNA-RNA -PROTEIN).

With respect to number of strands, four types of nucleic acids have been found in viruses:

1. Double stranded DNA:

Double stranded DNA has been reported in pox viruses, the bacteriophages T 2, T 4, T 6, T 3, T 7 and lamda, herpes viruses, adeno viruses, polyoma virus SV-40 and papilloma viruses.

2. Single stranded DNA:

Single stranded DNA is found in the bacteriophages ph i X 174 and M-13 and is cyclic.

3. Double stranded RNA:

Double stranded RNA has been found within viral capsid in the reoviruses of animals and in the wound tumour virus and rice dwarf viruses of plants.

4. Single stranded RNA:

Single stranded RNA is found in most of RNA viruses e.g. Tobacco mosaic virus, influenza virus, poliomylitis bacteriophage MS – 2, F – 2, Coliophage R 17 and the avian leukemia virus.

Viral Diseases in Human Beings

Mumps	Encephalitis
Influenza	Rabies (hydrophobia)
Small pox	Yellow fever
Chicken pox	Dengue fever
Measles	AIDS
Cold	Poliomyelitis

Disease	Tissue affected	Transmission
1. Yellow fever	General infections	Mosquitoes
2. Dengue fever	General infections	Mosquitoes
3. Small pox	Skin & mucous membrane	Contact, sputum etc.
4. Chiken pox	" "	
5. Measles	"	Contact
6. Diarrhoea	Intestinal	Faeces
7. Influenza	Respiratory tract	Nasal & oral discharges
8. Cold	" "	
9. Viral peumonia	"	Sputum
10. Poliomyelitis	Nervous tissue	Faeces, Sputum
11. Rabies	"	Animal bites
12. AIDS	Immune system	Homosexuality

REPLICATION OF VIRUSES

Introduction:

Viruses are obligate intracellular parasites. Hence, they need the help of a host cell for their replication. All viruses have to penetrate, replicate and come out of a cell. The exact way by which different viruses do this varies widely. Infection of a cell by virus may result in any one of the following types of infection.

Productive infection: The cells allow viruses to replicate and the progeny virions are released from the infected cell.

Abortive infection: The cells do not allow viruses to replicate and as a result of this daugter virions are not produced.

Restrictive infection: The cells allow minimal replication of viruses and as a result only few daughters alone are produced. However, the viral genome persists and can lead to serious consequence for the host. Eg. Epstein Barr and herpes simplex

Steps in Viral Replication

The following steps take place during viral replication;-

1. Adsorption

2. Penetration
3. Uncoating
4. Viral genome replication
5. Maturation
6. Release

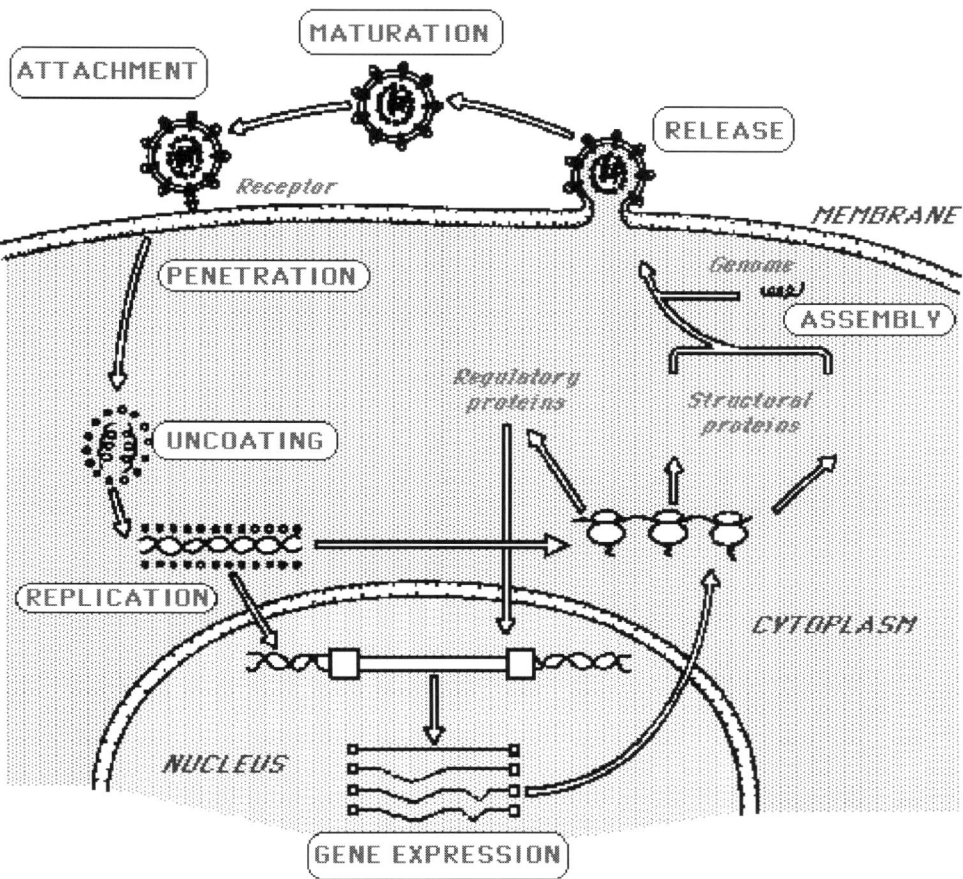

Fig.19.3 Steps in Viral Replication

1. Adsorption

The virus becomes attached to the cells, and at this stage, it can be recovered in the infectious form without cell lysis by procedures that either destroy the receptors or weaken their bonds to the virions. Animal viruses have specialized attachment sites distributed over the surface of the virion e.g. orthomyxoviruses and paramyxoviruses attach through glycoprotein spikes, and adenoviruses attach through the penton fibers. Adsorption occurs to specific cellular receptors. Some receptors are glycoproteins, others are phospholipids or glycolipids. These are usually macromolecules with specific physiological functions, such as complement receptors for EBV. Whether or not receptors for a certain virus are present on a cell depends on the species, the tissue and its physiological state. Cells lacking specific receptors are resistant. Attachment is blocked by antibodies that bind to the viral or cellular sites involved.

2. Penetration

Penetration rapidly follows adsorption, and the virus can no longer be recovered from the intact cell. The most common mechanism is receptor mediated endocytosis, the process by which many hormones and toxins enter cells. The virion is endocytosed and contained within a cytoplasmic vacuole.

3. Uncoating

A key step in uncoating is the acidification of the content of the endosome to a pH of about 5, owing to the activity of a proton pump present in the membrane. The low pH causes rearrangement of coat components, which then expose normally hidden hydrophobic sites. They bind to the lipid bilayer of the membrane, causing the extrusion of the viral core into the cytosol. For influenza virus, the acid-sensitive component is the core HA_2 unit of the haemagglutinin, for adenoviruses, it is the penton base.

4. Viral Nucleic Acid Replication

Virulent viruses, either DNA and RNA, shut off cellular protein synthesis and disaggregate cellular polyribosomes, favouring a shift to viral synthesis. The mechanism of protein synthesis shut-off varies even within the same viral family. Poliovirus, using a viral protease, causes cleavage of a 200 Kd cap-binding protein, which is required for initiation of translation of capped cellular messengers. In contrast to virulent viruses, moderate viruses e.g. polyomaviruses may stimulate the synthesis of host DNA, mRNA, and protein. This phenomenon is of considerable interest for viral carcinogenesis.

DNA Viruses

With animal DNA viruses, transcription and translation are not coupled. Except for poxviruses, transcription occurs in the nucleus and translation in the cytoplasm. Generally, the primary transcripts, generated by RNA polymerase II, are larger than the mRNAs found on ribosomes, and in some cases, as much as 30% of the transcribed RNA remains untranslated in the nucleus. The viral messengers, however, like those of animal cells, are monocistronic. Transcription has a temporal organization, with most DNA viruses only a small fraction of the genome is transcribed into early messengers. The synthesis of early proteins is the key initial step in viral DNA replication. After DNA synthesis, the remainder of the genome is transcribed into late messengers. The complex viruses have immediate early genes, which are expressed in the presence of inhibitors of protein synthesis, and delayed early genes, which require protein synthesis for expression. Regulation is carried out by proteins present in the virions, or specified by viral or cellular genes, interacting with regulatory sequences at the 5' end of the genes. These sequences may respond in *trans* to products produced by other genes and act in *cis* on the associated genes. Different classes of genes may be transcribed from different DNA strands and therefore in opposite directions e.g. polyomaviruses. The transcripts may undergo post- transcriptional processing so that nonessential intervening sequences are removed.

DNA replication

The mode of replication is semiconservative but the nature of the replicative intermediates depends on the manner of replication. Several methods of replication can be recognized.

A. Adenoviruses - Adenoviruses show asymmetric replication, which initiates at the 3' end of one of the strands using a protein primer. The growing strand displaces the preexisting strand of the same polarity and builds a complete duplex molecule. The displaced strand in turn replicates in a similar manner after generating a panhandle structure by pairing the inverted terminal repetitions.

B. Herpesviruses - Herpesviruses have linear genomes with terminal repeats. On reaching the nucleus, the terminal ends undergo limited exonucleotic digestion and then pair to form circles. Replication is thought to take place via a rolling circle mechanism, where concatemers are formed. During maturation, unit-length molecules are cut from the concatemers.

C. Papovaviruses - The DNA of papovaviruses are circular and the replication is bidirectional and symmetrical, via cyclic intermediates.

D. Parvoviruses - The replication of single stranded parvoviruses is initiated when +ve and -ve stranded DNA from different parvovirus particles come together to form a double stranded DNA molecule from which transcription and replication takes place.

E. Poxviruses - The striking feature of poxvirus DNA is that the two complementary strands are joined. The replicative intermediates, present in the cytoplasm, are special concatemers containing pairs of genomes connected either head to head or tail to tail.

F. Hepadnaviruses - Hepatitis B virus employs reverse transcription for replication. The genome consists of a partially double-stranded circular DNA with a complete negative strand and an incomplete positive strand. Upon entering the cell, the positive strand is completed and transcribed. RNA transcripts are in turn reverse-transcribed into DNA by a viral enzyme in several steps, following closely the model of retroviruses, including a jump of the nascent positive strand from one direct repeat (DR) to another.

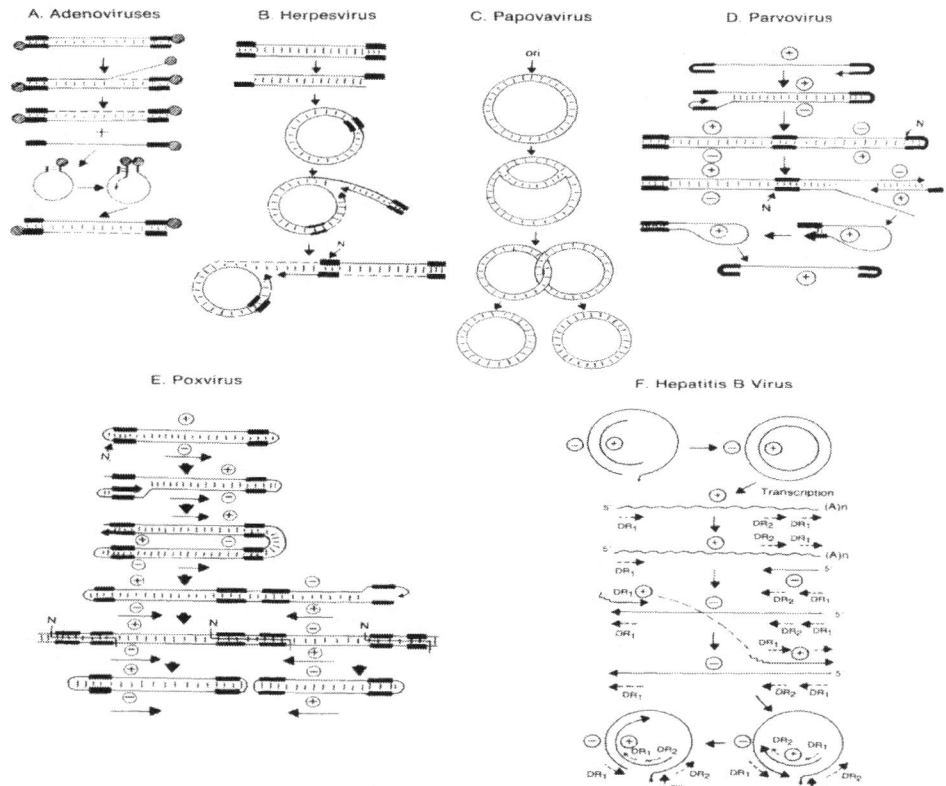

Fig. 19.4.DNA replication: The mode of replication is semiconservative but the nature of the replicative intermediates depends on the manner of replication.

RNA Viruses

The replication of RNA viral genomes is dictated by the absence of multiple translation units within the same messenger, a characteristic of all animal cell messengers. To overcome this difficulty, 3 main strategies have developed.

1. The viral mRNA acts directly as the messenger and is translated monocistronically, followed by cleavage to form different proteins.
2. The virion RNA is transcribed to yield various monocistronic mRNAs by initiating transcription at various places.
3. The genome itself is a collection of separate RNA fragments that are transcribed into monocistronic mRNAs.

RNA viruses can be placed into 7 classes, according to the nature of the viral RNA and its relation to the messenger.

Class I (e.g. picornaviruses, flaviviruses.) The genome, having +ve polarity, itself act as the messenger, specifying information for the synthesis of both structural and nonstructural proteins. The same RNA molecule also initiate replication that requires the expression of proteins first. This format allows little control over replication e.g. Poliovirus has no independent mechanism of controlling the numbers of structural proteins made.

Class II (e.g. coronaviruses, togaviruses.) Many +stranded RNA viruses have subgenomic RNA as part of their cycle. This would allow a certain amount of control. The subgenomic mRNA cannot be recognized by the RNA polymerase. It can be used solely for the synthesis of structural proteins etc. A second way to get round the problem is to make a nested set of RNAs. The nested set of RNA is the most efficient form of control. They can control which part of their genome to express.

	subgenomic	PTC	Nested	Splicing
Picornaviridae	N	Y	N	N
Togaviridae	Y	Y	N	N
Coronaviridae	Y	N	Y	N

With togaviruses, the 49S genome RNA is first translated into polyprotein that is processed into the nonstructural proteins. The subgenomic 26S mRNA, which is transcribed from the full length -ve RNA, is translated into a smaller polyprotein that is processed into viral structural proteins. With coronaviruses, a nested set of mRNAs is generated in the following manner: the -ve transcript is first generated form the genome, which is then transcribed into monocistronic mRNAs of different sizes. Each begins with an identical short 5' leader sequence that is joined to the transcripts at the start of the various genes and continues to the 3' end of the genome.

These mRNAs are not produced by splicing a genomic-size transcript because the virus is able to replicate in enucleated cells.

Class III (e.g. paramyxoviruses, rhabdoviruses.) The genome is of -ve polarity to the messenger. A virion RNA-dependent RNA transcriptase first transcribes the genomes into separate monocistronic messengers initiating at a single promoter. The transcriptase stops and restarts at each juncture between different genes.

Class IV (e.g. orthomyxoviruses, most bunyaviruses.) The -ve genome is in several distinct nonoverlapping pieces of ssRNA. The virion transcriptase generate a messenger from each piece. With orthomyxoviruses, most genomic segments contain a single gene but 2 fragments contain 2 overlapping genes: one is expressed by a full-length messenger, the other by a shorter messenger obtained from the former by splicing. The replication of orthomyxoviruses is unusual amongst RNA viruses in that it takes place within the nucleus. The nuclear function it requires is the 5' cap of cellular messengers, which it "pinches" after endonucleotic cleavage of the host messengers. The 5' cap is then used as primers in the synthesis of viral messengers.

Class V (e.g. arenaviruses, phleboviruses.) Arenaviruses have an ambisense genome in that half the genome is of -ve polarity and is transcribed into a messenger by a virion transcriptase, but the other half, which is of +ve polarity is transcribed twice: first a complete transcript of the genome is made, then the mRNA is transcribed form this transcript. This strategy is seen in the S (small) segment of the genome of phleboviruses. Ambisense genomes are unusual for RNA viruses but not for dsDNA viruses.

Class VI (e.g. reoviruses. Reoviruses.) contain distinct nonoverlapping segments of dsRNA, each is transcribed into an independent mRNA by the virion transcriptase. Most messengers are monocistronic, but one is bicistronic and expresses a second protein by initiating at an internal AUG in a different reading frame. Each segment of reovirus RNA is replicated independently. A nascent mRNA strand is first generated by the virion transcriptase, which then serves as the template for the replicase to make the negative strand. The two strands remain associated in a dsRNA molecule that ends up in a virion. This replication is asymmetric and conservative because (1) the -ve strand of the virion RNA servers as the initial template and (2) the parental RNA does not end up in the progeny.

Class VII (e.g. retroviruses.) Retroviruses are unique in that their genomes are transcribed into DNA and not RNA. They contain two identical ssRNA of +ve polarity, with a poly A tail at the 3' end and a cap at the 5' end. Each is transcribed into DNA by reverse transcriptase that then integrates into the cellular DNA as provirus. Transcription of the provirus by the cellular transcriptase yields the viral molecules that end up in virions.

Since RNA viruses of classes III to VII require a virion transcriptase for synthesizing a messenger, their purified viral RNAs are not infectious. Only those of classes I and II are infectious. With RNA viruses, there is no differentiation between early and late messengers.

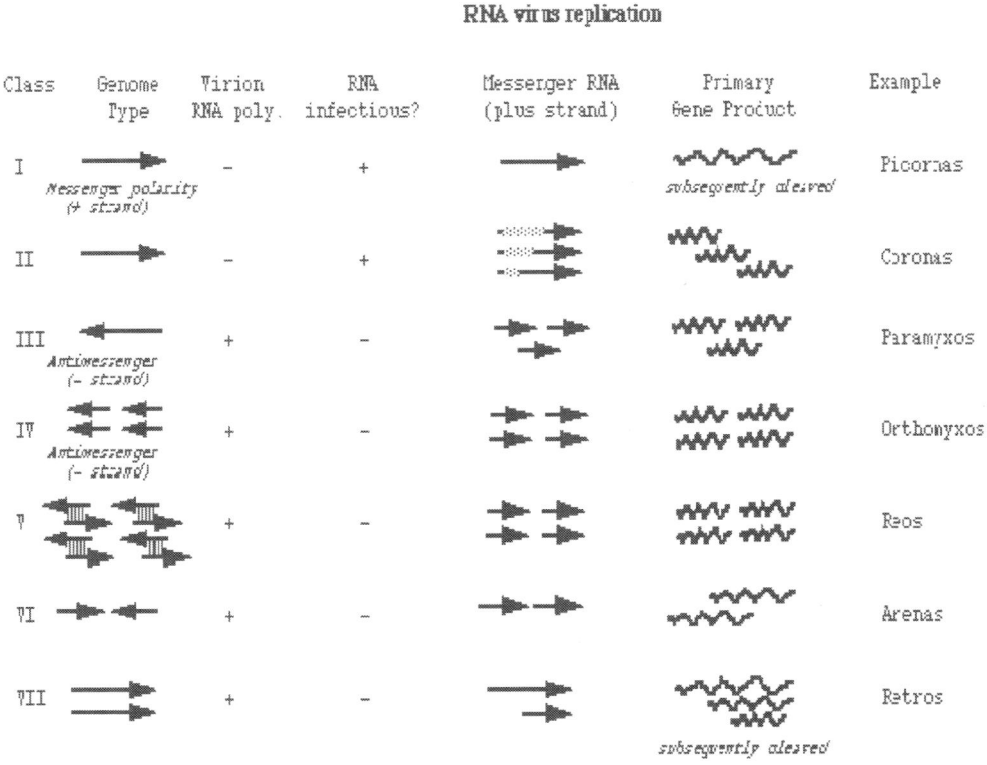

Fig.19.5.

Replication of Single-Stranded RNA Viruses (Classes I to V)

In all cases, replication consists of building a template strand complementary to the viral strand of the same length, which then servers as the template for progeny viral strands. These steps are carried out by a collection of enzymes of both viral and cellular origin, in association with the nucleocapsids of the infecting virions. In many instances, replication and transcription interfere with each other: with -ve stranded viruses, both template and transcripts are made from viral strands: with +ve viruses, a viral strand can be used as a messenger or replication template. Initially in the infection, there is no interference as the messenger function is needed to provide proteins needed for replication. Later the supply of these proteins regulate the rate of replication. e.g. with poliovirus, replication is initiated when the pVg protein becomes covalently linked to at the 5' ends of the RNA, apparently initiating the formation of a replication complex. Messenger and progeny often differed structurally. e.g., the messengers of influenza virus have capped leader sequences derived from cellular messengers. In addition, they lack 17 to 22 nucleotides at the 3' end. Moreover, replication requires ongoing protein synthesis to provide the required proteins, whereas transcription does not.

In RNA replication, the newly made template strand remains associated with the viral strand on which it is made, forming a double-stranded structure the length of the viral genome, known as the replicative form (RF). Synthesis of new strands occurs by conservative asymmetric synthesis, similar to adenoviruses. An RF with a nascent viral strand is known as RI (replicative intermediate) RF molecules are fairly abundant during replication because after the completion of a new strand, the replicase appear to remain associated for some time with the template before reinitiating synthesis. RFs accumulate at the end of replication, when no more RIs are formed. With the exception of orthomyxoviruses, the viral RNAs replicate in the cytoplasm. The replicase present in infected cells synthesize new viral RNA strands of both polarities. Transcription occurs at the same site as replication. It is unclear whether replication and transcription are carried out by different enzymes or by the same enzyme.

Maturation and Release

Maturation proceeds differently for naked, enveloped, and complex viruses.

Naked icosahedral viruses - Preassembled capsomers are joined to form empty capsids (procapsid) which are the precursors of virions. The assembly of capsomers to form the procapsid is often accompanied by extensive reorganization, which is revealed by changes in serological specificity and isoelectric point. eg. picornaviruses and adenoviruses. Naked icosahedral viruses are released from infected cells in different ways. Poliovirus is rapidly released, with death and lysis of infected cells. in contrast, the virions of DNA viruses that tend to mature in the nucleus tend to accumulate within infected cells over a long period and are released when the cell undergoes autolysis, and in some cases, may be extruded without lysis.

Enveloped Viruses - Viral proteins are first associated with the nucleic acid to form the nucleocapsid, which is then surrounded by an envelope. In nucleocapsid formation, the proteins are all synthesized on cytoplasmic polysomes and are rapidly assembled into capsid components. In envelope assembly, virus-specified envelope proteins go directly to the appropriate cell membrane (the plasma membrane, the ER, the Golgi apparatus), displacing host proteins. In contrast, the carbohydrates and the lipids are produced by the host cell. The viral envelope has the lipid constitution of the membrane where its assembly takes place. (eg. the plasma membrane for orthomyxoviruses and paramyxoviruses, the nuclear membrane for herpesviruses) A given virus will differ in its lipids and carbohydrates when grown in different cells, with consequent differences in physical, biological, and antigenic properties.

The envelope glycoproteins are synthesized in the following manner: the polypeptide backbone is first formed on polysomes bound to the ER, which then moves via transport vesicles to the Golgi apparatus where it attains it full glycosylation and fatty acid acylation. The matrix proteins that are present in viral envelope are usually not glycosylated and stick to the cytoplasmic side of the plasma membrane through hydrophobic domains. Matrix proteins connect the cytoplasmic domains of the envelope glycoproteins with the cell's cytoskeleton, and they gather the viral glycoproteins to form the virions. The selection of viral glycoproteins is efficient but not exclusive. eg. rhabdovirus virions contain 10 to 15% of nonviral glycoproteins. They may also contain glycoproteins specified by another virus infecting the same cell. Envelopes are formed around the nucleocapsids by budding of cellular membranes.

With orthomyxoviruses and paramyxoviruses, the viral glycoproteins incorporated in the membranes confer on the cell some properties of a giant virion. Thus cells infected by these viruses may bind RBCs (haemadsoption), and paramyxovirus-infected cells may fuse with uninfected cells to form multinucleated syncytia by the fusion of their membranes. This fusion is equivalent to the fusion of the virion's envelope with the plasma membrane of the host cell at the onset of infection.

Complex Viruses

Maturation of the highly organized poxviruses takes place in cytoplasmic foci called "factories" In contrast to simpler viruses, the poxvirus membrane contain newly synthesized lipids that differ in composition to the cellular lipids. The maturation of poxviruses after the precursors have been enclosed within the primitive membranes suggest that poxviruses may be transitional forms towards a cellular organization.

****

20.

BACTERIOPHAGES

Bacteriophages are bacterial viruses. They are the viruses that infect bacteria. They are obligate intracellular parasites that multiply inside bacteria by making use of some or all of the host biosynthesis machinery. They are also called ***phages***. These extrachromosomal genetic elements usually survive outside a host cell due to the presence of a nucleic acid genome surrounded by a protein coat. Phages occur widely in nature in close association with bacteria and are distributed widely in the soil, feces, and in other substances in the environment. They are associated with transmission of genetic material from one bacterium to another.

F. Twort (1915) and F. d 'Herelle (1917) independently discovered bacteriophages. Bacteriophages attacking Escherichia coli are called coliphages or T-phages. Max Delbruck (1938) numbered coliphages as T-even phages (T2, T4, T6 etc.) and T-odd phages (T1, T3, T5 etc.).

1. Structure:

T-even phages are spermatozoid or tadpole-like structure with a head and a tail. The head is a hexagonal, bipyramidal prismoid structure, while the tail is contractile and cylindrical.

The tail is attached to the lower side of the head. The extended portion between the head and tail is called disc or collar. The head of T_2 phage is approximately 950 Å x 650 Å in size.

The wall of the head is a proteinaceous membrane, consists of 2,000 similar protein subunits, enclosing a single, double-stranded DNA thread of about 50 nm in length.

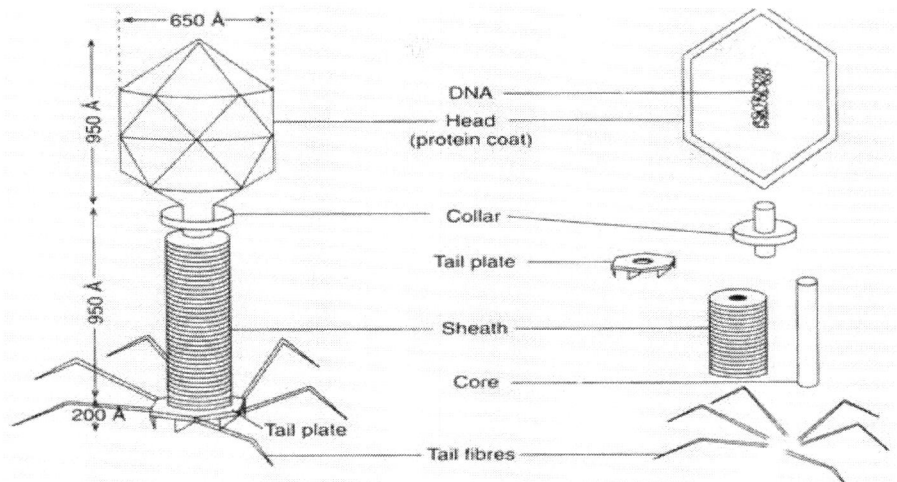

Fig.20. Structure of bacteriophage and its component

The molecular weight of DNA is 2,500,000 and the amount of nucleic acid is approximately 6×10^3 micrograms. The length of the tail is almost equal to the length of the head i.e., 950 Å and its diameter is 80 Å.

The tail consist of four parts (Fig.20):

(i) A central hollow tube or core (through which DNA passes during infection),

(ii) The proteinaceous sheath surrounding the core can contract longitudinally,

(iii) Hexagonal tail plate or basal plate (200 A thick), has a pin at every corner which is attached with the distal end of the sheath,

(iv) Six long thin tail fibres are attached to the tail plate, one at each corner. The tail fibres are about 1,500 Å in length. The tail fibres are the organ of attachment to the host cell wall.

Multiplication or Life Cycle of Phages

Phages multiply through two different types of life cycle. a. Lytic or virulent cycle b. Lysogenic or Avirulent life cycle

a. Lytic Cycle

During lytic cycle of phage, disintegration of host bacterial cell occurs and the progeny virions are released (Figure 1.5a). The steps involved in the lytic cycle are as follows:

(i) Adsorption

Phage (T_4) particles interact with cell wall of host (*E. coli*). The phage tail makes contact between the two, and tail fibres recognize the specific receptor sites present on bacterial cell surface. The lipopolysaccharides of tail fibres act as receptor in phages. The process involving the recognition of phage to bacterium is called **landing**. Once the contact is established between tail fibres and bacterial cell, tail fibres bend to anchor the pins and base plate to the cell surface. This step is called **pinning.**

ii) Penetration

The penetration process involves mechani-cal and enzymatic digestion of the cell wall of the host. At the recognition site phage digests certain cell wall structure by viral enzyme (lysozyme). After pinning the tail sheath contracts (using ATP) and appears shorter and thicker. After contraction of the base plate enlarges through which DNA is injected into the cell wall without using metabolic energy. The step involving injection of DNA particle alone into the bacterial cell is called **Transfection**. The empty protein coat leaving outside the cell is known as **'ghost'**.

(iii) Synthesis

This step involves the degradation of bacterial chromosome, protein synthesis and DNA replication. The phage nucleic acid takes over the host biosynthetic machinery. Host DNA gets inactivated and breaks down. Phage DNA suppresses the synthesis of bacterial protein and directs the metabolism of the cell to synthesis the proteins of the phage particles and simultaneously replication of Phage DNA also takes place.

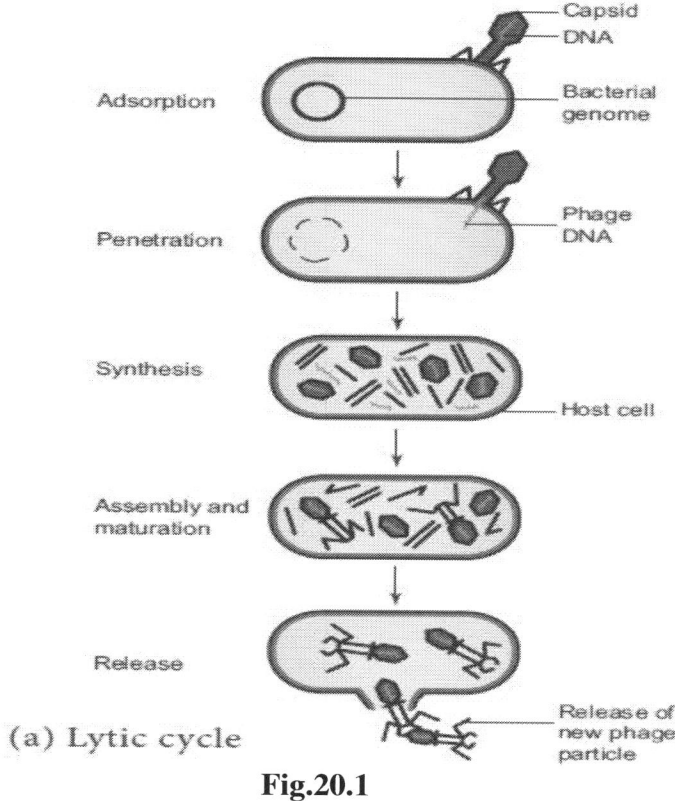

Fig.20.1

(iv) Assembly and Maturation

The DNA of the phage and protein coat are synthesized separately and are assembled to form phage particles. The process of assembling the phage particles is known as **maturation**. After 20 minutes of infection about 300 new phages are assembled.

(v) Release

The phage particle gets accumulated inside the host cell and are released by the lysis of host cell wall.

b. Lysogenic Cycle

In the lysogenic cycle the phage DNA gets integrated into host DNA and gets multiplied along with nucleic acid of the host. No independent viral particle is formed.

As soon as the phage injects its linear DNA into the host cell, it becomes circular and integrates into the bacterial chromosome by recombination. The integrated phage DNA is now called **prophage.** The activity of the prophage gene is repressed by two repressor proteins which are synthesized by phage genes. This checks the synthesis of new phages within the host cell. However, each time the bacterial cell divides, the prophage multiplies along with the bacterial chromosome. On exposure to UV radiation and chemicals the excision of phage DNA may occur and results in lytic cycle.

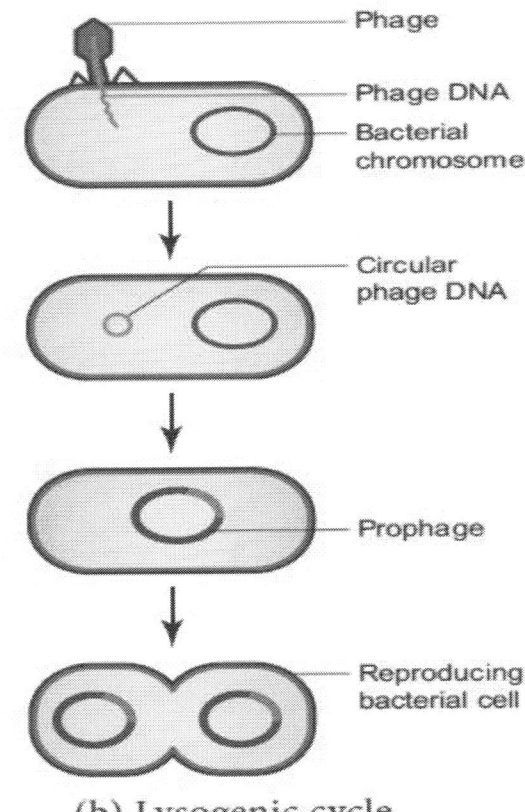

(b) Lysogenic cycle

Fig.20.2: Multiplication of Phages

Types of Bacteriophage.

Different strains of a serologically or otherwise identical species of bacteria are susceptible to one or more different bacteriophages. When a suspension of phages is deposited on the lawn culture of a susceptible bacterium, an area of clearing occurs after incubation due to lysis of the susceptible bacteria by the phages. These zones of lysis are called *plaques*. The shape, size, and nature of plaques are characteristic for different phages. Since a single phage particle is capable of producing one plaque, plaque assay can be used for titrating the number of viable phages in preparation. On the basis of this phenomenon, many bacterial species can be divided into various phage types. Phage typing has been used in epidemiological study of infections or outbreaks caused by *Staphylococcus aureus*, *Salmonella* spp., *Vibrio cholerae*, and many other bacteria. Different phages are available, which show difference in their specificity for genus, species, or strains.

Examples are (*i*) genus-specific bacteriophages for *Salmonella*, (*ii*) specific bacteriophages for all members or strains of *Bacillus anthracis*, and (*iii*) for all members of *V. cholerae* biotype classical (e.g., Mukherjee's phage IV). Mukherjee's phag IV lyses all strains of *V. cholerae* biotype classical, but not *V. cholerae* biotype Eltor. Phage typing of *S. aureus* is a pattern method in which a set of standard phages is employed for intraspecies typing of staphylococci. A strain of *Staphylococcus* may be lysed by a number of phages. Hence, the phage type of a strain is designated by the number of the different phages that lyse it. Phage typing of *Salmonella* Typhi is carried out by using prophage, which is active against only fresh isolates of *S. Typhi* possessing the Viantigen.

Comparison of multiplication of bacteriophages and viruses

Stage	Bacteriophages	Animal viruses
Attachment	Tail fibers attach to cell wall proteins	Attachment sites are plasma membrane proteins and glycoproteins
Penetration	Viral DNA is injected into host cell	Capsid enters by endocytosis or fusion
Uncoating	Not required	Enzymatic removal of capsid proteins
Biosynthesis	In cytoplasm	In nucleus (DNA viruses) or cytoplasm (RNA viruses)
Genome integration	Lysogeny	Latency; slow viral infections; cancer
Release	Host cell lysed	Enveloped viruses bud out; nonenveloped viruses rupture plasma membrane

****

21.

TOBACCO MOSAIC VIRUS (TMV)

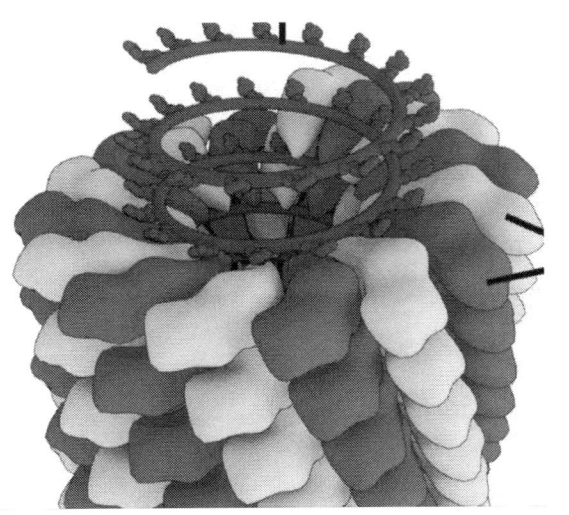

In 1886, Adolf Mayer first described the tobacco mosaic disease that could be transferred between plants, similar to bacterial infections. In 1892, Dmitri Ivanovsky gave the first concrete evidence for the existence of a non-bacterial infectious agent, showing that infected sap remained infectious even after filtering through the finest Chamberland filters. Later, in 1903, Ivanovsky published a paper describing abnormal crystal intracellular inclusions in the host cells of the affected tobacco plants and argued the connection between these inclusions and the infectious agent. However, Ivanovsky remained rather convinced, despite repeated failures to produce evidence, that the causal agent was an unculturable bacterium, too small to be retained on the employed Chamberland filters and to be detected in the light microscope. In 1898, Martinus Beijerinck independently replicated Ivanovsky's filtration experiments and then showed that the infectious agent was able to reproduce and multiply in the host cells of the tobacco plant. Beijerinck coined the term of "virus" to indicate that the causal agent of tobacco mosaic disease was of non-bacterial nature. *Tobacco mosaic virus* was the first virus to be crystallized. It was achieved by Wendell Meredith Stanley in 1935 who also showed that TMV remains active even after crystallization. For his work, he was awarded 1/4 of the Nobel Prize in Chemistry in 1946, even though it was later shown some of his conclusions (in particular, that the crystals were pure protein, and assembled by autocatalysis) were incorrect. The first electron microscopical images of TMV were made in 1939 by Gustav Kausche, Edgar Pfankuch and Helmut Ruska – the brother of Nobel Prize winner Ernst Ruska. In 1955, Heinz Fraenkel-Conrat and Robley Williams showed that purified TMV RNA and its capsid (coat) protein assemble by themselves to functional viruses, indicating that this is the most stable structure (the one with the lowest free energy). The crystallographer Rosalind Franklin worked for Stanley for about a month at Berkeley, and later designed and built a model of TMV for the 1958 World's Fair at Brussels. In 1958, she speculated that the virus was hollow, not solid, and hypothesized that the RNA of TMV is single-stranded. This conjecture was proven to be correct after her death and is now known to be the + strand. The investigations of tobacco mosaic disease and subsequent discovery of its viral nature were instrumental in the establishment of the general concepts of virology.

Symptoms and Signs

Symptoms induced by *Tobacco mosaic virus* (TMV) are somewhat dependent on the host plant and can include mosaic, mottling (Figures 20 and 20.1), necrosis (Figures 20.2 and 20.3), stunting, leaf curling, and yellowing of plant tissues. The symptoms are very dependent on the age of the infected plant, the environmental conditions, the virus strain, and the genetic background of the host plant. Strains of TMV also infect tomato, sometime causing poor yield or distorted fruits, delayed fruit ripening, and nonuniform fruit color (Figure 20.4).

Fig. 20   Fig. 20.1

Fig. 20.2   Fig.20.3

Fig.20.4

Structure of Tobacco Mosaic Virus (TMV):

TMV is a simple rod-shaped helical virus (Fig. 20.5) consisting of centrally located single-stranded RNA (5.6%) enveloped by a protein coat (94.4%). The rod is considered to be 3,000 Å in length and about 180 Å in diameter.

The protein coat is technically called 'capsid'. R. Franklin estimated 2,130 sub-units, namely, capsomeres in a complete helical rod and 49 capsomeres on every three turns of the helix; thus there would be about 130 turns per rod of TMV.

The diameter of RNA helix is about 80 Å and the RNA molecule lies about 50 Å inward from the outer-most surface of the rod. The central core of the rod is about 40 Å in diameter. Each capsomere is a grape like structure containing about 158 amino acids and having a molecular weight of 17,000 dalton as determined by Knight.

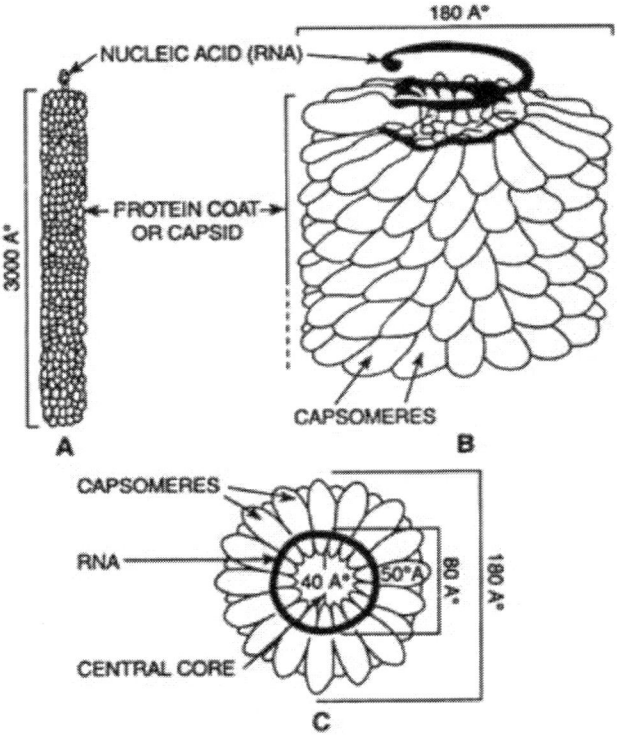

Tobacco mosaic virus (TMV). A. surface view; B. an enlarged portion showing RNA-capsomere arrangement; C. view in section.

Fig.20.5

The ssRNA is little more in length (about 3300 Å) slightly protruding from one end of the rod. The RNA molecule consists of about 7300 nucleotides; the molecular weight of the RNA molecule being about 25,000 dalton.

Life-Cycle (Replication) of Tobacco Mosaic Virus (TMV):

Plant viruses like TMV penetrate and enter the host cells in toto and their replication completes within such infected host cells (Fig. 13.21). Inside the host cell, the protein coat dissociates and viral nucleic acid becomes free in the cell cytoplasm.

Although the sites for different steps of the viral multiplication and formation of new viruses have not yet been determined with absolute certainty, the studies suggest ha alter becoming free in the cell cytoplasm the viral-RNA moves into the nucleus (possibly into the nucleolus).

The viral-RNA first induces the formation of specific enzymes called 'RNA polymerases' the single-stranded viral-RNA synthesizes an additional RNA strand called replicative RNA.

This RNA strand is complementary to the viral genome and serves as 'template' for producing new RNA single strands which is the copies of the parental viral-RNA. The new viral-RNAs are released from the nucleus into die cytoplasm and serve as messenger-RNAs (mRNAs). Each mRNA, in cooperation with ribosomes and t-RNA of the host cell directs the synthesis of protein subunits.

After the desired protein sub-units (capsomeres) have been produced, the new viral nucleic acid is considered to organize the protein subunit around it resulting in the formation of complete virus particle, the virion.

No 'lysis' of the host cell, as seen in case of virulent bacteriophages, takes place. The host ells remain alive and viruses move from one cell to the other causing systemic infection. When transmitted by some means the viruses infect other healthy plants.

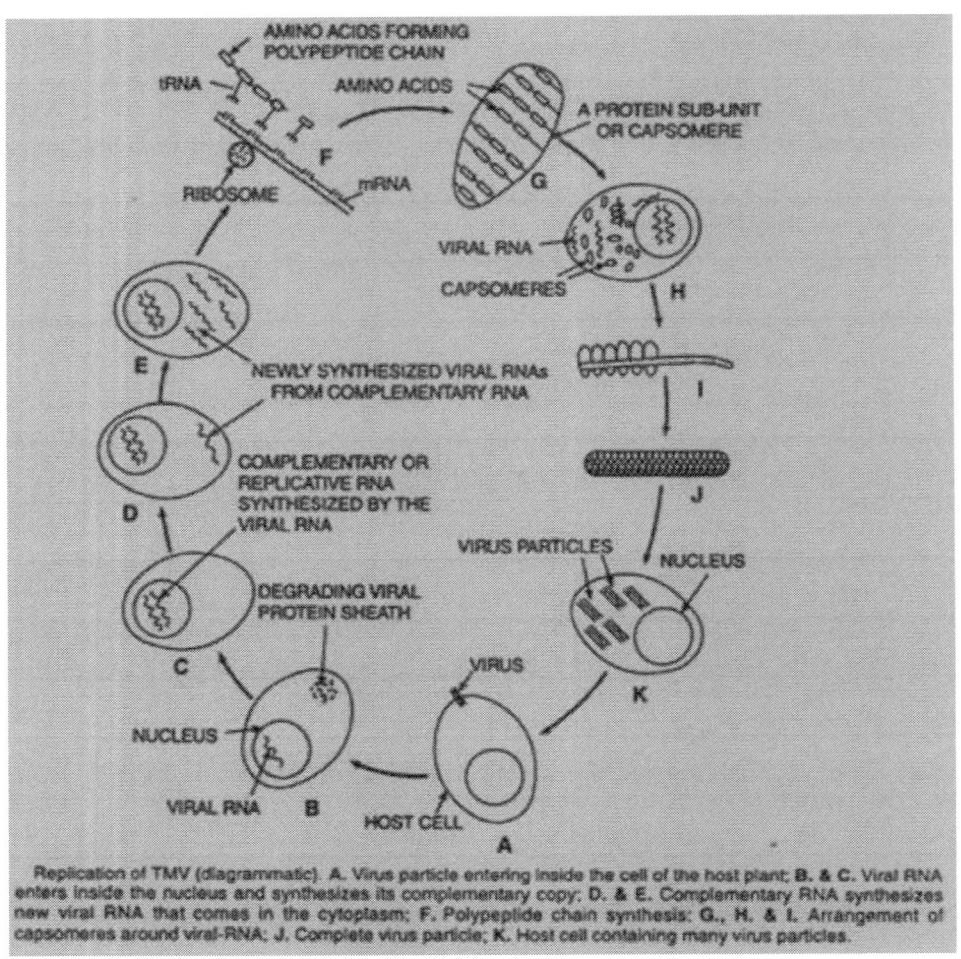

Fig.20.6.

Disease Cycle and Epidemiology

Transmission from plant to plant

TMV is very easily transmitted when an infected leaf rubs against a leaf of a healthy plant, by contaminated tools, and occasionally by workers whose hands become contaminated with TMV after smoking cigarettes. A wounded plant cell provides a site of entry for TMV. The virus can also contaminate seed coats, and the plants germinating from these seeds can become infected. TMV is extraordinarily stable. Purified TMV (Fig.20.7.) has been reported to be infectious after 50 years storage in the laboratory at 4°C/40°F.

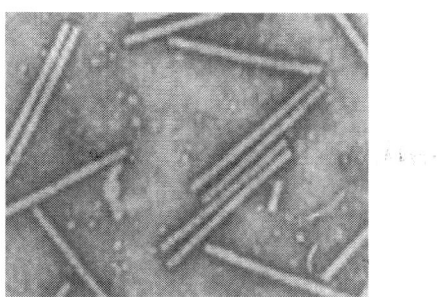

Fig.20.7

Movement in the infected plant

TMV uses its movement protein to spread from cell-to-cell through plasmodesmata, which connect plant cells (figure 10). Normally, the plasmodesmata are too small for passage of intact TMV particles.

The movement protein (probably with the assistance of as yet unidentified host proteins) enlarges the plasmodesmatal openings so that TMV RNA can move to the adjacent cells, release the movement protein and host proteins, and initiate a new round of infection. As the virus moves from cell to cell, it eventually reaches the plant's vascular system (veins) for rapid systemic spread through the phloem to the roots and tips of the growing plant.

Epidemiology

The TMV disease cycle and its epidemiology are intimately related because the virus is completely dependent on the host for replication and spread. There is wide variation in disease incidence, depending on the time of disease onset in the field and on cropping practices. For example, a few plants could become infected early in the season, either from TMV on the seed coat or by workers contaminating plants. The disease could then spread rapidly throughout the field or greenhouse by TMV-infected plants contacting healthy plants, or by equipment or workers. TMV can also survive or overwinter in infected plant debris or perennial (weedy) hosts and, perhaps, in the soil. Agricultural practices, such as continuous cropping, have the potential to be a particular problem, especially in greenhouse facilities, where TMV inoculum may increase in more than one plant species.

Disease Management

Greenhouse management

Horticultural practices. To reduce infection of plants with TMV all tools should be washed with soap or a 10% solution of household bleach to inactivate the virus. TMV-contaminated soil should be discarded. To avoid transmitting the virus from an infected plant to healthy plants, the watering hose or watering can should not be allowed to make contact with the plants. Care should be taken to dispose of dead leaves and old plants, because dry, TMV-infected leaves can be blown around the greenhouse as 'dust' which can subsequently infect healthy plants if they are wounded.

Cross protection. Inoculation of a mild strain of the virus onto young plants can protect them from subsequent infection by more severe strains of TMV. This is a well-documented control strategy, called "cross protection," that is successfully applied in greenhouse operations. Transgenic plants also offer alternative strategies for virus control (see Biotechnology) (Fig.20.8).

Figure 20.8.

****

22.

POTATO VIRUS Y, POTATO VIRUS X (PVX) AND WILD POTATO MOSAIC VIRUS (WPMV)

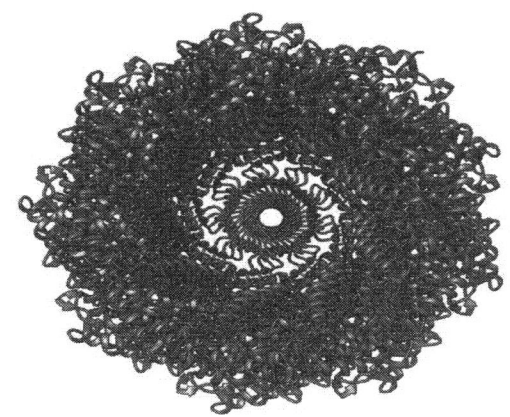

Potato virus Y (PVY) is an aphid-borne virus that causes yield losses and tuber quality defects in commercial potato crops.

In seed crops PVY infection increases the risk of the seed lot being downgraded or rejected from certification. PVY infects other solanaceous crops including tomato and capsicum. The infection occurs in most potato growing areas Australia and overseas.

Strains and symptoms

A number of different strains of PVY can occur. In Western Australia, the 'ordinary' or 'o' strain has been detected in potatoes, however in Victoria and South Australia the 'NTN' strain is present. Other strains occur overseas.

Symptoms of infection vary with cultivar, plant age, environmental conditions and PVY strain. Leaf symptoms can range from mild to severe mosaic or mottling. In severe cases leaf drop can occur. Infected plants are often stunted.

Fig.22. King Edward potato plant infected with PVY showing mild leaf mottle and distortion

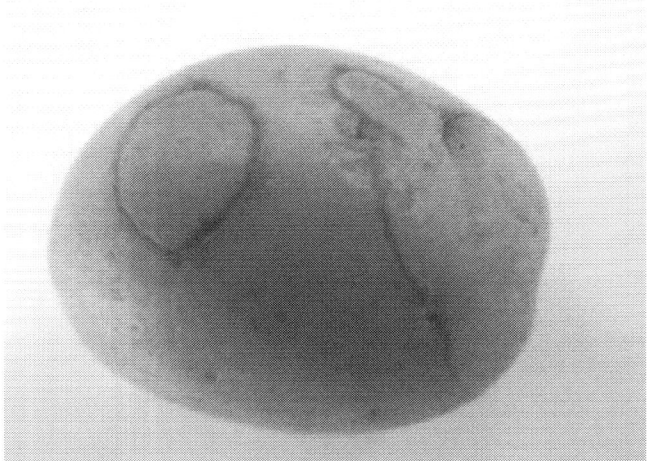

Fig. 22.1. Tubers from infected plants are often small and can have necrotic or dead rings on the skin.

Necrotic rings on potato tubers from PVY-infected Nadine plants

Sources of virus

Infected seed potato tubers are the principal source of PVY spread to other potato plants. Sources are old potato crops, volunteer or self-sown potatoes, crops including capsicum, tomato and tobacco, and weeds such as nightshade and cape gooseberry.

Infected old crop plants, self-sown potato plants and weeds allow the virus to survive between growing seasons. These plants are often found growing within crops, on roadside verges and along fencelines.

PVY is spread from these infected plants to potato crops by aphids. PVY has a wide host range including plants in the Solanaceae, Amaranthaceae and Chenopodiaceae families.

Fig. 22.2. Volunteer or self-sown potatoes can be a source of PVY

Aphid vectors and transmission

PVY is primarily spread by aphids, including green peach (*Myzus persicae*) and potato (*Macrosiphum euphorbiae*) aphids, as well as other species that migrate through the crop.

PVY is transmitted non-persistently. This means when an aphid feeds on a PVY-infected plant, it picks up the virus within 1-2 seconds of feeding. If the aphid then moves to a healthy plant and begins to feed (for 1-2 seconds) the virus is released and transferred to the healthy plant leading to infection. After the aphid has probed one or two healthy plants the virus is lost until it probes another infected plant.

A small number of aphids can spread the virus to a large number of plants quickly as they search for a suitable host plant to colonise.

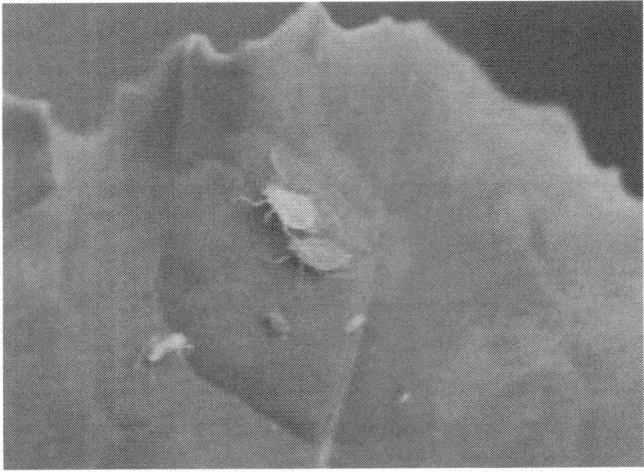

Fig.22.3. PVY is spread by green peach aphids

Contact transmission

PVY is readily spread between plants when machinery damages infected leaves and the infective sap then brushes or rubs onto healthy plants. Such virus spread can be reduced by cleaning cutting implements and machinery with a 1:4 dilution of household bleach followed by water rinse; 1% Virkon; or a 20% solution of skim milk powder.

Yield and quality losses

When potato plants become infected early, yield losses of 10-80% have been reported. Losses can be due to small and few tubers being produced. The severity of symptoms and magnitude of yield depends on the time of infection, number of plants infected, strain of the virus and the variety grown.

Management

To minimise PVY spread in potato crops an integrated disease management approach is necessary, however the most important management practice is to use seed lots with zero levels of PVY. The addition of other practices in combination will provide added control.

- Plant seed potatoes with zero level of PVY. WA certified and registered seed potatoes with Rating 1 and Rating 2 have a zero tolerance for PVY.

- Use PVY resistant potato varieties when available. Commercial varieties used in WA that are resistant to PVY are Royal Blue, FL1867 and FL2195.
- Plant a non-host border crop around the potato crop about four weeks before planting, for example, wheat, oats, sorghum. This acts as a cleansing barrier for aphids. If PVY is entering the crop from an outside source then infective aphids may feed on the barrier crop, lose the virus and will no longer be infective when they land on the potato crop.
- Plant new crops upwind from older crops – there is less infection upwind from infection sources as aphids can be blown along with the wind.
- Remove any potato plants showing virus symptoms – removing virus sources within the crop may help to slow down the spread of the virus to nearby plants especially early in the season.
- Employ good hygiene practices. Use 1:4 dilution of household bleach or 1% Virkon to wash equipment and machinery.
- Avoid moving machinery, equipment and workers from old crops to new ones to minimise spread from older to young crops.
- Remove and destroy old potato crops immediately after the final harvest to minimise virus spread to new crops.
- Destroy volunteer potato plants and weeds before planting to reduce any potential virus and aphid sources for new crops. (www.agric.wa.gov.au)

POTATO VIRUS X (PVX)

Potato virus X (PVX) is a plant pathogenic virus of the family *Alphaflexiviridae* and the order *Tymovirales*. It is the type species of the genus *Potexvirus*.

PVX is found mainly in potatoes and is only transmitted mechanically. There are no insect or fungal vectors for this virus. This virus causes mild or no symptoms in most potato varieties, but when Potato virus Y is present, synergy between these two viruses causes severe symptoms in potatoes. The virion has helical symmetry and a deeply grooved, highly hydrated surface and is made of a single-stranded positive-sense RNA genome of approximately 6.4 kb. This is wrapped in approximately 1300 units of a single coat protein (CP) type, with 8.9 CP units per helix turn. The genome is capped at the 5′-end and poly-adenylated at the 3′-terminus. It contains five open reading frames (ORFs) encoding five proteins: the RNA-dependent RNA Polymerase

(RdRP), the movement proteins encoded by three overlapping ORFs that form the Triple Gene Block module (TGBp1, TGBp2, and TGBp3), and the CP (coat protein).

Virus indexing and limited generation production of potato, which starts from disease-free tissue culture plantlets, has nearly eliminated this virus from many countries' potato supply.

WILD POTATO MOSAIC VIRUS

Wild potato mosaic virus (WPMV) is a plant pathogenic virus of the family Potyviridae. The **Potyviridae** are a family of plant viruses. They are flexuous filamentous rod-shaped particles. Their genome is composed of positive-sense RNA which is surrounded by a protein coat made up of a single viral encoded protein called a capsid. All induce the formation of virus inclusion bodies called cylindrical inclusions ('pinwheels') in their hosts. These are composed of a single protein (70 kDa) made in their hosts from a single viral genome product.

Symptoms:

1. **Latent mosaic:** Standard varieties of potato are symptomless carriers of the latent mosaic virus. Some other varieties, when infected, respond with an acute systemic reaction commonly known as top necrosis killing the top of the main stem or a branch and processing downwards until the plant is killed. Tuber production may be prevented or if smaller tubers are formed they commonly show internal necrosis. Still other varieties may show mottling, stunting, necrotic spots on leaves and in some cases necrotic spots in the tuber.
2. **Mild Mosaic:** This disease is caused by one of the several strains of mild mosaic virus. Symptom on susceptible varieties is mottling and stunting which is favoured by cold temperature. Affected plants produce undersized tubers. When some strain of latent mosaic virus is present in mild mosaic affected plant; synergistic effect of both viruses give strong symptoms.
3. **Rugose Mosaic:** This disease is incited by the vein-banding virus. Mottling is inconspicuous in most cases. Rugosity and mottling of the surface of the leaflets together with some stunting are the most prominent early symptoms. Necrosis usually follows, starting as individual spots on leaflets and progressing until the entire leaf dies, sometimes dropping but often clinging to the stem. Necrosis is most severe on the lowest leaves and progresses upward slowly; a tuft of rugose leaves often persisting at the tip until the top dies prematurely. The tubers ordinarily show no symptoms other

than the reduction in size induced by the damage to the top. If some strain of latent mosaic virus is also present, synergistic effect produce strong symptoms in this case also.

4. **Yellows or Leaf Curl:** It cause curling or rolling of leaves and top leaf becomes yellow in color. Plants do not grow properly and remain stunted. Tubers are also smaller in size. The loss in yield is up to 50% transmitted by aphid

Causal agents:

- Latent mosaic virus (Virus A)
- Mild mosaic Virus (Virus X)
- Vein banding virus (Virus Y)
- Yellows or Leaf curl virus.

Disease Cycle: The chief means of perpetuation of the three potato viruses is by way of seed tuber. Local dissemination of the latent mosaic virus is by mechanical contact. Mild mosaic and vein banding viruses are aphid transmitted.

Epidimiology: Air temperature has a marked effect upon the expression of symptoms. Latent mosaic and mild mosaic are favored by cool temperatures in symptom expression, while rugose mosaic is enhanced by warm weather and suppressed by cool weather

Control:

1. Use of disease free certified seed.
2. Rouging of diseased plants and burying them deep in soil.
3. Insect control in case of Mild and Rugose mosaic.
4. Avoid working of labour and animals from diseased to health crop in case of latent mosaic virus.
5. Resistant varieties (like chippewa & Irish cobs).
6. Early harvesting of the crop.

****

23. MYCOVIRUSES, KURU VIRUS, MEASLES (RUBEOLA) VIRUS, ONCOGENIC OR CANCERCAUSING VIRUSES AND VIROIDS

Mycoviruses (Ancient Greek: *mykes* ("fungus") + Latin *virus*), also known as **mycophages**, are viruses that infect fungi. The majority of mycoviruses have double-stranded RNA (dsRNA) genomes and isometric particles, but approximately 30% have positive-sense, single-stranded RNA (+ssRNA) genomes.

True mycoviruses demonstrate an ability to be transmitted to infect other healthy fungi. Many double-stranded RNA elements that have been described in fungi do not fit this description, and in these cases they are referred to as virus-like particles or VLPs. Preliminary results indicate that most mycoviruses co-diverge with their hosts, i.e. their phylogeny is largely congruent with that of their primary hosts. However, many virus families containing mycoviruses have only sparsely been sampled. Mycovirology[4] is the study of mycoviruses. It is a special subdivision of virology and seeks to understand and describe the taxonomy, host range, origin and evolution, transmission and movement of mycoviruses and their impact on host phenotype.

Host range and incidence

Mycoviruses are common in fungi (Herrero et al., 2009) and are found in all four phyla of the true fungi: Chytridiomycota, Zygomycota, Ascomycota and

Basidiomycota. Fungi are frequently infected with two or more unrelated viruses and also with defective dsRNA and/or satellite dsRNA.[12][13] There are also viruses that simply use fungi as vectors and are distinct from mycoviruses because they cannot reproduce in the fungal cytoplasm.

It is generally assumed that the natural host range of mycoviruses is confined to closely related vegetability compatibility groups or VCGs which allow for cytoplasmic fusion, but some mycoviruses can replicate in taxonomically different fungal hosts. Good examples are mitoviruses found in the two fungal species *Sclerotinia homoeocarpa* and *Ophiostoma novo-ulmi*. Nuss et al. (2005) described that it is possible to extend the natural host range of *Cryphonectria parasitica* hypovirus 1 (CHV1) to several fungal species that are closely related to *Cryphonectria parasitica* using *in vitro* virus transfection techniques. CHV1 can also propagate in the genera *Endothia* and *Valsa*, which belong to the two distinct families Cryphonectriaceae and Diaporthaceae, respectively. Furthermore, some human pathogenic fungi are also found to be naturally infected with mycoviruses, including AfuPmV-1 of *Aspergillus fumigatus* and TmPV1 of *Talaromyces marneffei* (formerly *Penicillium marneffei*).

Transmission

A significant difference between the genomes of mycoviruses to other viruses is the absence of genes for 'cell-to-cell movement' proteins. It is therefore assumed that mycoviruses only move intercellularly during cell division (e.g. sporogenesis) or via hyphal fusion. Mycoviruses might simply not need an external route of infection as they have many means of transmission and spread due to their fungal host's life style:

- Plasmogamy and cytoplasmic exchange over extended periods of time
- Production of vast amounts of asexual spores
- Overwintering via sclerotia
- More or less effective transmission into sexual spores

However, there are potential barriers to mycovirus spread due to vegetative incompatibility and variable transmission to sexual spores. Transmission to sexually produced spores can range from 0% to 100% depending on the virus-host combination. Transmission between species of the same genus sharing the same habitat has also been reported including *Cryphonectria* (*C. parasitica* and *C.* sp), *Sclerotinia* (*Sclerotinia sclerotiorum* and *S. minor*), and *Ophiostoma* (*O. ulmi* and *O. novo-ulmi*). Intraspecies transmission has also been reported between *Fusarium poae* and black *Aspergillus* isolates. However, it is not known how fungi overcome the genetic barrier; whether there is some form of recognition process during physical contact or some other means of exchange, such as vectors. Research using *Aspergillus* species indicated that transmission efficiencies might depend on the hosts viral infection status (infected with no, different, or same virus), and that mycoviruses might play a role in the regulation of secondary mycoviral infection. Whether this is also true for other fungi is not yet known. In contrast to acquiring mycoviruses spontaneously, the loss of mycoviruses seems very infrequent and suggests that either viruses actively moved into spores and new hyphal tips, or the fungus might facilitate the mycoviral transport in some other way.

KURU VIRUS

Kuru is a slow virus disease primarily of women reported from Eastern- highlands of New Guinea. This is a degenerative disease of central nervous system which was found in those women who participated in the cannibalistic consumption as a ritual of their dead relatives as a mark of respect and mourning and tasted the brain tissue of the dead. The disease started with trembling and uncontrollable shivering lasting for a year. Infected persons were called "kuru" that means dancers. The disease finally culminated into the "dance of death". The men were seldom infected from the disease since they did not take part inthis tribal ritual. However, infection passedon to children from their mothers who did not wash their hands before washing their children or while wiping the nose or eyes. Dr. D. Carlton Gajdusek who was awarded Nobel Prize discovered that "Kuru" is an infectious disease, as proved by him by chimpanzee inoculation. Now this cannibalistic ritual has been banned by the Government of New Guinea consequently the disease has almost disappeared from that country.

MEASLES (RUBEOLA) VIRUS

There exista only one serotype of virus which causos measles or rubeola. The measles virus possesses a short survival time outside the body but it can survive or remain viable for a long timein droplet nuclei which are the air-borne particles. These droplet nuclei can spread the disease by the aerosol route through upper respiratory route or conjunctiva. This virus can be adapted to grow in cell cultures of human and monkey; and embryonated eggs. The measles virusis found is respiratory tract secretions of patients who are in early stages of measles, The measles virus nfter initial multiplication is disseminated through blood to the mucous membranes of the intestinal tract and urinary tract to the skin and to the central nervous system.

The virus on passing into the blood is deposited throughout the body. The incubation period is of 9 to 12 days followed by fever, cough, runny nose (coryza) and conductivitis. At this stage red macules or ulcer type lesions called 'Koplik spots' appearon the inside of the cheek. The rash appears on the skin after about 3 days, It is believed that the red rash is formed as a result of the immune reaction with viral antigens on the surface of infected blood capillary cells, therefore, causing the dilation ofthe capillaries accompanied by leakage of blood from the infected capillary tissue. There is no antiviral treatment however, the antibiotics are administered to check the secondary infection. At present the control of the measles depends on the use of attenuated vaccine to the children between 12 and 15 months of age. As determined by antibody test those who have not acquired inadequate protection may be vaccinated for the second time before the child goes to school at the age of 5 or 6 years.

ONCOGENIC OR CANCER CAUSING VIRUSES

The tumour inducing viruses of animals are called oncogenic viruses. They include both RNA and DNA viruses which can produce cancer in animals finally forming tumours. The formation of tumours is induced as a result of modification of genome of the cell because of incorporation of DNA from other cells or viruses. The transformed cancerous cells possess changed phenotypic, biochemical and other properties which make them distinct from other infected cells or the cells which have been infected but have not produced the tumour.

VIROIDS

Viroids are the smallest infectious pathogens known. They are composed solely of a short strand of circular, single-stranded RNA that has no protein coating. All known viroids are inhabitants of higher plants, and most cause diseases, whose respective economic importance on humans vary widely.

The first discoveries of viroids in the 1970s triggered the historically third major extension of the biosphere—to include smaller lifelike entities —after the discoveries, in 1675 by Antonie van Leeuwenhoek (of the "subvisible" microorganisms) and in 1892 by Dmitri Iosifovich Ivanovsky (of the "submicroscopic" viruses). The unique properties of viroids have been recognized by the International Committee on Taxonomy of Viruses, in creating a new order of subviral agents.

The first recognized viroid, the pathogenic agent of the potato spindle tuber disease, was discovered, initially molecularly characterized, and named by Theodor Otto Diener, plant pathologist at the U.S Department of Agriculture's Research Center in Beltsville, Maryland, in 1971. This viroid is now called Potato spindle tuber viroid, abbreviated PSTVd.

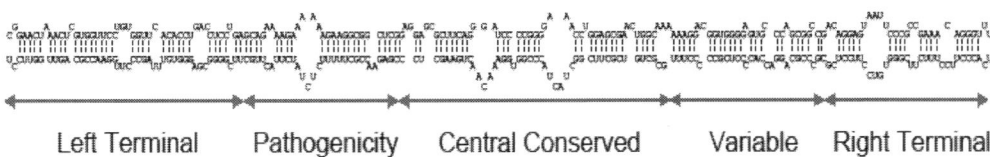

Fig. 23.

Although viroids are composed of nucleic acid, they do not code for any protein. The viroid's replication mechanism uses RNA polymerase II, a host cell enzyme normally associated with synthesis of messenger RNA from DNA, which instead catalyzes "rolling circle" synthesis of new RNA using the viroid's RNA as a template. Some viroids are ribozymes, having catalytic properties that allow self-cleavage and ligation of unit-size genomes from larger replication intermediates.

With Diener's 1989 hypothesis that viroids may represent "living relics" from the widely assumed, ancient, and non-cellular RNA world—extant before the evolution of DNA or proteins—viroids have assumed significance beyond plant pathology to evolutionary science, by representing the most plausible RNAs capable of performing crucial steps in abiogenesis, the evolution of life from inanimate matter.

The human pathogen hepatitis D virus is a subviral agent similar to a viroid.

Transmission

The reproduction mechanism of a typical viroid. Leaf contact transmits the viroid. The viroid enters the cell via its plasmodesmata. RNA polymerase II catalyzes rolling-circle synthesis of new viroids.

Viroid infections can be transmitted by aphids, by cross contamination following mechanical damage to plants as a result of horticultural or agricultural practices, or from plant to plant by leaf contact.[9][10]

Replication

Viroids replicate in the nucleus (*Pospiviroidae*) or chloroplasts (*Avsunviroidae*) of plant cells in three steps through an RNA-based mechanism. They require RNA polymerase II, a host cell enzyme normally associated with synthesis of messenger RNA from DNA, which instead catalyzes "rolling circle" synthesis of new RNA using the viroid as template.[11][12]

RNA silencing

There has long been uncertainty over how viroids induce symptoms in plants without encoding any protein products within their sequences. Evidence suggests that RNA silencing is involved in the process. First, changes to the viroid genome can dramatically alter its virulence. This reflects the fact that any siRNAs produced would have less complementary base pairing with target messenger RNA. Secondly, siRNAs corresponding to sequences from viroid genomes have been isolated from infected plants. Finally, transgenic expression of the noninfectious hpRNA of potato spindle tuber viroid develops all the corresponding viroid-like symptoms. This indicates that when viroids replicate via a double stranded intermediate RNA, they are targeted by a dicer enzyme and cleaved into siRNAs that are then loaded onto the RNA-induced silencing complex. The viroid siRNAs contain sequences capable of complementary base pairing with the plant's own messenger RNAs, and induction of degradation or inhibition of translation causes the classic viroid symptoms.

RNA world hypothesis

Diener's 1989 hypothesis had proposed that the unique properties of viroids make them more plausible macromolecules than introns, or other RNAs considered in the past as possible "living relics" of a hypothetical, pre-cellular RNA world. If so, viroids have assumed significance beyond plant virology for evolutionary theory, because their properties make them more plausible candidates than other RNAs to perform crucial steps in the evolution of life from inanimate matter (abiogenesis). Diener's hypothesis was mostly forgotten until 2014, when it was resurrected in a review article by Flores et al., in which the authors summarized Diener's evidence supporting his hypothesis as:

1. Viroids' small size, imposed by error-prone replication.
2. Their high guanine and cytosine content, which increases stability and replication fidelity.
3. Their circular structure, which assures complete replication without genomic tags.
4. Existence of structural periodicity, which permits modular assembly into enlarged genomes.
5. Their lack of protein-coding ability, consistent with a ribosome-free habitat.
6. Replication mediated in some by ribozymes—the fingerprint of the RNA world.

The presence, in extant cells, of RNAs with molecular properties predicted for RNAs of the RNA World constitutes another powerful argument supporting the RNA World hypothesis.

****

24.

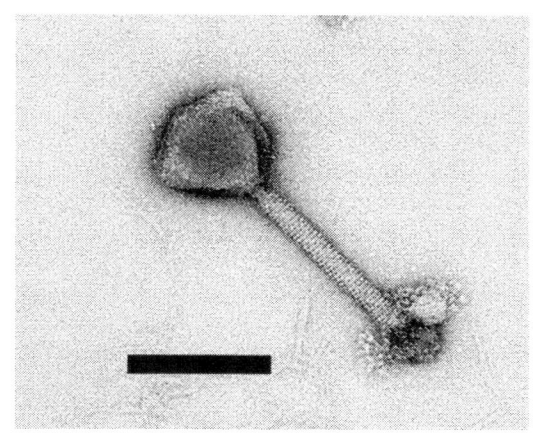

CYANOPHAGES

Discovery of Cyanophages:

Cyanophages are the phages (Fig. 24.) that attack cyanobacteria. Cyanophages were first discovered by Safferman and Morris (1963) from a waste stabilization pond of Indiana University (USA). The first cyanophage studied by Safferman and Morris was the cyanophage attacking Lyngbya, Plectonema and Phormidium.

They named the virus as LPP-1 using the first letter of the three genera. Thereafter, several serological strains of LPP were isolated from different parts of world and named LPP-1, LPP-2, LPP-3, LPP-4 and LPP-5. Besides LPP groups of cyanophages, a large number of other cynophages such as SM-1, AS-1, N-1, C-1, AR-1. A-1, etc. have been reported in recent years (Table 13.1).

Waste stabilization ponds, eutrophic lakes and polluted water support the luxurient growth of cyanobacteria. These can be obnoxious bloom in water reservoirs like lakes and result in fish mortality. Therefore, the cyanophages can play a significant role in control of blooms. So far the problems with them that they are specific to genus and difficult to isolate.

Table.1. Some important Cyanophages known to us

Cynophages	Sources	Host
(i) LPP Group		
LPP-1	Waste stabilization ponds, Indiana, U.S.A.	Lyngbya, Plectonema and Phormidium
LPP-2	Waste stabilization ponds Indiana, U.S.A.	Same but serologically different
LPP-3	Russia	Same but serologically different.
LPP-4*	Banaras Hindu University, India	Same but serologically different.
LPP-5	Banaras Hindu University, India	Same but serologically different
(ii) G-II Group		
Long tailed**	Polluted water, B.H.U., India	Plectonema borryanum
Unnamed	Stream, Japan	Oscillatoria princeps
D-1	Scotland	Same as LPP-1
(iii) N-Group		
C-1	Polluted water, B.H.U.	Cylindrospermum sp.
AR-1	Polluted water, B.H.U.	Anabaenopsis circulans, Raphidiopsis indica
N-1	Polluted water, B.H.U.	Nostoc muscorum
(iv) SM-Group		
SM-1	Waste stabilization ponds Indiana, U.S.A.	Synecococcus elongatus and Microcytis aeruginosa
SM-2	Fresh water	Synecococcus elongatus and Microcytis aeruginosa
AS-1	Polluted water	Anacystis nidulans and Synechococcus cedrorum
AS-1 M	Polluted water	Anacystis nidulans and Synechococcus cedrorum, also M. aeruginosa

Morphology of Cyanophages:

Morphology of LPP-1 has been studied in detail as compared to the other cyanophages. The cyanophages differ morphologically (Fig. 24.) as well as in physico-chemical properties.

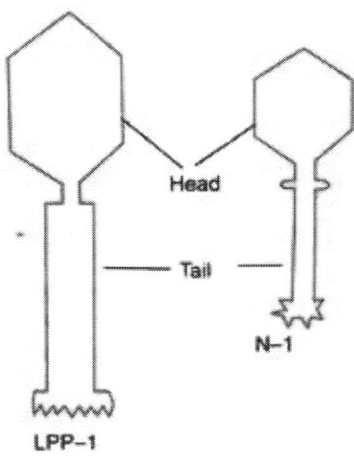

Fig.24. Structure of cyanophages

Sources of some of the cyanophages are given in Table 13.1. The LPP-1 group of cyanophages has an icosahedral head and a tail and are similar to T3 and T7 bacteriophages, whereas the N-1 group resembles with T2 and T4 phages.

Like T-even phages the tail may be contractile or non-contractile. In some groups the tail is absent. The AS-1 group has the largest cyanophages. The group G-III and D-1 are serologically related but do not show any relationship with T-phages.

Replication of Cyanophages:

Cyanophages resemble T-even bacteriophages in their growth cycle. However, cyanophage LPP-1 is much more studied and our discussion on growth cycle is based on this virus. After the LPP-1 is adsorbed on the host cyanobacterium, the viral-DNA is injected into the host cytoplasm; the injection mechanism is still obscure.

Unlike bacteriophages, it does not takes over the complete charge of host cell machinery immediately; the depression of protein synthesis starts soon after the injection of viral DNA but its complete blockage occurs by the end of 5th hours.

The synthesis of viral DNA starts immediately after injection and continues till the lysis of host. The host DNA is not completely degraded (about 50% of it is converted to acid soluble material by the end of 7th hour of infection) and most of the degraded material is further incorporated into viral DNA.

Therefore, it is considered that LPP-1 does not completely arrest the host DNA synthesis. It has been established further that synthesis of viral DNA takes place in the virogenic stoma or invaginated photosynthetic lamellae. However, finally, complete virus-like structures consisting of head and tail are formed and they are released upon the lysis of the cyanobacterial cell.

****

25.

TYPES OF VIRAL INFECTION

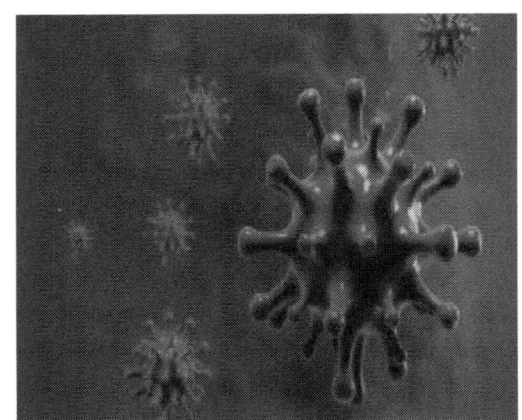

Respiratory Viral Infections

Respiratory viral infections affect the lungs, nose, and throat. These viruses are most commonly spread by inhaling droplets containing virus particles. Examples include:

- **Rhinovirus** is the virus that most often causes the common cold, but there are more than 200 different viruses that can cause colds. Cold symptoms like coughing, sneezing, mild headache, and sore throat typically last for up to 2 weeks.
- **Seasonal influenza** is an illness that affects about 5% to 20% of the population in the US every year. More than 200,000 people per year are hospitalized annually in the US due to complications of the flu. Flu symptoms are more severe than cold symptoms and often include body aches and severe fatigue. The flu also tends to come on more suddenly than a cold.
- **Respiratory Syncytial Virus (RSV)** is an infection that can cause both upper respiratory infections (like colds) and lower respiratory infections (like pneumonia and bronchiolitis). It can be very severe in infants, small children, and elderly adults.

Frequent hand-washing, covering the nose and mouth when coughing or sneezing, and avoiding contact with infected individuals can all reduce the spread of respiratory infections. Disinfecting hard surfaces and not touching the eyes, nose, and mouth can help reduce transmission as well.

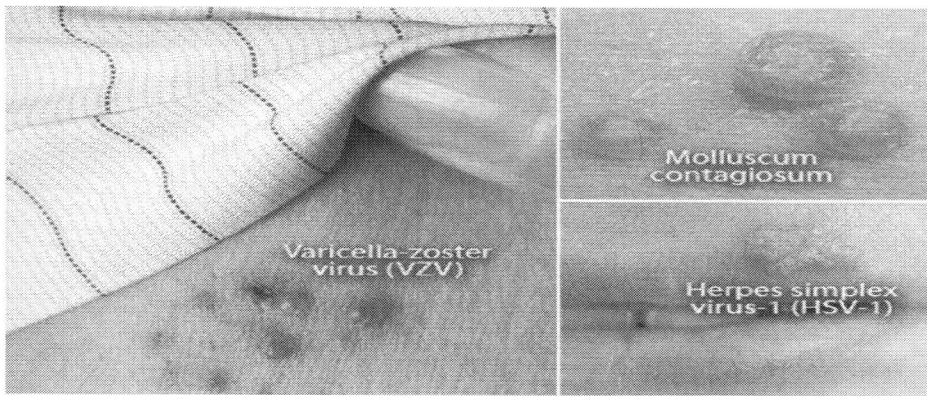

Fig.25.

Viral Skin Infections

Viral skin infections can range from mild to severe and often produce a rash. Examples of viral skin infections include:

- **Molluscum contagiosum** causes small, flesh-colored bumps most often in children ages 1 to 10 years old; however, people of any age can acquire the virus. The bumps usually disappear without treatment, usually in 6 to 12 months.
- **Herpes simplex virus-1 (HSV-1)** is the common virus that causes cold sores. It's transmitted through saliva by kissing or sharing food or drink with an infected individual. Sometimes, HSV-1 causes genital herpes. An estimated 85% of people in the US have HSV-1 by the time they are in their 60s.
- **Varicella-zoster virus (VZV)** causes itchy, oozing blisters, fatigue, and high fever characteristic of chickenpox. The chickenpox vaccine is 98% effective at preventing infection. People who have had chickenpox (or in extremely rare instances, people who have received the chickenpox vaccine) are at risk for developing shingles, an illness caused by the same virus. Shingles can occur at any age, but it occurs most often in people age 60 or older.

The best way to avoid viral skin infections is to avoid skin-to-skin contact (especially areas that have a rash or sores) with an infected individual. Some viral skin infections, such as varicella-zoster virus, are also transmitted by an airborne route. Communal showers, swimming pools, and contaminated towels can also potentially harbor certain viruses.

Foodborne Viral Infections

Viruses are one of the most common causes of food poisoning. The symptoms of these infections vary depending on the virus involved.

- **Hepatitis A** is a virus that affects the liver for a few weeks up to several months. Symptoms may include yellow skin, nausea, diarrhea, and vomiting. Up to 15% of infected individuals experience recurrent illness within 6 months of infection.
- **Norovirus** has been reported to be responsible for outbreaks of severe gastrointestinal illness that happen on cruise ships, but it causes disease in many situations and locations. About 20 million people in the U.S. become sick from these highly contagious viruses every year.
- **Rotavirus** causes severe, watery diarrhea that can lead to dehydration. Anyone can get rotavirus, but the illness occurs most often in babies and young children.

Rotaviruses and noroviruses are responsible for many (but not all) cases of viral gastroenteritis, which causes inflammation of the stomach and intestines. People may use the terms "stomach virus" or "stomach flu" to refer to viral gastroenteritis, which causes nausea, vomiting, diarrhea, and abdominal pain.

It's not pleasant to think about it, but foodborne viral illnesses are transmitted via the fecal-oral route. This means that a person gets the virus by ingesting virus particles that were shed through the feces of an infected person. Someone with this type of virus who doesn't wash their hands after using the restroom can transfer the virus to others by shaking hands, preparing food, or touching hard surfaces. Contaminated water is another potential source of infection.

Sexually Transmitted Viral Infections

Sexually transmitted viral infections spread through contact with bodily fluids. Some sexually transmitted infections can also be transmitted via the blood (blood-borne transmission).

- **Human papillomavirus (HPV)** is the most common sexually-transmitted infection in the US. There are many different types of HPV. Some cause genital warts while others increase the risk of cervical cancer. Vaccination can protect against cancer-causing strains of HPV.

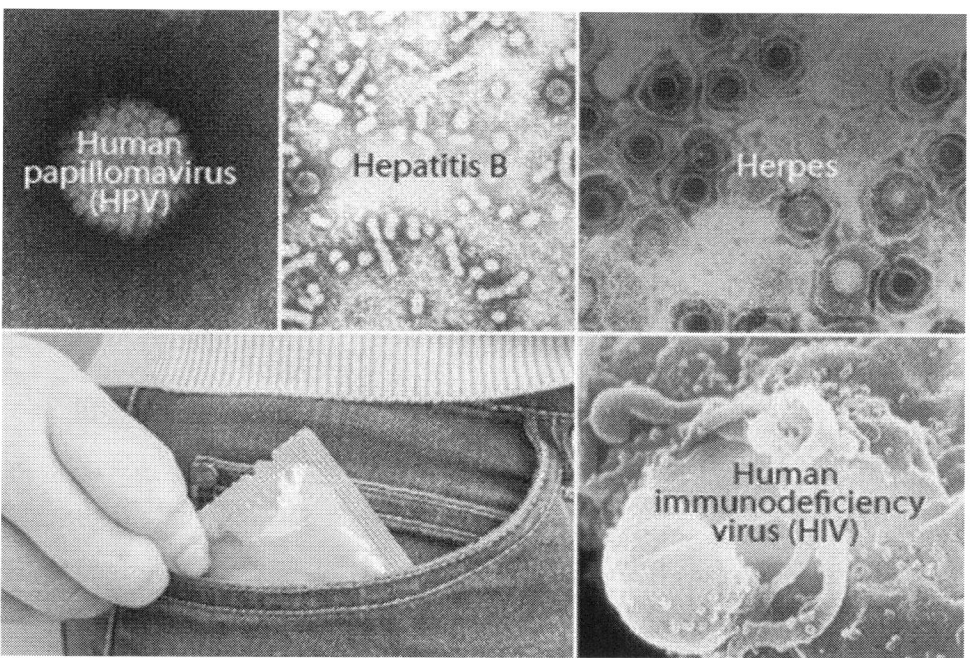

Fig.25.1

- **Hepatitis B** is a virus that causes inflammation in the liver. It's transmitted through contaminated blood and bodily fluids. Some people with the virus don't have any symptoms while others feel like they have the flu. The hepatitis B vaccine is more than 90% effective at preventing infection.
- **Genital herpes** is a common sexually-transmitted infection caused by herpes simplex virus-2 (HSV-2). Herpes simplex virus-1 (HSV-1), the virus responsible for cold sores, can also sometimes cause genital herpes. There's no cure for genital herpes. Painful sores often recur during outbreaks. Antiviral medications can decrease both the number and length of outbreaks.

- **Human immunodeficiency virus (HIV)** is a virus that affects certain types of T cells of the immune system. Progression of the infection decreases the body's ability to fight disease and infection, leading to acquired immune deficiency syndrome (AIDS). HIV is transmitted by coming into contact with blood or bodily fluids of an infected person.

People can reduce the risk of getting a sexually-transmitted viral infection by abstaining from sex or only having sex while in a monogamous relationship with someone who does not have a sexually-transmitted infection. Using a condom decreases, but doesn't entirely eliminate, the risk of acquiring a sexually-transmitted infection. Minimizing the number of sexual partners and avoiding intravenous drug use are other ways to reduce the risk of acquiring sexually-transmitted and bloodborne viral infections.

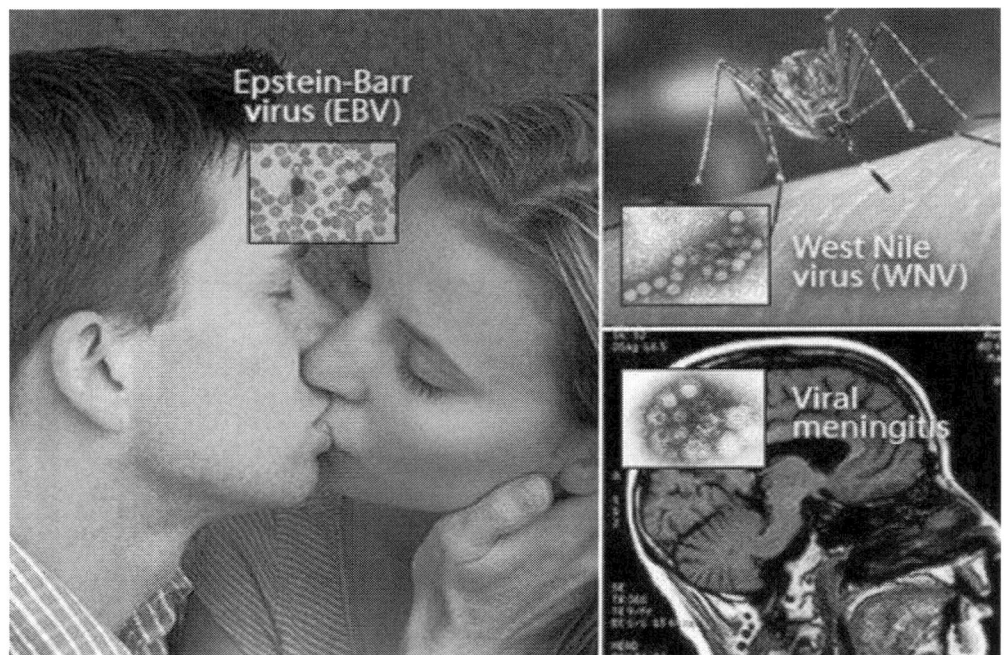

Fig.25.2

Other Viral Infections

Viruses are abundant in the world and cause many other infections ranging from mild to life-threatening.

- **Epstein-Barr virus (EBV)** is a type of herpes virus that's associated with fever, fatigue, swollen lymph nodes, and an enlarged spleen. EBV is a very common virus that causes mononucleosis ("mono"). More than 90% of adults have been infected with this "kissing disease" that is spread primarily through saliva.
- **West Nile virus (WNV)** is a virus that's most commonly transmitted by infected mosquitos. Most people (70% to 80%) with WNV don't have any symptoms while others develop a fever, headache, and other symptoms. Less than 1% of people with WNV develop inflammation of the brain (encephalitis) or inflammation of the tissue surrounding the brain and spinal cord (meningitis).
- **Viral meningitis** is an inflammation of the lining of the brain and spinal cord that causes headache, fever, stiff neck, and other symptoms. Many viruses can cause viral meningitis, but a group of viruses called enteroviruses are most often to blame.

Antiviral Medication and Other Treatment

Many viral infections resolve on their own without treatment. Other times, treatment of viral infections focuses on symptom relief, not fighting the virus. For example, cold medicine helps alleviate the pain and congestion associated with the cold, but it doesn't act directly on the cold virus.

There are some medications that work directly on viruses. These are called antiviral medications. They work by inhibiting the production of virus particles. Some interfere with the production of viral DNA. Others prevent viruses from entering host cells. There are other ways in which these medications work. In general, antiviral medications are most effective when they're taken early on in the course of an initial viral infection or a recurrent outbreak. Different kinds of antiviral medications may be used to treat chickenpox, shingles, herpes simplex virus-1 (HSV-1), herpes simplex virus-2 (HSV-2), HIV, hepatitis B, hepatitis C, and influenza.

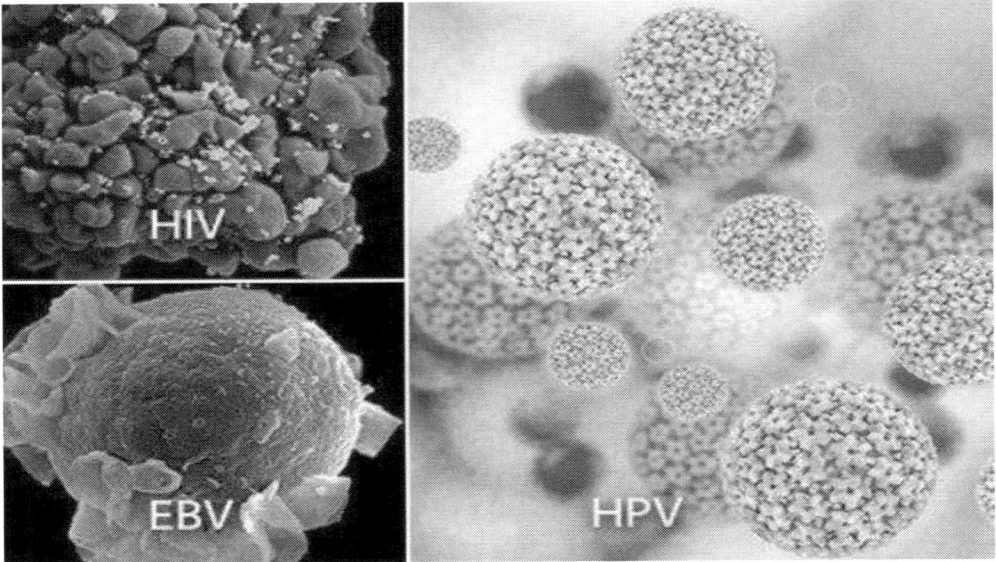

Fig.25.3

Viruses and Cancer

Viruses insert themselves into host cell DNA in order to make more virus particles. Cancer is a disease that occurs as the result of mutations or alterations to DNA. Because viruses affect the DNA of host cells, viruses are known to contribute to several different types of cancer. Viruses known to increase the risk of cancer include:

- Epstein-Barr virus (EBV) for nasopharyngeal cancer, Burkitt lymphoma, Hodgkin's lymphoma, and stomach cancer
- Hepatitis B and hepatitis C for liver cancer
- Human immunodeficiency virus (HIV) for Kaposi sarcoma, invasive cervical cancer, lymphomas, and other cancers
- Human T-lymphotrophic virus-1 (HTLV-1) for T-cell leukemia/lymphoma (ATL)
- Human papilloma virus (HPV) for cervical cancer
- Merkel cell polyomavirus (MCV) for a rare skin cancer called Merkel cell carcinoma

Viral Illness Prevention

Vaccines can reduce the risk of acquiring some viral illnesses. Vaccines are available to help protect against the flu, hepatitis A, hepatitis B, chickenpox, herpes zoster (shingles), cancer-causing strains of human papillomavirus (HPV), measles/mumps/rubella (MMR), polio, rabies, rotavirus, and other viruses.

Vaccines vary in effectiveness and in the number of doses required to confer protection. Some vaccines require booster shots to maintain immunity. (www.onhealth.com)

****

26. REOVIRUSES

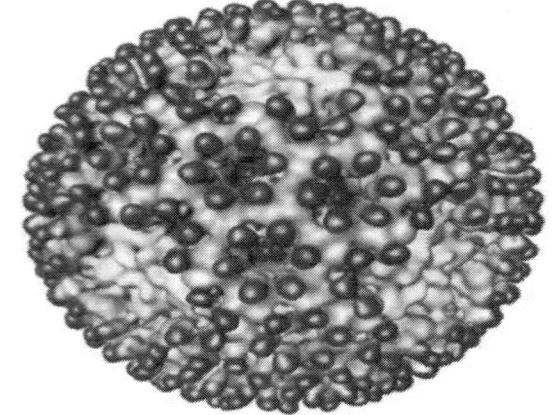

The family **Reoviridae** contains viruses which are most complex in nature. The term Reo stands for "respiratory enteric orphan," which was named because the first member was identified in the respiratory and the enteric tract of animals and humans and was not associated with any type of disease. They are generally spherical in shape and have icosahedral symmetry. They do not contain any envelope. The different viruses belonging to this family have their names which are indicative of their unique morphological features. For example in case of rotaviruses, rota stands for wheel like capsid with spikes, similarly in orbiviruses, orbi stands for ring shaped capsid. Some of the reoviruses are transmitted through the bite of female culicoides (blue tongue virus) or through tick bite (Colorado Tick fever).

The family *Reoviridae* contains six genera based on group specific antigen present on VP6 capsid protein

- **Orthoreovirus**
- **Orbivirus**
- **Coltivirus**
- **Rotavirus**
- **Seadornavirus**
- **Aquareovirus**

Table .1 Diseases caused by Reoviruses:

VIRUS	DISEASE
Orthoreovirus	Respiratory and enteric diseases
Orbivirus	Blue tongue fever in cattle and Sheep
	African Horse Sickness disease
Coltivirus	Colorado Tick fever
Rotavirus	Gastroenteritis and Diarrhoea
Aquareovirus	Diseases of Fish

1. Virion Property

Reovirus particles are non-enveloped, spherical having a diameter of approximately 85 nm. Genome contains a linear dsRNA divided into 10-12 segments. The overall genome size varies from 16 to 29 Kbp. They contain a cap at their 5' end while poly A tails are absent from 3' end.

2. Virus Replication

Virus replication takes place in the cytoplasm of the cell. Because of the segmented RNA genome chances of reassortment of genomic segments between different strains is very high. This results in genetic drift and shift leading to diversity among viruses which is reflected by numerous serotypes within each genus.

Virus enters the cell by receptor mediated endocytosis. The coated vesicle uncoats and fuses to the lysosomes under low pH condition. Virions are disrupted and viral inner core is released into the cytoplasm . Virus associated RNA polymerase utilizes the negative strand of each dsRNA segment as template to transcribe viral mRNA. These viral mRNA's are translated to form viral structural proteins that eventually assemble to form the infectious virion. Progeny viruses remain cell associated but are often released by lysis of the infected cell. Genomic RNA replication takes place within sub-viral particles in the cytoplasm of infected cells.

3. Important Reoviruses

ROTAVIRUS

Rotaviruses are the most important human pathogens which lead to life threatening diarrhoea in young ones. It was first isolated from a children hospital in Australia. Rotavirus infection leads to destruction of intestinal villi. Transmission of the virus mainly takes place by fecal-oral route. Destruction of enterocyte (intestinal cells) causes mal-digestion and poor absorption of food. Intestinal epithelium of young ones has a slow turnover rate and that is the reason that rotavirus infection is more severe in case of infants. Rotavirus infection is characterized by severe diarrhoea, dehydration, weight loss and fatigue. Virus can be isolated in faeces in high amount. Enzyme immunoassay is commonly used to detect viral infection in infected individuals. Genetic reassortment vaccines are available for rotavirus infection.

AFRICAN HORSE SICKNESS

African horse sickness is caused by a member of genus orb virus . It is a pantropic and fatal disease of horses, predominantly infecting endothelial cells and myocardium. Acute form of the disease is characterized by pneumonia, interlobular pulmonary oedema, pericarditis, haemorrhages and oedema of the visceral organs. The death can occur within 5 days in highly acute form of the disease. A more prolonged form of the sickness involves the **cardiac vascular system**. Mortality can be more than 80% depending on the immune status of the animal and virulence of the isolate. **Subclinical disease** can occur in donkeys and vaccinated horses. Diagnosis is usually carried out by using complement fixation tests and haemagglutination inhibition tests.

BLUETONGUE VIRUS

It is another important disease of livestock caused by a member of genus orbivirus . It causes high mortality in sheep and decrease in productivity of other farm animals. The virus is transmitted to the animals by the bite of Culicoides mosquitoes. The chances of outbreaks are more during the breeding season of Culicoides. Major signs of the disease include high fever, swelling of the face, hyperemia around the coronary band and **cyanosis of the tongue** (blue colour of the tongue). The disease can be restricted by controlling mosquitoes while polyvalent live vaccines are available for the animals.

COLORADO TICK FEVER

Colorado Tick fever is a disease caused by bite of tick (*Dermacentor andersoni*) infected with Coltivirus. Transmission of this disease is reported to be through blood transfusion from an infected individual. The virus usually infects and replicates in the erythrocyte. The symptoms of the disease include fever, headache, vomiting, abdominal pain, and encephalitis.

****

27. RETROVIRUS

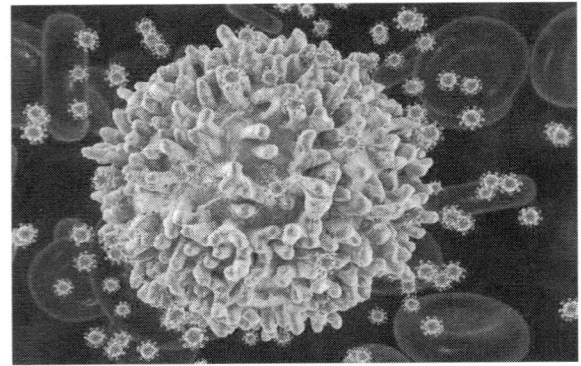

A **retrovirus** is a type of RNA virus that inserts a copy of its genome into the DNA of a host cell that it invades, thus changing the genome of that cell. Once inside the host cell's cytoplasm, the virus uses its own reverse transcriptase enzyme to produce DNA from its RNA genome, the reverse of the usual pattern, thus *retro* (backwards). The new DNA is then incorporated into the host cell genome by an integrase enzyme, at which point the retroviral DNA is referred to as a provirus. The host cell then treats the viral DNA as part of its own genome, transcribing and translating the viral genes along with the cell's own genes, producing the proteins required to assemble new copies of the virus.

Although retroviruses have different subfamilies, they have three basic groups. The oncoretroviruses (oncogenic retroviruses), the lentiviruses (slow retroviruses) and the spumaviruses (foamy viruses). The oncoretroviruses are able to cause cancer in some species, the lentiviruses able to cause severe immunodeficiency and death in humans and other animals, and the spumaviruses being benign and not linked to any disease in humans or animals.

Many retroviruses cause serious diseases in humans, other mammals, and birds. Human retroviruses include HIV-1 and HIV-2, the cause of the disease AIDS. Also the Human T-lymphotropic virus (HTLV) causes disease in humans. The murine leukemia viruses (MLVs) cause cancer in mouse hosts. Retroviruses are valuable research tools in molecular biology, and they have been used successfully in gene delivery systems.

Multiplication cycle of a typical retrovirus

The first stage in the infection process is to copy the RNA in the virus particle to give a complementary single strand of DNA. This is done by the most characteristic enzyme in the virus particle: reverse transcriptase. The same enzyme then goes on to eliminate the RNA strand from the hybrid molecule and synthesize a second complementary strand of DNA. This double-stranded DNA form is called the provirus and is integrated into the chromosomal DNA in the host *cell*. Viral mRNA can then be transcribed from the provirus template using the usual transcription machinery of the host *cell* (Figure 10.25).

The retrovirus genome can sometimes carry genes that cause cancer (so-called oncogenes) if integrated into the genome of the host cell. The introduction of an oncogene right into the heart of the host-cell genome can transform a normal host cell to a tumour *cell*. For example, Rous sarcoma virus, which naturally infects chickens and causes solid tumours in connective tissue ('sarcomas'), carries a *src* gene that codes for an enzyme that phosphorylates cellular proteins. This is thought to affect

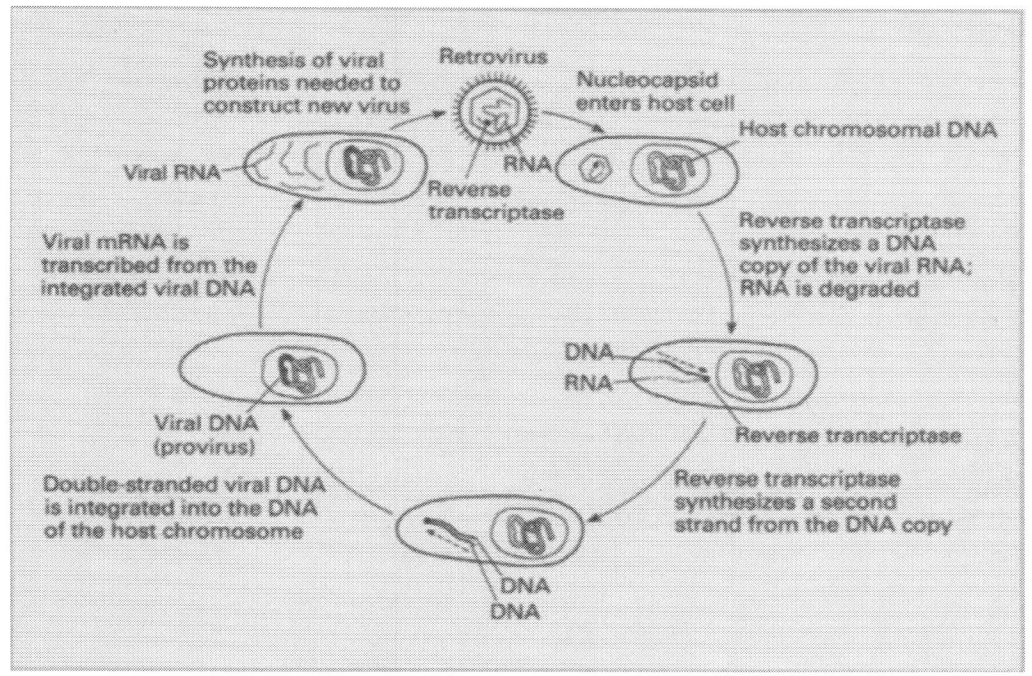

Fig.27. The multiplication cycle of a typical retrovirus.

cell growth and behaviour and so cause transformation. Other retroviruses are known which can cause certain types of leukaemia. Human T *cell* lymphotropic virus (HTL V) causes adult T *cell* leukaemia in patients, especially in south western Japan and in the Caribbean. However, by far the most is known about another retrovirus that does not transform its host cells but kills them instead. This is human immunodeficiency virus (HIV) which causes AIDS. The Baltimore classification is useful because it can be applied to all viruses infecting animals, bacteria and plants. However, it is clear that some of the classes contain very wide assemblages of different sorts of viruses. Class I is a good example. It puts together smallpox and phage T4 which are totally different in their structure, function and biology.

Human immunodeficiency viruses (HIV)

The human immunodeficiency viruses 1 and 2 (HIV-1, HIV-2) originated from the simian immunodeficiency viruses (SIVs) of primates. Thus, HIV-1 and HIV-2 each had a zoonotic origin but now spread directly from human to human. HIV-1 was first isolated in 1983 and HIV-2 in 1986 and they represent two different epidemics. The SIV of chimpanzees (SIV_{cpz}) gave rise to HIV-1 in humans, and the SIV of the sooty mangabey monkey (SIV_{sm}) to HIV-2 in humans.[1] It is still uncertain exactly how the transmission of these SIVs to humans occurred, but it may have been during the hunting and preparation of these primates for food, by the indigenous people of these areas in Central and Western Africa, where these primate species live.[2] Studies using molecular clock evolutionary assumptions have suggested that the ancestor virus for HIV-1 appeared in around 1931[3] and that of HIV-2 in around 1940.[4] After this initial transmission event, it is likely individuals infected with these primate SIVs then transmitted the human form of the viruses (HIV-1, HIV-2) to other people in their communities, from where it spread, world-wide.

Classification of HIV

The two human immunodeficiency viruses, HIV-1 and HIV-2, are members of the family of Retroviruses, in the genus of Lentiviruses. Retroviruses have been found in various vertebrate species, associated with a wide variety of diseases, in both animals and humans. In particular, retroviruses have been found to be associated with malignancies, autoimmune diseases, immunodeficiency syndromes, aplastic and haemolytic anaemias, bone and joint disease and diseases of the nervous system

The many different strains of HIV-1 have been separated into major (M), new (N) and outlier (O) groups, which may represent three separate zoonotic transfers from chimpanzees. Groups N and O are mainly confined to West and Central Africa (Gabon and Cameroon), though cases of Group O have been found world-wide due to international travel, after contact with infected individuals from these areas. The HIV strains in Group M are the ones mainly responsible for the HIV/AIDS pandemic, and they are so diverse that they have been subclassified into subtypes (or clades) A-K. This huge diversity of HIV-1 is important when diagnostic testing, treatment and monitoring are applied as the results may differ between different subtypes or clades (see **HIV Global Genetic Diversity and Epidemiology** below).[1] The diversity of HIV-2 is much less, but subtypes A-H have been proposed.[5]

HIV structure

The human immunodeficiency viruses are approximately 100 nm in diameter. It has a lipid envelope, in which are embedded the trimeric transmembrane glycoprotein gp41 to which the surface glycoprotein gp120 is attached (Box 1.1). These two viral proteins are responsible for attachment to the host cell and are encoded by the *env* gene of the viral RNA genome. Beneath the envelope, is the matrix protein p17, the core proteins p24 and p6 and the nucleocapsid protein p7 (bound to the RNA), all encoded by the viral *gag* gene. Within the viral core, lies 2 copies of the ~10 kilobase (kb) positive-sense, viral RNA genome (i.e. it has a diploid RNA genome), together with the protease, integrase and reverse transcriptase enzymes. These three enzymes are encoded by the viral *pol* gene. There are several other proteins coded for by both HIV-1 and HIV-2, with various regulatory or immuno-modulatory functions, including *vif* (viral infectivity protein), *vpr* (viral protein R), *tat* (transactivator of transcription), *rev* (regulator of viral protein expression) and *nef* (negative regulatory factor). An additional protein found in HIV-1 but not HIV-2 is *vpu* (viral protein U). Similarly, *vpx* (viral protein X) is found in HIV-2 and not HIV-1.

HIV replication

The main attachment receptor for HIV is the CD4 molecule that is present on the CD4 positive T (helper) lymphocyte, macrophages, and microglial cells. The viral gp120 binds initially to this CD4 molecule, which then triggers a conformational change in the host-cell envelope that allows binding of the co-receptor (either CCR5 or CXCR4) which is required for fusion between virus envelope and cell membrane.

Macrophages carry the CCR5 co-receptor, hence HIV strains requiring the CCR5 co-receptor for entry are also referred to as 'macrophage-tropic' although they also infect lymphocytes. These HIV strains are also known, phenotypically, as R5 or non-syncytium inducing (NSI) strains as they do not form syncytia (cell-fusion) when cultured with CD4 lymphocytes *in vitro*. Primary HIV-1 infections tend to involve this R5 NSI macrophage-tropic phenotype. Uncommonly, individuals may have a homozygous deletion mutation in the CCR5 gene (CCR5Δ32) resulting in the absence of the CCR5 molecule on their macrophages. Therefore, these individuals cannot be infected by this R5 phenotype. The 'lymphotrophic' HIV strains use CXCR4 as the co-receptor. These viruses are also known as X4 viruses and do produce syncytia (i.e. are phenotypically syncytium-inducing, SI) when cultured *in vitro* with CD4 lymphocytes. X4 viruses tend to appear later in about 50% of HIV-1 subtype B-infected individuals, but seldom with other subtypes, as they progress to AIDS. So far,

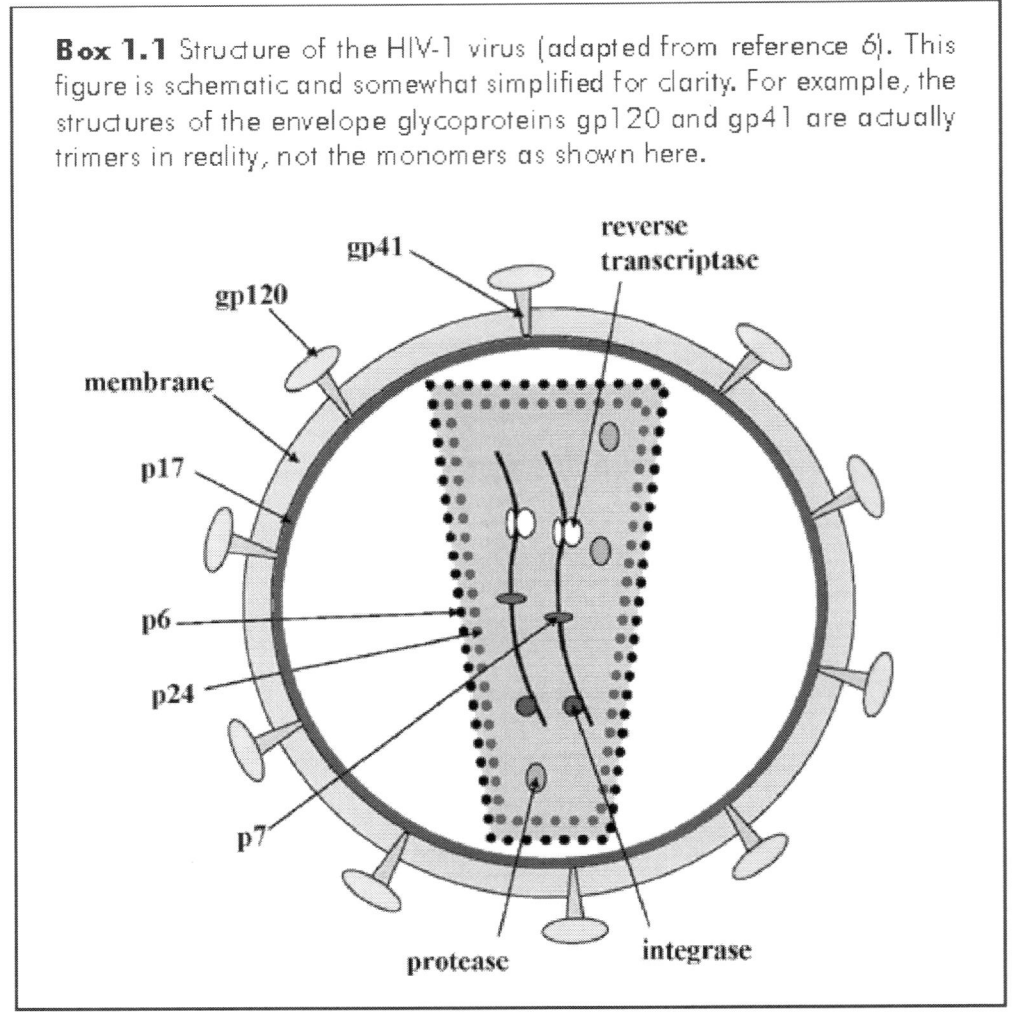

Fig.27.1.

CXCR4 deficient individuals have not been found. This attachment and fusion process allows the HIV viral core to enter the host-cell.

All retroviruses encode a reverse transcriptase enzyme that transcribes its viral RNA into double-stranded DNA (dsDNA), which is then integrated, via the action of the integrase enzyme into the host-cell genome (Box 1.2). The viral integrated dsDNA or 'provirus' then acts as a template for viral genomic and messenger RNA transcription by the host cell's nucleic acid replicating machinery. Recombination between these two RNA strands during viral replication, coupled with the extremely error-prone action of the RT enzyme, give rise to the extreme genetic diversity of HIV.

Integration of the linear provirus dsDNA into the genome of the host-cell establishes an infection that lasts for the lifespan of the cell, and all its progeny, which usually means life-long infection for the organism, in this case the human host. Viral replication occurs along with cellular replication and is enhanced by various factors, including coinfection with other organisms, the presence of inflammatory cytokines and cellular activation. During cellular replication, the provirus is transcribed by the host-cell RNA polymerase II enzyme, and the viral messenger RNA (vmRNA) and genomic RNA, are carried with the cellular mRNAs, to be translated into proteins. This vmRNA codes for a *gag-pol* precursor polypeptide that is ultimately cleaved by the viral-encoded protease enzyme to produce the *gag* and *pol* viral proteins. In addition, the vmRNA is also spliced to produce other vmRNAs coding for the viral proteins *tat, rev, vif, vpr, vpu* (for HIV-1), as well as the *env* precursor polypeptide. Ultimately, the *env* precursor polypeptide is cleaved by cellular (not viral) proteases, producing the envelope glycoproteins gp41 and gp120. These viral proteins, together with the replicated diploid viral genomic RNA, are assembled and enveloped by budding through the host-cell membrane, producing complete HIV virions.

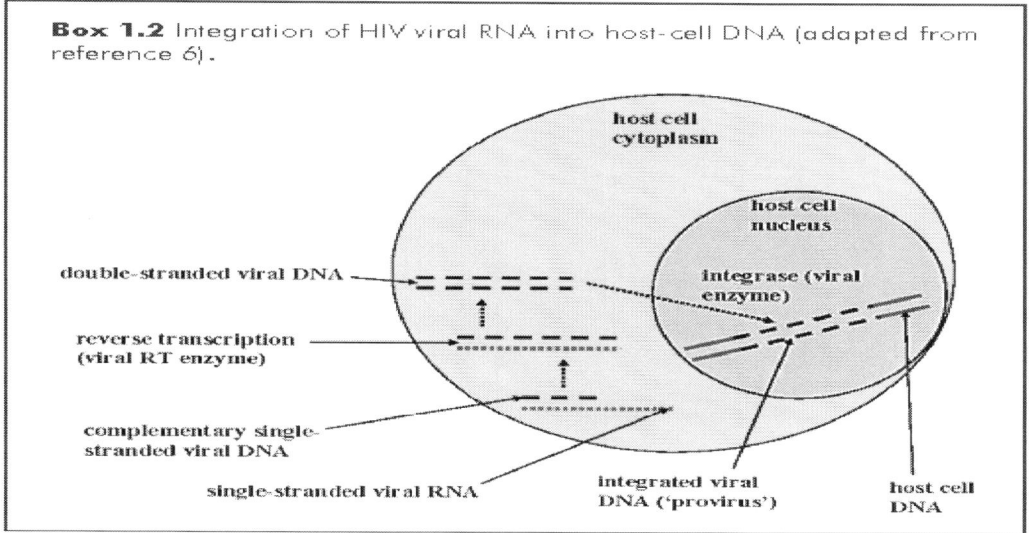

Box 1.2 Integration of HIV viral RNA into host-cell DNA (adapted from reference 6).

Fig.27.2.

AIDS

World AIDS day is celebrated every year on 1st of December.

Approximately 35 million people are affected with HIV worldwide. In India around 2.5 million people are suffering from HIV while in US more than one million people are suffering. There is decline in HIV infectivity in India as well as other parts of the world during the past 10 years. About 80 percent of the AIDS cases are reported in men between the age group of 20 to 44. In general males, percentage varies from 78 to 80 percent and the female around 20 to 22 percent. Nearly 1.8 percent of the total AIDS cases are children born to HIV infected mothers. In HIV infection an infectious doze means presence of ten thousand or more particles in the body. The level of HIV particle in different body fluids varies.

"AIDS is defined as HIV positive patient with CD4+ count less than 200cell/mm^3 or CD4+ count less than 14% of total lymphocyte population".

34.1 Transmission of HIV –

There are three major ways of HIV transmission

1) Sexual interaction

2) Blood transfusion

3) Perinatal infection

1) Sexual interaction - Risk of transmission of HIV is higher if person is infected with other sexually transmitted disease (STD) eg. herpes, syphyllis, Gonorrhoea etc. Passive partner is always in the higher risk side. This is more established way of HIV transmission.

2) Blood transfusion – HIV transmission is also possible by the way of infected blood, syringes, needles and other body fluids including saliva. Since now blood donors are screened first for HIV, the number of HIV cases decreased to a greater extent.

3) Perinatal infection – Young born children get infected if the mother is infected with HIV. There are several theories established regarding transmission of HIV through placenta or during delivery. Many studies have suggested involvement of breast milk for the transmission of HIV.

Early Phase of HIV Infection

Decrease in the number of circulating CD4+ lymphocyte is the hallmark of AIDS. CD4+ T helper cell and the macrophages are the major reservoir of HIV. Replication of HIV in macrophages results in budding of progeny virions through the membranes of endoplasmic reticulum (ER) which means that it acquires its envelope from ER similar to that of Coronaviruses. After entry into the body the virus travels to circulating lymph nodes and starts its replication. Initial infection of about 1 to 3 weeks results in fever, high virus titer in blood and high depletion of CD4+ T helper cell. After one month of infection the virus titer gets reduced in the blood circulation because of cytotoxic T cells, natural killer cells and antibody dependent cellular cytotoxicity

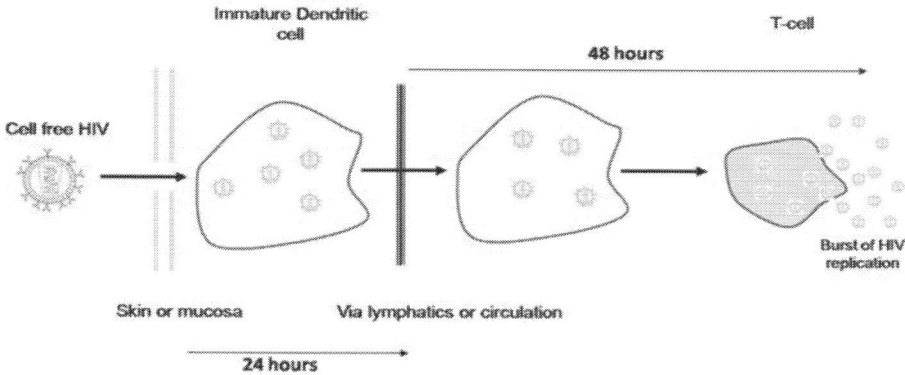

Fig.27.3.Mucosal entry of HIV and its path of circulation:

HIV-infected person during the latent period

Cells are called latently infected if they are infected by HIV and are not active in producing virus but can be activated by various signals during the course of infection. Since they are not producing the virus they would not be detected by the immune system and acts as a perfect reservoir of the virus in the body. Only a small amount of cells are latently infected with HIV (around 1%). During the asymptomatic stage, the virus appears in low titers in the blood and it is counter balanced by the host immune system.

AIDS associated disease conditions

Several specific syndromes are associated with infection by HIV. These include:

1. **Fever and Lymphadenopathy** It is characterized by loss of weight and malaise.

2. **Opportunistic infections:** HIV infected patients have lower immune response because of depletion of CD4+ cells. Several diseases that rarely affect normal individuals may occur in HIV positive patient. The organisms are: *Pneumocystis carinii*, a causative agent of pneumonia, tuberculosis, fungal infection such as candidiasis, herpesvirus, *Salmonella* , *Shigella* and *Campylobacter* .

3. **Cancer:** Kaposi's sarcoma is a rare type of cancer that occurs in HIV-infected persons. These normally benign lesions become malignant and disseminate to involve visceral organs.

4. **Wasting disease:** Disease is characterized by hide bound condition.

5. **AIDS dementia:** Sometime HIV infection of the brain leads to condition that mimics Alzheimer's disease.

Antiviral drug against AIDS and some facts

a) Around 95% reduction of virus within 14 days after treatment with essentially any one of the following drug (nucleoside RT inhibitors, nonnucleoside RT inhibitiors, protease inhibitors).

b) Average half life of an HIV infected cell is about 24-48 hrs.

c) Approximately 5% of total CD4+ T lymphocytes are productively infected in an infected individual at any given time during the latent period.

d) The CD4+ T lymphocytes that are dying each day are being replaced nearly as fast as they die. This means that bone marrow, spleen and other reservoirs of T lymphocytes must be producing the cells at an exponential rate.

e) Nearly within 7 -25 days after treatment with any one drug there is emergence of resistant virus.

f) The rapid emergence of mutant virus suggests that the resistant virus was already present in the population at the time drug treatment was started.

g) Resistant viruses do not grow quite as well as the wild-type viruses.

h) Upon removal of the anti HIV drug, the wild type virus once again becomes predominant over the course of infection.

i) One potential factor in the development of AIDS may be excessive stress on the immune system due to rapid turnover of T lymphocytes.

Control

    I. The use of condoms during sexual intercourse can reduce the chance of infection.

    II. Avoiding intravenous drugs (Needle sharing).

    III. Monitoring the blood for HIV before transfusion.

    IV. Educating people regarding the cause, severity, and preventive measure of HIV.

    V. Anti HIV drugs: These drugs are often given in combination of two or three. Nucleoside and Non-nucleoside analogues are called as reverse transcriptase inhibitors.

    VI. Protease Inhibitors- saquinavir

    VII. Nucleoside analogues- Zidovudine and AZT

    VIII. Fusion inhibitors- Enfuvirtide

    IX. Non- nucleoside analogues- Nevirapine

****

28.

ISOLATION AND PURIFICATION OF VIRUSES AND COMPONENTS

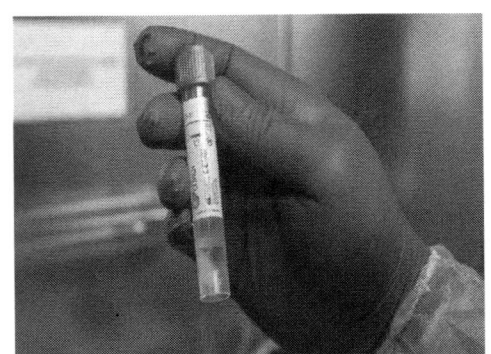

Isolation of Viruses:

Viruses are isolated from infected host cells containing mature virions. The cells are mechanically disrupted and the cell contents are released in a suitable buffer solution. The suspension containing the virions and cell ingredients is then subjected to centrifugation for several times at different speeds to fractionate the virions from other cell components. Such procedure is known as differential centrifugation.

A more refined technique is the density gradient centrifugation which is usually applied for getting a more purified sample of viruses. Before subjecting the sample of virus to density gradient centrifugation, it is initially purified by differential centrifugation. A density gradient is prepared in a centrifuge tube. For example, a sucrose gradient contains a linearly increasing concentration of sucrose from top to bottom of the centrifuge tube.

The partially purified virus sample is poured on the top and the tube is subjected to high-speed centrifugation for several hours in an ultracentrifuge. The centrifugal force drives the viral particles towards the bottom until they settle at a density gradient of sucrose which equals that of the virions, forming a concentrated zone or band.

The suspension of virions can then be removed from this band with the help of a pipette. The bacteriophages causing lytic infections can be isolated by a more or less similar method by differential centrifugation of the lysate to eliminate cell debris and the non-lysed intact cells of the host bacteria.

Assay of Viruses:

Viral assay means determination of number of viral particles per unit volume of a sample. Several methods are available for this purpose. The total number of viral particles in a sample including both viable and non-viable virions can be counted directly by means of electron microscopy.

For this purpose, a known volume of a purified sample of the virus is mixed with a known volume of a suspension of minute polystyrene latex beads of known concentration. The mixture is sprayed in droplets on a supporting membrane, dried and shadowed.

From the ratio of the number of beads and that of viral particles the number of virions per unit volume can be calculated, when the preparation is examined under an electron microscope. For example, if in a preparation, a droplet reveals presence of 200 viral particles and 20 latex beads, the concentration of virions / ml would be 200/20 multiplied by the number of beads per ml in the suspension.

If the bead concentration in the original suspension is 5×10^{10} per ml. The viral count would be $200/20 \times 5 \times 10^{10}$/ml = 5×10^{11}/ml. The concentration of viruses in a sample is also known as its titre. Among the other methods for assay of viral titres in a sample, the plaque method is a very important one.

It makes use of the infectivity of the viruses and their capacity to destroy the infected cells. The plaque method was first developed for detection and counting of bacteriophages and later it was extended to the study of animal viruses. The plaque method is widely used for its simplicity, accuracy and reproducibility.

For assay of bacteriophages, a sample is diluted many-fold (~10^{-20} to 10^{-25}) and mixed with a drop of a dense liquid culture of the host bacterium and a few milliliters of molten soft agar medium at 44°C. The mixture is poured on the surface of an already solidified hard nutrient agar in a plate and uniformly spread to allow the phages and bacteria to be evenly distributed.

Overnight incubation of the plates reveals the presence of a number of clear areas, known as plaques, on a lawn of continuous growth of the host bacterium. Plaques are formed due to infection and destruction of the host cells producing the clear areas. The number of plaques is proportional to the concentration of the virus.

Each plaque is produced by a plaque-forming unit (PFU). Thus, if an aliquot of 0.1 ml of a 10^{-20} dilution of the phage sample is plated and produce an average of 40 plaques per plate, the titre is $40/0.1 \times 10^{20}$ PFU/ml. It should be noted that the number of PFUs cannot be taken as equal to the number of phages, but the two are proportional to each other.

The plaque method with necessary modification has also been used for assay of animal viruses. In place of bacteria, a suspension of cultured animal cells is used as host. The bacteriological nutrient medium is replaced by appropriate nutrient medium for animal cell culture.

As in the case of bacteriophage assay, the viral sample is serially diluted and aliquots of appropriate dilutions are spread on monolayers of host culture cells growing on a solid support e.g. in a petridish. The virions are allowed to get attached to the host cells for an hour or so, and then the monolayer is covered with a soft agar or some other gelling medium to prevent free horizontal diffusion of the viral particles.

The infectious progeny particles released by lysis of the infected cells remain more or less localized to produce foci or plaques. The plaques can be counted to determine the infectivity titre of the virus sample. Development of visible plaques may require from 1 to 2 days or even several weeks depending on the virus. For facilitating detection and counting of plaques, it may be necessary to stain the cell layer with a dye like neutral red which stains only the living cells, or a stain like trypan blue which stains only the dead cells. The accuracy of plaque assay depends on the number of plaques counted.

If too many plaques develop on a plate, some of these tend to fuse. As a result, the count becomes lower than what it should be. Many viruses produce sharp plaques, while others produce plaques with a diffuse margin.

Again some viruses produce large plaques and others small ones. Depending on the nature of the virus to be assayed, the dilution is to be accordingly determined to get reliable results. In case of bacteriophages, the number of plaques in plates is proportional to the concentration of the virus i.e. the relationship between plaque number and the viral concentration is linear.

Another method of assay of animal viruses which was used previously and now practically abandoned and replaced by the plaque assay is the pock assay. In this method the appropriately diluted viral sample is inoculated into the epithelial layer of the chrioallantoic membrane (CAM) of a chick embryo in an embryonated chicken egg. Characteristic infection lesions, called pocks appear after an incubation period of 36 to 72 hr. The pocks are opaque areas — usually white — and can be located on the transparent CAM.

The plaque method may also be suitably modified for enumerating plant viruses. For this purpose, a known volume of a properly diluted sample of the plant virus is applied on the leaf surface of a susceptible host plant after the leaf has been mechanically injured by mildly rubbing the surface with an abrasive like carborandum.

After incubation for several days necrotic lesions appear on the leaf. From the number of lesions per unit area, the dilution factor of the applied viral sample and the inoculum volume, the concentration (titre) of the virus can be calculated.

Cultivation of Viruses:

viruses are obligate intracellular parasites so they depend on host for their survival. They cannot be grown in non-living culture media or on agar plates alone, they must require living cells to support their replication.

The primary purpose of virus cultivation is:

1. To isolate and identify viruses in clinical samples.
2. To do research on viral structure, replication, genetics and effects on host cell.
3. To prepare viruses for vaccine production.

Cultivation of viruses can be discussed under following headings:

1. Animal Inoculation
2. Inoculation into embryonated egg
3. Cell Culture

1. Animal Inoculation

- Viruses which are not cultivated in embryonated egg and tissue culture are cultivated in laboratory animals such as mice, guinea pig, hamster, rabbits and primates are used.
- The selected animals should be healthy and free from any communicable diseases.
- Suckling mice(less than 48 hours old) are most commonly used.
- Suckling mice are susceptible to togavirus and coxsackie viruses, which are inoculated by intracerebral and intranasal route.
- Viruses can also be inoculated by intraperitoneal and subcutaneous route.
- After inoculation, virus multiply in host and develops disease. The animals are observed for symptoms of disease and death.
- Then the virus is isolated and purified from the tissue of these animals.
- Live inoculation was first used on human volunteers for the study of yellow fever virus.

Advantages of Animal Inoculation

1. Diagnosis, Pathogenesis and clinical symptoms are determined.
2. Production of antibodies can be identified.
3. Primary isolation of certain viruses.
4. Mice provide a reliable model for studying viral replication.
5. Used for the study of immune responses, epidemiology and oncogenesis.

Disadvantages of Animal Inoculation

1. Expensive and difficulties in maintenance of animals.
2. Difficulty in choosing of animals for particular virus
3. Some human viruses cannot be grown in animals,or can be grown but do not cause disease.
4. Mice do not provide models for vaccine development.
5. It will lead to generation of escape mutants
6. Issues related to animal welfare systems.

2. Inoculation into embryonated egg

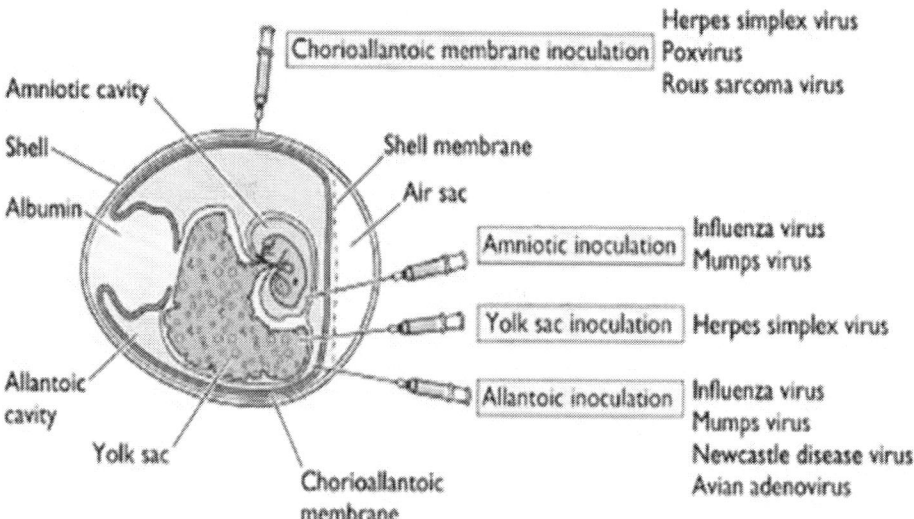

Fig. 28. Inoculation into embryonated egg

- Good pasture in 1931 first used the embryonated hen's egg for the cultivation of virus.
- The process of cultivation of viruses in embryonated eggs depends on the type of egg which is used.
- Viruses are inoculated into chick embryo of 7-12 days old.
- For inoculation, eggs are first prepared for cultivation, the shell surface is first disinfected with iodine and penetrated with a small sterile drill.
- After inoculation, the opening is sealed with gelatin or paraffin and incubated at 36°c for 2-3 days.
- After incubation, the egg is broken and virus is isolated from tissue of egg.
- Viral growth and multiplication in the egg embryo is indicated by the death of the embryo, by embryo cell damage, or by the formation of typical pocks or lesions on the egg membranes
- Viruses can be cultivated in various parts of egg like chorioallantoic membrane, allantoic cavity, amniotic sac and yolk sac.

1. **Chorioallantoic Membrane (CAM):**

- Inoculation is mainly for growing poxvirus.
- After incubation and incubation, visible lesions called pocks are observed, which is grey white area in transparent CAM.

- Herpes simplex virus is also grown.
- Single virus gives single pocks
- This method is suitable for plaque studies.

2. Allantoic cavity:

- Inoculation is mainly done for production of vaccine of influenza virus, yellow fever, rabies.
- Most of avian viruses can be isolated using this method.

3. Amniotic sac:

- Inoculation is mainly done for primary isolation of influenza virus and the mumps virus.
- Growth and replication of virus in egg embryo can be detected by haemagglutination assay.

4. Yolk sac inoculation:

- It is also a simplest method for growth and multiplication of virus.
- It is inoculated for cultivation of some viruses and some bacteria (Chlamydia, Rickettsiae)
- Immune interference mechanism can be detected in most of avian viruses.

Advantages of Inoculation into embryonated egg

1. Widely used method for the isolation of virus and growth.
2. Ideal substrate for the viral growth and replication.
3. Isolation and cultivation of many avian and few mammalian viruses.
4. Cost effective and maintenance is much easier.
5. Less labor is needed.
6. The embryonated eggs are readily available.
7. Sterile and wide range of tissues and fluids
8. They are free from contaminating bacteria and many latent viruses.
9. Specific and non specific factors of defense are not involved in embryonated eggs.
10. Widely used method to grow virus for some vaccine production.

Disadvantages of Inoculation into embryonated egg

1. The site of inoculation for varies with different virus. That is, each virus have different sites for their growth and replication.

3. Cell Culture (Tissue Culture)

There are three types of tissue culture; organ culture, explant culture and cell culture.

Organ cultures are mainly done for highly specialized parasites of certain organs e.g. tracheal ring culture is done for isolation of coronavirus.

Explant culture is rarely done.

Cell culture is mostly used for identification and cultivation of viruses.

- Cell culture is the process by which cells are grown under controlled conditions.
- Cells are grown in vitro on glass or a treated plastic surface in a suitable growth medium.
- At first growth medium, usually balanced salt solution containing 13 amino acids, sugar, proteins, salts, calf serum, buffer, antibiotics and phenol red are taken and the host tissue or cell is inoculated.
- On incubation the cell divide and spread out on the glass surface to form a confluent monolayer.

Types of cell culture

1. Primary cell culture:

- These are normal cells derived from animal or human cells.
- They are able to grow only for limited time and cannot be maintained in serial culture.
- They are used for the primary isolation of viruses and production of vaccine.
- Examples: Monkey kidney cell culture, Human amnion cell culture

2. Diploid cell culture (Semi-continuous cell lines):

- They are diploid and contain the same number of chromosomes as the parent cells.
- They can be sub-cultured up to 50 times by serial transfer following senescence and the cell strain is lost.
- They are used for the isolation of some fastidious viruses and production of viral vaccines.
- Examples: Human embryonic lung strain, Rhesus embryo cell strain

3. Heteroploid cultures (Continuous cell lines):

- They are derived from cancer cells.
- They can be serially cultured indefinitely so named as continuous cell lines
- They can be maintained either by serial subculture or by storing in deep freeze at -70°c.
- Due to derivation from cancer cells they are not useful for vaccine production.
- Examples: HeLa (Human Carcinoma of cervix cell line), HEP-2 (Humman Epithelioma of larynx cell line), Vero (Vervet monkey) kidney cell lines, BHK-21 (Baby Hamster Kidney cell line).

Susceptible Cell Lines

1. **Herpes Simplex** Vero Hep-2, human diploid (HEK and HEL), human amnion
2. **VZV** human diploid (HEL, HEK)
3. **CMV** human diploid fibroblasts
4. **Adenovirus** Hep2, HEK,
5. **Poliovirus** MK, BGM, LLC-MK2, human diploid, Vero
6. **Coxsackie B** MK, BGM, LLC-MK2, vero, hep-2
7. **Echo** MK, BGM, LLC-MK2, human diploid, Rd
8. **Influenza A** MK, LLC-MK2, MDCK
9. **Influenza B** MK, LLC-MK2, MDCK
10. **Parainfluenza** MK, LLC-MK2
11. **Mumps** MK, LLC-MK2, HEK, Vero
12. **RSV** Hep-2, Vero
13. **Rhinovirus** human diploid (HEK, HEL)
14. **Measles** MK, HEK
15. **Rubella** Vero, RK13

Advantages of cell culture

1. Relative ease, broad spectrum, cheaper and sensitivity

Disadvantage of cell culture

1. The process requires trained technicians with experience in working on a full time basis.
2. State health laboratories and hospital laboratories do not isolate and identify viruses in clinical work.
3. Tissue or serum for analysis is sent to central laboratories to identify virus.

Cultivation of plant viruses and bacteriophages

Cultivation of plant viruses

There are some methods of Cultivation of plant viruses such as plant tissue cultures, cultures of separated cells, or cultures of protoplasts, etc. viruses can be grown in whole plants.

Leaves are mechanically inoculated by rubbing with a mixture of viruses and an abrasive. When the cell wall is broken by the abrasive, the viruses directly contact the plasma membrane and infect the exposed host cells. A localized necrotic lesion often develops due to the rapid death of cells in the infected area. Some plant viruses can be transmitted only if a diseased part is grafted onto a healthy plant.

Cultivation of bacteriophages

Bacteriophages are cultivated in either broth or agar cultures of young, actively growing bacterial cells.

****

29.

THE MYCOSES (MYCOSIS)

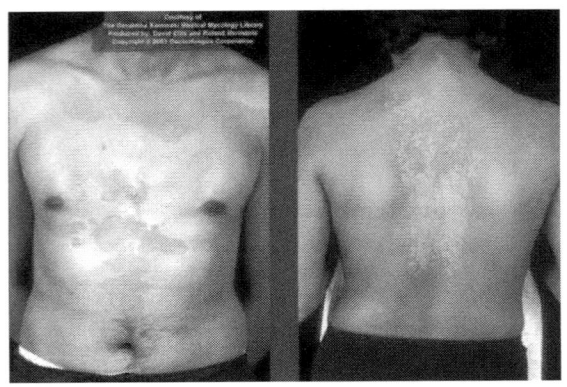

There are about 50,000 to 200,000 species of fungi estimated by the mycologists and 50 of them have been recognised as human pathogens. Greek mykes = fungus, osis condition, therefore, the term mycosis (mycoses) stands for any disease caused by the fungus. There are 15 to 20 main types of mycoses of human besides the diseases which are not severe. It is generally seen that the mycoses develop in persons whose susceptibility to fungal infections has been increased by administration of immunosuppressive drugs or corticosteroids or broad spectrum antibiotics while treating non mycotic diseases. The old persons are found to be more susceptible to mycosis in comparison to the young. Keeping in view the terminology observed in 1968 in the "Ciba Foundation Symposium, Systemic Mycosis" they have been classified into (1) superficial mycoses, (2) deep-seated or systemic mycoces and (3) opportunistic infections.

The fungal infection at first cause acute inflammation because of accumulation of polymorphonuclear leucocytes. This is followed by the allergic necrosis because of hypersensitivity of the immune system of the host as a result of fungal attack. Sometimes the mycosis is caused by the fungi forming part of normal microflora of the host as in case of candidiasis caused by Candida albicans. Such fungi that cause mycosis from the normal microflora of the host are referred to as endogenous in origin. Exogenous fungi i.e., the fungi which come from outside as from soil or droppings of the birds may also cause infection eg., the disease histoplasmosis. As a result of movement of air the exogenous fungi are inhaled and infection is caused in lungs which later on spread to other parts of the body.

Common principles of therapy of the mycosis

The therapy of mycosis is based on the following three principles, which may be applicable for other types of diseases also.

1. General care

The mycosis may result any of the symptom generally seen due to any type of infection, as fever, chills, night sweats, lack of appetite, weight-loss and depression. Sometimes dehydration and depletion of electrolytes requires rehydration and replacement of electrolytes. An appetizing, high protein vitamin enriched diet is considered reasonable. The patients require rest, proper oxygen therapy in respiratory problems, and ventilatory conditions. The toxicity of fungi due to eating of poisonous mushrooms should be removed by inducing vomiting and washing out body cavity, the replacement of electrolyte and correction of hypoglycemia (decreased blood sugar attended by anxiety perspiration and delirium or comma), is required. The delirium which is an abnormal mental condition based on hallucitation or illusion may occur in high fever or may be toxic in origin particularly due to consumption of poisonous mushrooms, which occasionally turn to coma i.e., complete loss of consciousness, The acidosis in which the depletion of body's alkali reserve take place disturbing the acid base balance of the body, therefore, needs to be corrected to avoid hepatic necrosis.

2. Surgery

The surgery may be required in some infectious diseases. Proper attention should be paid to incision and drainage of localized infection as it is considered that the pus is not properly treated by chemotherapy so must be removed. The disfiguring skin lesions in some patients of blastomycosis may be improved by plastic surgical repair. The blastomycosis is a granulomatous condition caused by budding yeast-like organisms, It may affect skin, viscera and bones.

Generally two types of blastomycosis have been identified. (1)North American blastomycosis caused by the fungus Blastomyces dermatitidis and (2) the South American blastomycosis is caused by *B. brasiliensis* (syn. Para coccidioides brasiliensis), The patients with endocarditis (inflammation of inner lining ok heart called endocardium) have sometimes been cured by the removal of vegetations and replacement of the diseased valve with a prosthesis Le, an artificial substitute for a missing part.

3. Chemotherapy

The discoveries of the drugs salvarsan in 1909, sulfa drugs in 1936 and the penicillin in 1929 ushered the mankind into the modern era of chemotherapy, The chemicals that have been used from time to time include iodides, hydroxystilbamidine isethionate, antibiotics and several other drugs Above all, the patient should take the treatment under the supervision of a propery qualified physician.

****

30. SUPERFICIAL MYCOSES OR DERMATOPHYTOSIS

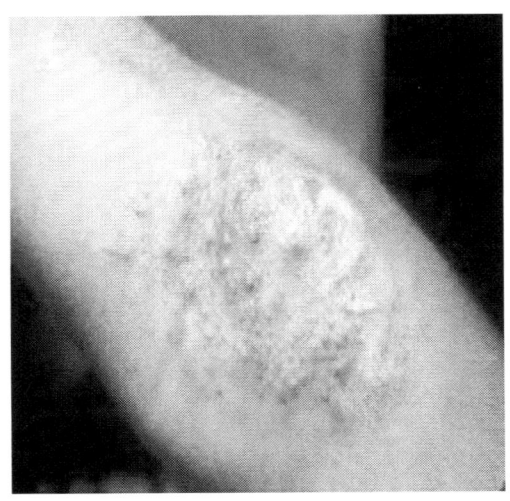

Superficial mycoses. The term "superficial mycosis" applies to diseases affecting the outermost layer of the skin (stratum corneum), or growing along hair shafts. The most common superficial mycosis is pityriasis versicolor, causing patches of hypo- or hyper-pigmentation of the neck, shoulders, chest, and back, and caused by lipophilic basidiomycete yeasts of the genus Malassezia.

Other superficial mycoses:

White piedra: soft, beige nodules on the distal ends of hair shafts (Trichosporon species),
Black piedra: small firm black nodules on the hair shaft (Piedraia hortae)
Tinea nigra: brown to black stain on the palm of the hand or sole of the foot (Hortaea werneckii.) Clinical descriptions and microscopic morphology of these agents may be found in Veasey et al., 2017

Disease	Causative organisms	Incidence
Pityriasis versicolor Seborrhoeic dermatitis including Dandruff and Follicular pityriasis	*Malassezia* spp. (a lipophilic yeast)	Common
Tinea nigra	*Hortaea werneckii*	Rare
White piedra	*Trichosporon* spp.	Common
Black piedra	*Piedraia hortae*	Rare

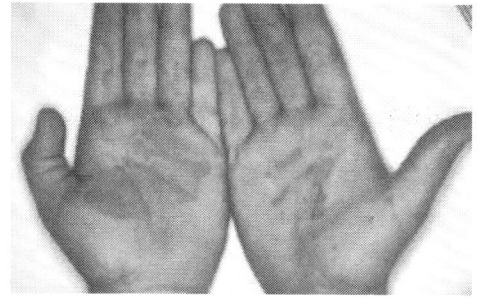

Fig.30. Superficial mycoses

Cutaneous mycoses, the dermatophytes (def: "skin fungi".) The dermatophytes are confined to grow on the non-living outer layers of the skin (stratum corneum), and only exceptionally invade living tissue. Unlike superficial mycoses, dermatophyte infections can be itchy, inflammatory, and affect the smooth skin, hair, and nails. There are three genera of dermatophytes: Microsporum, Trichophyton and, less commonly, Epidermophyton. Dermatophytes produce extracellular enzymes (keratinases) capable of hydrolyzing keratin.

Athlete's Foot. This is general term, usually applied to dermatophytosis, but also to other infections and non-infectious conditions, singly or in combination, as follows (alphabetic order, not order of prevalence):

- Bacterial infection (erythrasma, Pseudomonas, staphylococci, streptococci.)
- Fungal infection (tinea pedis)
- Injury due to over-vigorous removal of peeling skin.
- Psoriasis, eczema, or keratolysis exfoliativa (excessive sweating and friction can cause the skin to exfoliate. Exposure to detergents and solvents also can cause exfoliation)
- Soft corns-- areas of white moist skin between the toes.

Dermatomycoses. Uncommon causes of cutaneous mycoses are fungi other than dermatophytes. Mold agents of nail infections, are Neoscytalidium, Scopulariopsis, Aspergillus spp., Fusarium spp. Candida albicans primary skin conditions (e.g.: diaper rash) also are considered dermatomycoses.

Skin manifestations of systemic or disseminated mycoses include blastomycosis, cryptococcosis, sporotrichosis, talaromycosis (formerly penicilliosis.)

DERMATOPHYTES AND DERMATOPHYTOSIS

Dermatophytosis (syn: ringworm, tinea) is a communicable skin disease of the non-living outer layer of the epidermis, the stratum corneum, and may invade hair and nails, caused by a group of hyaline molds in three genera: *Microsporum*, *Trichophyton*, and, less commonly, *Epidermophyton*.

These fungi break down keratin in skin, hair, and nails. They do not invade living tissue or, in hair, beyond the keratogenous zone. The classic skin lesion is ringworm -- circular with an active border, inflammation, scaling, and pruritus.

"Ringworm." This term originated in the 15th century to describe any skin rash patterned in a ring formation. It is a misnomer because the etiology is fungal, not parasitic. In 1841, David Gruby, a Paris-based physician, described a fungus that is a cause of scalp ringworm and, from the 1850's, ringworm was considered a fungal disease. This discovery was made even before dermatology was a recognized medical specialty. Scalp ringworm in children caused by *Trichophyton tonusurans* is sometimes called Gruby's disease.

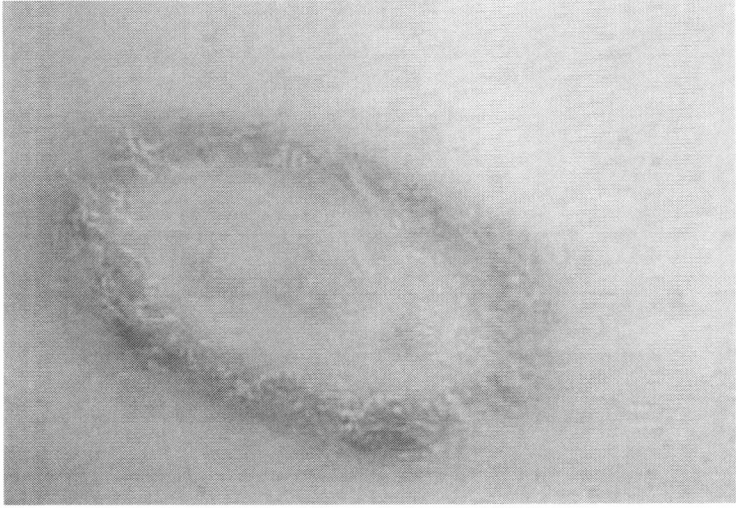

Fig.30.1. Ringworm

"Tinea" (def.: Latin, "larva" or "worm,") This refers to dermatophyte infections on particular body sites, not to genus and species of the causative agent! (Table 1).

The ancient Romans associated skin lesions on humans with holes in wool blankets caused by the larvae of moths, the Tineidae.

The dermatophytoses are characterized by anatomic site-specificity according to genera. e.g.: *Epidermophyton floccosum* infects only skin and nails, but does not invade hair shafts and follicles. *Microsporum* spp. infect hair and skin, but do not involve nails. *Trichophyton* spp. may infect hair, skin, and nails.

ETIOLOGIC AGENTS

Trichophyton **species** (14 species.)

As a group, these species infect skin, hair, and nails, but single species have anatomic site restrictions. They rarely cause subcutaneous infections and those occur in immuno-compromised individuals. *Trichophyton* spp. take 2 to 3 wks to grow in culture. Microconidia are the characteristic structures of the genus *Trichophyton*. Macroconidia, when present, are smooth, with thin-walls, septate (0-10 septa), and pencil-shaped. Colonies consist of loose aerial mycelium secreting pigments in a variety of colors.

Trichophyton rubrum

This most common cause of tinea corporis does not infect hair. It can rarely cause subcutaneous infections in immunocompromised individuals, such as patients with chronic myelogenous leukemia.

- Colony morphology (Figure 1). Flat to slightly raised, white to cream, suede-like with a pinkish-red reverse. The downy type has a yellow-brown to wine-red reverse.
- Microscopic morphology (Figure 2). Granular type: numerous clavate to pyriform microconidia formed evenly on hyphae, resembling "birds on a wire." Moderate numbers of smooth, thin-walled multiseptate, pencil shaped macroconidia. It is the parent strain of downy type. The downy type has fewer microconidia, macroconidia usually absent

Trichophyton tonsurans

A major causative agent of tinea capitis in the U.S.

Colony morphology.

Vary in texture and color. Suede-like to powdery, flat with a raised center or folded, often with radial grooves. Color varies from buff to yellow, (the so called "sulfureum" form) to dark-brown. The reverse varies from yellow-brown to red-brown to mahogany.

Microscopic morphology.

Numerous microconidia along hyphae or on short conidiophores. Vary in size and shape from long clavate to broad pyriform, at right angles to the hyphae ("match stick" shape) and may enlarge into balloon forms. Occasional smooth, thin-walled, irregular, clavate macroconidia in some cultures.

Trichophyton interdigitale

This anthropomorphic species is a common cause of tinea pedis.

Colony morphology.

Flat, white to cream color, powdery to suede-like surface. Reverse is yellowish and pinkish brown, becoming red-brown with age.

Microscopic morphology.

Sub-spherical to pyriform microconidia, occasional spiral hyphae and spherical chlamydoconidia. Occasional slender, clavate, smooth-walled, multiseptate macroconidia

Microsporum **species** (3 species: *M.canis, M. audouinii, M. ferrugineum*). The genus *Microsporum* is identified by its macroconidia.

Microsporum canis is a common zoophilic dermatophyte species infecting smooth skin and hair of humans, nails rarely are affected. Scalp infections with this species are identified because infected hairs fluoresce bright green when illuminated with UV-emitting Wood's light.

Colony morphology (Figure 3). Surface is loose, cottony, white to cream color; reverse usually a bright yellow, but may be non-pigmented.

Microscopic morphology. Macroconidia large, spindle-shaped, multicelled, thick walls, and echinulate (spiny) often with a terminal knob (Figure 4); microconidia are uncommon.

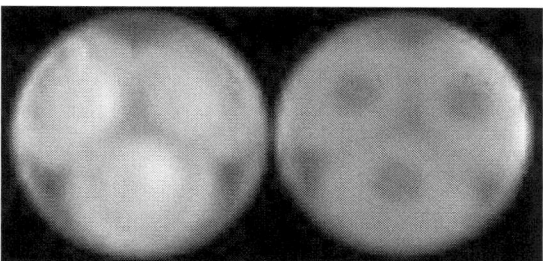

Fig.30.2. 3*Microsporum canis*. Colony Morphology, (S) surface; (R) reverse.
Photo credit: Jim Gathany, CDC

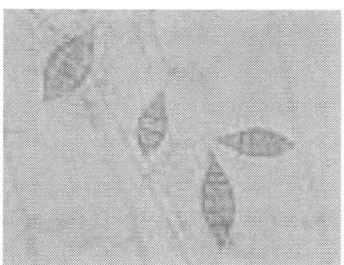

Fig.30.3. 4*M. canis*. Microscopic morphology: spindle-shaped thick-walled, multi-celled macroconidia (microconidia uncommon), lactofuchsin stain. Photo credit: E. Reiss, CDC

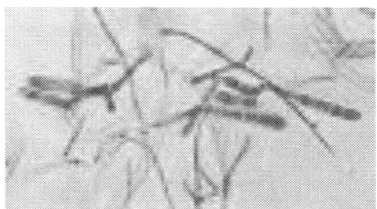

Fig.30.4. 5*Epidermophyton floccosum*. Microscopic morphology: macroconidia are oval to club-shaped, most with 2-4 septa, walls smooth, often in clusters of 2 or 3, microconidia absent. Photo credit: Geraldine Kaminski Library, #163

Epidermophyton floccosum

This species infects skin and nails and rarely hair.

Colony morphology. Suede-like surface, greenish-brown or khaki, often raised and folded in the center. Reverse is yellowish-brown.

Microscopic morphology. No microconidia. Multiple smooth, club-shaped macroconidia, 2-6 cells, often in clusters of 2 or 3 (Figure 5)

GEOGRAPHIC DISTRIBUTION/ECOLOGIC NICHE ECOLOGY

Geographic distribution: Dermatophytes occur worldwide, but some species are geographically limited. Dermatophytes causing human infections may have different natural sources and modes of transmission. Knowledge of the species of dermatophyte and source of infection are important for proper treatment control of the source. Invasion by zoophilic or geophilic species may cause inflammatory disease in humans (Table 2.)

- **Anthropophilic** – These spp. are usually associated with humans only; transmission from human-to-human is by close contact or through contaminated objects, e.g.: sharing caps and combs, contact by high school wrestlers (tinea gladiatorum.)
- **Zoophilic -** Usually associated with animals; transmission to humans is by close contact with animals (cats, dogs, cattle) or with contaminated products. Sometimes transmission is from human to human in outbreaks, or from human to animal, e.g.: child with tinea capitis or corporis spreading the infection to pet kitten.
- **Geophilic** – Species are found in the soil and are transmitted to humans by direct exposure.

Table 2. Classification of dermatophytes based on ecologic source and host response			
	Geophilic	Zoophilic	Anthropophilic
Infection	Acute inflammation	Moderate inflammation	Non-inflammatory

Transmission	From environment	Animal -> Human Human <- Animal	Human-> Human
Resolution	Rapid	Self-limited outbreaks	Chronic

TRANSMISSION

Dermatophytosis is a communicable disease with fungal elements and/or conidia persisting on fomites like towels, bedsheets, caps, combs; and on surfaces like gym shower rooms. Outbreaks of tinea capitis, once very prevalent among poor children in early decades of the 20th century (Fox 1909) are seen today in pediatric refugees (Mashiah et al., 2016.) Asymptomatic adult carriers may be a source of tinea capitis in children. Another example is transmission between pets or livestock and humans.

Clinical syndromes

The skin infections caused by dermatophytes are chronic infections of the skin often found in the warm humid areas of the body, such as athlete's foot and jock itch. Typical ringworm lesions are circular, which have an inflamed border containing papules and vesicles surrounding a clear area of relatively normal skin. These lesions are associated with variable degrees of scaling and inflammation. Broken hair and thickened broken nails are often seen in this lesion. Clinically, ringworm can be classified depending on the site affected.

These are (*a*) *Tinea capitis* involving scalp,
(*b*) *Tineacorporis* involving nonhairy skin of the body,
(*c*) *Tinea cruris* affecting groin,
(*d*) *Tinea pedis* affecting foot, and
(*e*) *Tinea barbae* affecting beard areas of face and neck.

Favus is a chronic ringworm infection affecting hair follicle.I t leads to alopecia and scarring.

Laboratory diagnosis

Laboratory diagnosis is based on demonstration of fungal element in clinical specimen by microscopy and confirmation by culture. The specimens include skin scrapings and nail clippings or hair taken from the areas suspected to be infected by dermatophytes. These entire specimens are treated with alkali solution to clear epithelial cells and other debris. Direct microscopy is useful only for diagnosis, while culture is always carried out to identify the specific causative fungal agent.

Treatment

Treatment of dermatophyte infection is carried out by use of local antifungal drugs, such as miconazole, clotrimazole, econazole, etc., or by treatment orally with griseofulvin.

****

31.

CANDIDIASIS

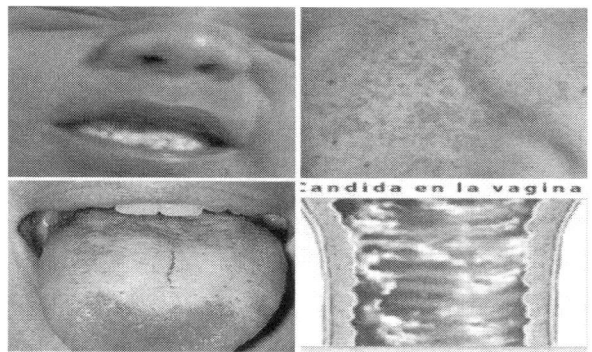

Candidiasis is a fungal infection caused by a yeast (a type of fungus) called *Candida*. Some species of *Candida* can cause infection in people; the most common is *Candida albicans*. *Candida* normally lives on the skin and inside the body, in places such as the mouth, throat, gut, and vagina, without causing any problems. *Candida* can cause infections if it grows out of control or if it enters deep into the body (for example, the bloodstream or internal organs like the kidney, heart, or brain). Some types of *Candida* are resistant to the antifungals used to treat them.

Fig.31. *C. albicans*

Candida species are the most common fungal pathogens that affect humans. These species are true opportunistic pathogens that take advantage of the host's debilitated condition and gain access to the circulation and deep tissues. The genus *Candida* includes more than 100 species, of which only few cause disease in humans. *C.*

albicans and occasionally other species cause candidiasis, a major infection in immunocompromised hosts.

C. albicans is the most common *Candida* species, which causes opportunistic infections in immunocompromised hosts. It forms the part of the normal flora of the mucous membrane of the gastrointestinal, genitourinary, and respiratory tract.

Common opportunistic infections

Disease	Causative fungus
Candidiasis	*Candida albicans, Candida tropicalis,* and other species
Aspergillosis	*Aspergillus fumigates, Aspergillus flavus, Aspergillus niger,* and other species
Zygomycosis	*Mucor, Rhizopus,* and *Absidia* species
Pneumocystis carinii pneumonia (PCP)	*Pneumocystis jiroveci*
Penicilliosis	*Penicillium marneffei*
Pseudoallescheria boydii infection	*Pseudoallescheria boydii*
Fusarium solani infection	*Fusarium solani*
Meningitis	*Cryptococcus neoformans*

Clinical illnesses

Candida causes a wide spectrum of clinical illnesses as follows:

Cutaneous candidiasis: *Candida* species in immunocompetent host can cause infection of any warm and moist part of the body exposed to environment. It causes infection of the nail, rectum, and other skin folds.

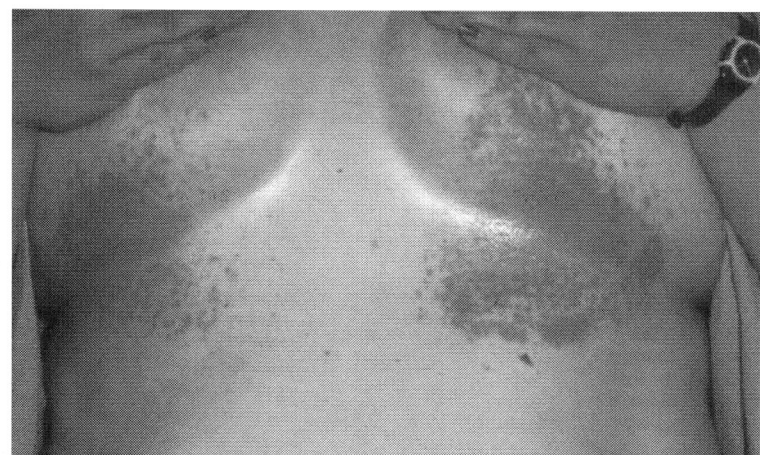
Fig. 31.1. Cutaneous candidiasis

Mucocutaneous candidiasis: Mucocutaneous candidiasis (thrush, perianal disease, etc.) is the most common manifestation of candidiasis, but usually does not cause any mortality. In patients with advanced immunodeficiency due to HIV infection, *Candida* species can cause severe oropharyngeal and esophageal candidiasis that result in poor intake of food, leading to malnutrition, wasting, and early death. These patients are also usually resistant to treatment with antifungal therapy.

Chronic mucocutaneous candidiasis: This is a heterogeneous group of clinical syndromes. This syndrome is characterized by chronic, treatment-resistant, superficial *Candida* infection of the skin, nails, and oropharynx. However, these patients do not show any evidence of disseminated candidiasis.

Systemic candidiasis: These include endocarditis, gastrointestinal tract candidiasis, respiratory tract candidiasis, genitourinary candidiasis, and hepatosplenic candidiasis. Systemic candidiasis may be candidemia and disseminated candidiasis. In patients with AIDS, oral thrush and *Candida* esophagitis are more common but not candidemia and disseminated candidiasis. *Candida* endophthalmitis and central nervous infection (CNS) infection due to *Candida* species are other complications of *Candida* infection.

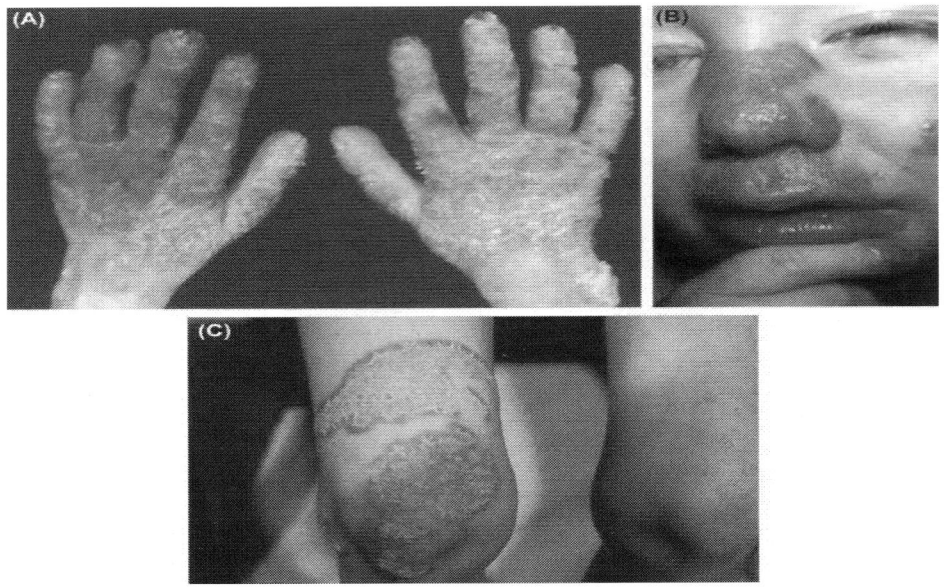

Fig. 31.2. Chronic mucocutaneous candidiasis

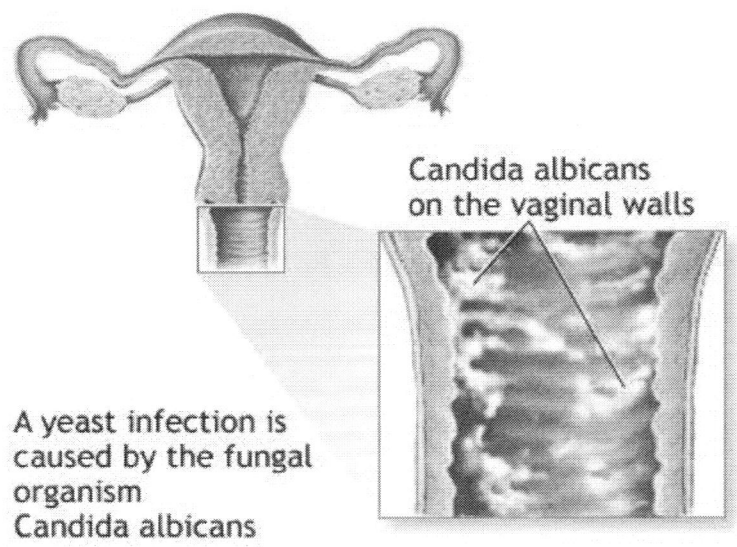

Fig.31.3. Genitourinary candidiasis

Disseminated candidiasis: This is increasingly becoming a problem in patients with serious hematologic malignancies that are treated with immunosuppressive drugs for over a long period of time. Severe neutropenia in these patients is the most important predisposing condition for life-threatening infection caused by *Candida*. In this condition, *Candida* usually

spreads through the circulation and involves many organs, such as lungs, spleen, kidney, liver, heart, and brain. However, disseminated candidiasis is not a major problem in patients with AIDS. In such patients, serious infection of the oropharynx and the upper gastrointestinal tract is the major problem. The development of these conditions in previously healthy individuals not receiving broad-spectrum antibiotic therapy should be strongly suspected for possibility ofinfection with HIV.

Diagnosis

In oral candidiasis, simply inspecting the person's mouth for white patches and irritation may make the diagnosis. They may also take a sample of the infected area to determine what organism is causing the infection.

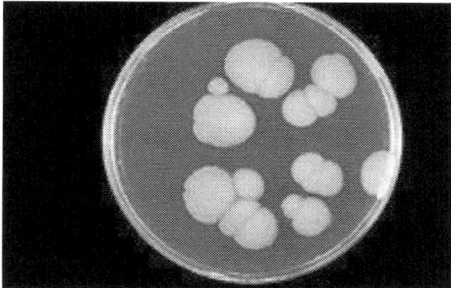

Fig. 31.4. Agar plate culture of *C. albicans*

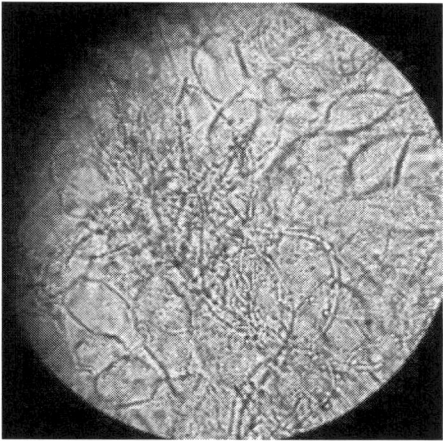

Fig.31.5. KOH test on a vaginal wet mount, showing slings of pseudohyphae of *Candida albicans* surrounded by round vaginal epithelial cells, conferring a diagnosis of candidal vulvovaginitis

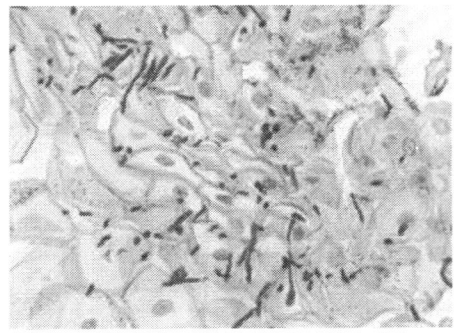

Fig.31.6.Micrograph of esophageal candidiasis showing hyphae, biopsy specimen, PAS stain

Symptoms of vaginal candidiasis are also present in the more common bacterial vaginosis;[aerobic vaginitis is distinct and should be excluded in the differential diagnosis. In a 2002 study, only 33% of women who were self-treating for a yeast infection actually had such an infection, while most had either bacterial vaginosis or a mixed-type infection.

Diagnosis of a yeast infection is done either via microscopic examination or culturing. For identification by light microscopy, a scraping or swab of the affected area is placed on a microscope slide. A single drop of 10% potassium hydroxide (KOH) solution is then added to the specimen. The KOH dissolves the skin cells, but leaves the *Candida* cells intact, permitting visualization of pseudohyphae and budding yeast cells typical of many *Candida* species.

For the culturing method, a sterile swab is rubbed on the infected skin surface. The swab is then streaked on a culture medium. The culture is incubated at 37 °C (98.6 °F) for several days, to allow development of yeast or bacterial colonies. The characteristics (such as morphology and colour) of the colonies may allow initial diagnosis of the organism causing disease symptoms. Respiratory, gastrointestinal, and esophageal candidiasis require an endoscopy to diagnose. For gastrointestinal candidiasis, it is necessary to obtain a 3–5 milliliter sample of fluid from the duodenum for fungal culture. The diagnosis of gastrointestinal candidiasis is based upon the culture containing in excess of 1,000 colony-forming units per milliliter.

Epidemiology

Candida species is distributed worldwide. In recent times, *Candida* species have replaced *Cryptococcus* species as the most common fungi affecting the CNS of immunocompromised patients worldwide. *C. albicans* and *Candida glabrata* are responsible for causing infection in 70–80% of patients with invasive candidiasis. *Candida tropicalis* is an important cause of candidemia in patients with leukemia and in those who have undergone bone marrow transplantation. *Candida parapsilosis* is an important pathogen associated with the use of vascular catheters. Since *Candida* is present as a part of normal flora already in the skin and mucous membrane of the host, it causes infection in the infected host; it is therefore not transmitted.

Treatment

Treatment for *Candida* infections consist of the administration of antifungal drugs. Examples of the drugs of choice include amphotericin B, fluconazole, ketoconazole, and nystatin. The real possibility of the development of irritative side effects makes monitoring during therapy a prudent precaution.

****

32.

MUCORMYCOSIS

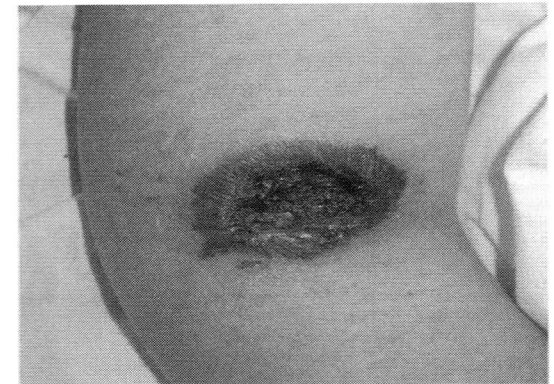

Mucormycosis is the general term that indicates any fungal infection caused by various genera of the class Zygomycetes. Another term used in medical and lay publications that means the same is phycomycosis. Mucormycosis can result in an acute, rapidly advancing, and occasionally fatal disease caused by different fungi commonly found in the soil or environment. These fungal infections are diagnosed relatively infrequently; however, they occur in individual people who are debilitated in some major way (uncontrolled diabetics, immunocompromised patients) and occasionally in groups of people that are injured (often multiple injuries and penetrating injuries that are contaminated with soil and water from the environment). Such groups of people are those that are injured in disasters such as tsunamis, hurricanes, earthquakes or tornadoes, where otherwise healthy people can have contaminated soil and water inhaled, embedded in wounds, or simply forced into skin, mouth, eyes, and nose by the force of water, soil, or wind pressure. The disease is not passed person to person.

A cluster of mucormycosis infections occurred in people who initially survived devastating tornadoes which struck Joplin, Missouri, on May 23, 2011. Thirteen cases were confirmed, all in persons with severe wounds, including fractures, multiple wounds, penetrating injuries, and blunt trauma. Ten patients required intensive care and five died.

Because the majority of mucormycosis infections are caused by one family member in the class of Zygomycetes (family member Mucoraceae), many clinicians now term the disease mucormycosis instead of zygomycosis, the more "general" term. The lay press has used terms like "Black Death" and "Zombie disease" to describe this fungal infection but such terms seldom help people to understand this disease. Such terms may cause misunderstandings between the patients, their families, and the public; many clinicians think these potentially harmful or cruel terms should not be used by responsible individuals.

Causes mucormycosis

Zygomycetes represent the general class of fungi that cause mucormycosis. *Rhizopus arrhizus* species from the Mucoraceae family are the most commonly identified cause of mucormycosis in humans. Other fungal causes may include Mucor species, *Cunninghamella bertholletiae*, *Apophysomyces elegans*, Absidia species, Saksenaea species, *Rhizomucor pusillus*, Entomophthora species, Conidiobolus species, and Basidiobolus species. Mucoraceae are found worldwide and in the ecosystem are responsible for initiating and decaying most organic material in the environment. Most fungi are identified by their unique morphological appearance (see Figure 1) viewed microscopically and determined by a professional practiced in fungal identification (microbiologist or pathologist).

In general, mucormycosis is an infection not often seen by many doctors because the fungal causes are not readily infectious. Usually an infection develops because of some unusual circumstance that places the fungi in contact with compromised or injured animal or human tissue. However, once established, the fungi can rapidly multiply in blood vessel walls where it effectively reduces and cuts off blood to tissues, thereby creating its own decaying organic food source resulting in widespread tissue destruction. If this fulminant spread of fungi is not stopped, death is the outcome.

Factors for mucormycosis

A risk factor is any debilitating disease process, especially diseases that can yield compromised blood flow to tissue. The classic example is the patient with uncontrolled diabetes and foot ulcers where dirt or debris can easily reach compromised tissue. Patients with burns, malignancies, immunocompromised patients, patients with a splenectomy, and people with wounds (usually severe) that have been contaminated with soil or environmental water are at higher risk to get mucormycosis. Consequently, people injured in environmental disasters are, as a group, at high risk for this infection.

Symptoms

Most of the symptoms of mucormycosis do not differ to any major extent between the various fungal causes. Most authorities describe the signs and symptoms of the disease according to the predominant or initial body area that is infected. Some patients have more than one body area infected. The following is a list of signs and symptoms (note that many authors prefer the term mucormycosis instead of zygomycosis since the majority of fungi, when identified, are from the Mucoraceae family of fungi):

- Rhinocerebral mucormycosis: fever, headache, reddish and swollen skin over nose and sinuses, dark scabbing in the nose by the eye(s), visual problems, eye(s) swelling, facial pain

Rhinocerebral Mucormycosis

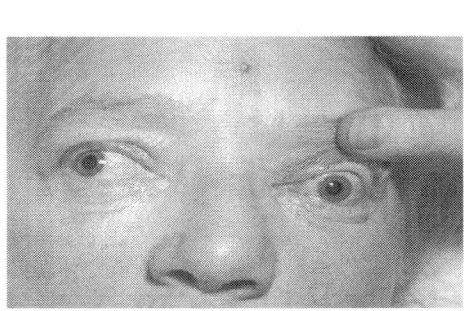
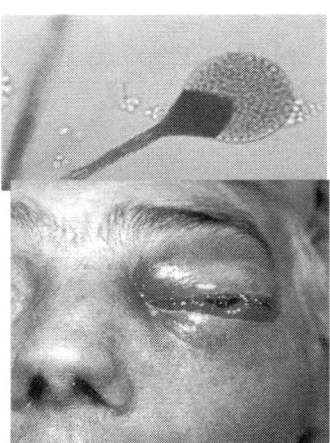

Fig.32. Rhinocerebral mucormycosis

- Pulmonary (lung) mucormycosis: fever, coughing sometimes with bloody or dark fluid production, shortness of breath
- GI mucormycosis: diffuse abdominal pain, bloody and sometimes dark vomitus, abdominal distension
- Renal mucormycosis: fever, flank pain
- Cutaneous mucormycosis: initially, reddish and swollen skin often adjacent to an area of skin trauma, that becomes an ulcer with a dark center and sharply defined edges.

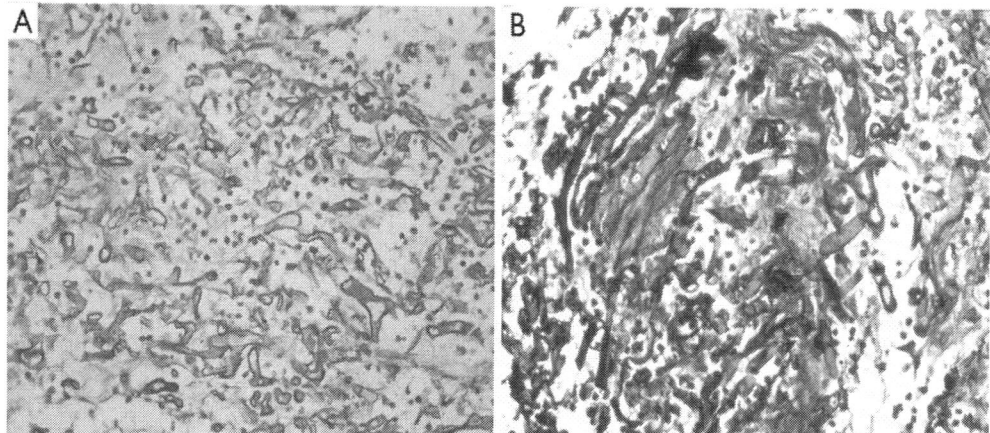

Fig.32.1. Chronic pulmonary mucormycosis

- Disseminated mucormycosis: initially may have any of the above symptoms; as the disease spreads to other organs, headaches, fever, and mental-status changes occur

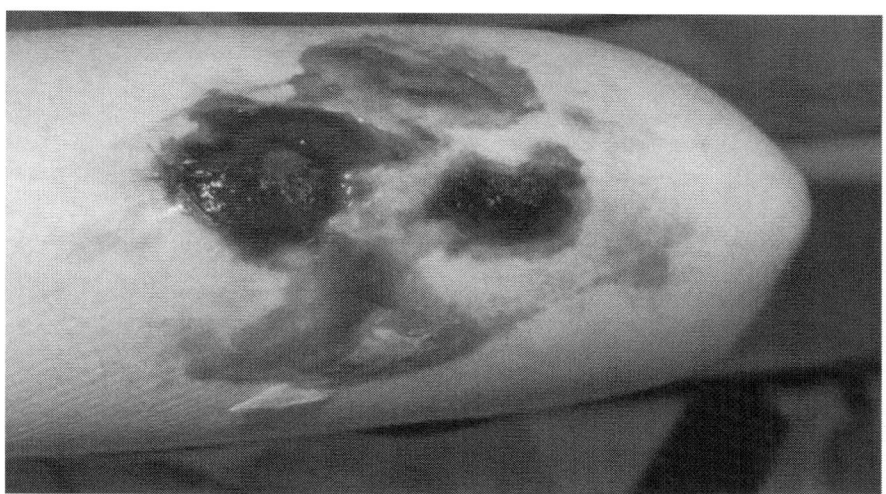

Fig .32.2. Disseminated mucormycosis

Although these symptoms suggest that a patient may have mucormycosis, they are not definitive. In addition, they may not develop very quickly because it may take a few days to over a week in many people before the symptoms develop. When they do initially develop, it is not unusual to ascribe the symptoms to causes other than fungi (often to secondary bacterial infections). Consequently, the fungal diagnosis may be delayed (see diagnosis section below).

Mucormycosis diagnosed

Presumptive diagnosis is based on the patient's history, physical exam, and the patient's risk factors for getting a fungal infection. A definitive diagnosis is difficult. Although tests such as CT or MRI may help define the extent of infections or tissue destruction, their findings are not specific for mucormycosis. There are no serological or blood tests that are helpful. Growth of the fungi from a biopsy (tissue obtained by surgical removal or endoscopes with biopsy tool) of infected tissue, accompanied by special tissue stains looking for unique structural components, may identify the fungus and help make the definitive diagnosis. This helps distinguish mucormycosis from other fungal diseases such as candidiasis and histoplasmosis. However, it is still sometimes difficult to determine the specific fungal genus and species infecting the patient. Consequently, mucormycosis is often a "working" diagnosis that clinicians use because the supportive care and treatments for the causative fungal agents are essentially the same. Figure 2 shows a periorbital eye infection eventually diagnosed as mucormycosis.

Treatment of mucormycosis

Treatments of mucormycosis need to be fast and aggressive. The need for speed is because by the time even the presumptive diagnosis is made, often the patient has suffered significant tissue damage that cannot be reversed. Most patients will require both surgical and medical treatments. Most infectious-disease experts say that without aggressive surgical debridement of the infected area, the patient is likely to die. Medications play an important role. Two main goals are sought at the same time: antifungal medications to slow or halt fungal spread and medications to treat any debilitating underlying diseases. Amphotericin B (initially intravenous) is the usual drug of choice for antifungal treatment. In addition, posaconazole or isavuconazole may treat mucormycosis. Patients with underlying diseases like diabetes need their diabetes optimally controlled. Patients normally on steroids or undergoing treatment with deferoxamine (Desferal; used to remove excess iron in the body) are likely to have these medications stopped because they can increase the survival of fungi in the body. Patients may need additional surgeries and usually need antifungal therapy for an extended time period (weeks to months) depending on the severity of the disease. Consultation with an infectious-disease expert is advised

33.

ASPERGILLOSIS

Aspergillosis is the name given to a wide variety of diseases caused by fungal infections from species of *Aspergillus*. Aspergillosis occurs in humans, birds and other animals.

Aspergillosis occurs in chronic or acute forms which are clinically very distinct. Most cases of acute aspergillosis occur in people with severely compromised immune systems, e.g. those undergoing bone marrow transplantation. Chronic colonization or infection can cause complications in people with underlying respiratory illnesses, such as asthma, cystic fibrosis, sarcoidosis,[4] tuberculosis, or chronic obstructive pulmonary disease. Most commonly, aspergillosis occurs in the form of chronic pulmonary aspergillosis (CPA), aspergilloma, or allergic bronchopulmonary aspergillosis (ABPA). Some forms are intertwined; for example ABPA and simple aspergilloma can progress to CPA. Other, noninvasive manifestations include fungal sinusitis (both allergic in nature and with established fungal balls), otomycosis (ear infection), keratitis (eye infection), and onychomycosis (nail infection). In most instances, these are less severe, and curable with effective antifungal treatment.

The most frequently identified pathogens are *Aspergillus fumigatus* and *Aspergillus flavus*, ubiquitous organisms capable of living under extensive environmental stress. Most people are thought to inhale thousands of *Aspergillus* spores daily but without effect due to an efficient immune response. Taken together, the major chronic, invasive, and allergic forms of aspergillosis account for around 600,000 deaths annually worldwide.

Aspergillosis has several forms:

- **Pulmonary aspergillosis:** Aspergillosis usually develops in open spaces in the body, such as cavities in the lungs caused by preexisting lung disorders. The infection may also develop in the ear canals and sinuses. In the sinuses and lungs, aspergillosis typically develops as a ball (aspergilloma) composed of a tangled mass of fungus fibers, blood clots, and white blood cells. The fungus ball gradually enlarges, destroying lung tissue in the process, but usually does not spread to other areas.

- **Invasive aspergillosis:** Less often, aspergillosis becomes very aggressive and rapidly spreads throughout the lungs and often through the bloodstream to the brain, heart, liver, and kidneys. This rapid spread occurs mainly in people with a very weakened immune system.
- **Allergic bronchopulmonary aspergillosis:** Some people who have asthma or cystic fibrosis develop a chronic allergic reaction with cough, wheezing, and fever if *Aspergillus* colonizes the lining of their airways.
- **Superficial aspergillosis:** This form is uncommon. It may develop in burns, under bandages, after damage to the eye, or in the sinuses, mouth, nose, or ear canal.

Symptoms

Pulmonary aspergillosis

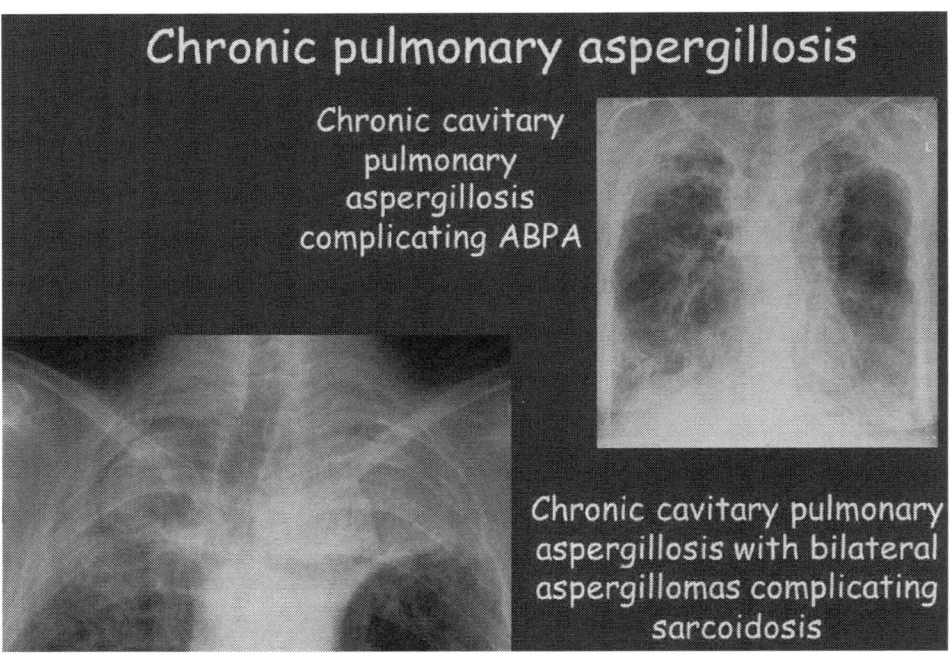

Fig.33.Pulmonary aspergillosis

A fungus ball in the lungs may cause no symptoms and may be discovered only when a chest x-ray is taken for other reasons. Or it may cause repeated coughing up of blood and, rarely, severe, even fatal bleeding.

Invasive aspergillosis

A rapidly invasive infection in the lungs often causes cough, fever, chest pain, and difficulty breathing. Without treatment, this form of invasive aspergillosis is fatal.

Aspergillosis that spreads to other organs makes people very ill. Symptoms include fever, chills, shock, delirium, and blood clots. Kidney failure, liver failure (causing jaundice), and breathing difficulties may develop. Death can occur quickly.

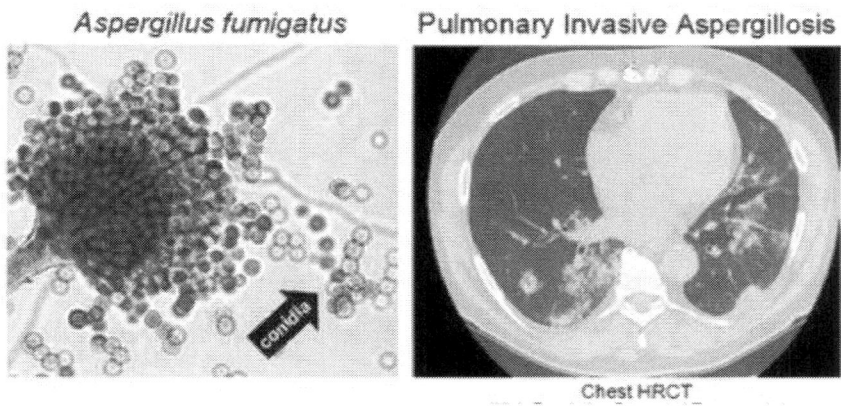

Fig.33.1. Invasive aspergillosis

Superficial aspergillosis

Aspergillosis of the ear canal causes itching and occasionally pain. Fluid draining overnight from the ear may leave a stain on the pillow.

Aspergillosis of the sinuses causes a feeling of congestion and sometimes pain or discharge or bleeding from the nose.

Risk Factors

People who are immunocompromised — such as patients undergoing hematopoietic stem cell transplantation, chemotherapy for leukaemia, or AIDS — are at an increased risk for invasive aspergillosis infections. These people may have neutropenia or corticoid-induced immunosuppression as a result of medical treatments. Neutropenia is often caused by extremely cytotoxic medications such as cyclophosphamide. Cyclophosphamide interferes with cellular replication including that of white blood cells such as neutrophils. A decreased neutrophil count inhibits the ability of the body to mount immune responses against pathogens. Although tumor necrosis factor alpha (TNF-α) — a signaling molecule related to acute inflammation responses is produced, the abnormally low number of neutrophils present in neutropenic patients leads to a depressed inflammatory response. If the underlying neutropenia is not fixed, rapid and uncontrolled hyphal growth of the invasive fungi will occur and result in negative health outcomes.

Epidemiology

Aspergillosis is thought to affect more than 14 million people worldwide,[24] with allergic bronchopulmonary aspergillosis (ABPA, >4 million), severe asthma with fungal sensitization (>6.5 million), and chronic pulmonary aspergillosis (CPA, ~3 million) being considerably more prevalent than invasive aspergillosis (IA, >300,000). Other common conditions include Aspergillus bronchitis, *Aspergillus* rhinosinusitis (many millions), otitis externa, and *Aspergillus* onychomycosis (10 million). Alterations in the composition and function of the lung microbiome and mycobiome have been associated with an increasing number of chronic pulmonary diseases such as COPD, cystic fibrosis, chronic rhinosinusitis and asthma.

Treatment

- Antifungal drugs
- Sometimes surgery to remove fungi

Aspergillosis that affects only a sinus or a single area in the lung requires treatment but does not pose an immediate danger because it progresses slowly. However, if the infection is widespread or if people appear seriously ill or have a weakened immune system, treatment is started immediately.

Invasive aspergillosis is treated with antifungal drugs, such as voriconazole, isavuconazole, or sometimes posaconazole or itraconazole. However, some forms of *Aspergillus* do not respond to these drugs and may need to be treated with amphotericin B or with a combination of drugs. Any condition that is weakening the immune system should be corrected if possible. For example, doctors may advise people who are taking corticosteroids, which suppress the immune system, to stop.

Aspergillosis in the ear canal is treated by scraping out the fungus and applying drops of antifungal drugs. Collections of fungi in the sinuses must usually be removed surgically.

Fungus balls in the lungs (aspergillomas) usually do not require treatment with drugs and do not usually respond to drugs. If these balls cause bleeding (causing people to cough up blood) or other symptoms, they may need to be removed surgically. Surgery cures the infection but is often risky because many of these people have other disorders.

****

34.
PREDACEOUS FUNGI

Carnivorous fungi or **predaceous fungi** are fungi that derive some or most of their nutrients from trapping and eating microscopic or other minute animals. More than 200 species have been described, belonging to the phyla Ascomycota, Mucoromycotina, and Basidiomycota. They usually live in soil and many species trap or stun nematodes (nematophagous fungus), while others attack amoebae or collembola.

Fungi that grow on the epidermis, hair, skin, nails, scales or feathers of living or dead animals are considered to be dermatophytes rather than carnivores. Similarly, fungi in orifices and the digestive tract of animals are not carnivorous, and neither are internal pathogens. Neither are insect pathogens that stun and colonize insects normally labelled carnivorous if the fungal thallus is mainly in the insect as does *Cordyceps*, or if it clings to the insect like the Laboulbeniales. All of these are examples of parasitism or scavenging.

Two basic trapping mechanisms have been observed in carnivorous fungi that are predatory on nematodes:

- constricting rings (active traps)
- adhesive structures (passive traps)

The Predaceous fungi are also termed as Nematophagous fungi. The Nematophagous fungi are of three main types on thebasis of ecological habitat:
   a. Nematode trapping fungi,
   b. Endoparasitic fungi,

c. Egg parasites.

Nematode trapping fungi:

Fungi capturing Nematodes are called Nematode trapping fungi. Such fungi are evolvedstructural adaptations to trap or penetrate their prey. They may be Predatory or Endoparasites. There are a varieties of ways by which fungi trap Nemetodes resulting in their death.

Adhesive Hyphae:

The fungal Hyphae form adhesive which capture Nematodes. These Hyphae produce adhesive any point in response to Nematode contact or the Hyphae are coated with adhesive along their entire surface. At the point of Hyphae where contact is made for capture, a thick and yellowish chemical material is secreted. An outgrowth of Hyphae similar to Appressorium develops. When the Nematode is trapped it becomes inactive first and killed in the last after penetration of hyphae.

Adhesive branches:

The Nematode trapping fungi produce the most primitive and simple organ of capture, the adhesive branches, which are a few cells in height. From the main prostrate Hyphae short laterals grows as erect branches on or below the substrate.

Adhesive nets:

Nets are formed by fungal Hyphae (coated with adhesive material) which are adhesive in nature. Nets may be in the form of a single hoop like loop to a complex multi-branched networks.

Adhesive knobs:

A distinct adhesive cell is produced at the apex of a slender non-adhesive stalk containing 1-3 cells. A thin film of adhesive material is produced over the surface of knob.

Non-Constricting rings:

From the prostrate creeping septate hyphae there arise erect and lateral branches which form non-constricting rings. Initially the branch is slender but widens subsequently and being curver to form a circular structure. At the point where tip of branch makes contact with supporting stalk, cell walls get fused. This results in formation of threecelled ring with a stalk.

Constricting rings:

The constricting rings are produced similar to non- constricting rings but the supporting stalk is shorter and stouter.

The body of Nematode consists of a low molecular weight peptide which is called Nemin. Nemin is water soluble and potential stimulant for trap formation. It causes morphological changes in Nematophagous fungi. The lectin of fungus *A. oligospora* binds especially to the sugar, N-acetyl-D-galactosamine, present on Nematode cuticle.

The trapped Nematodes secrete mucilage which has been identified under electron microscope. The Nematode cuticle is lysed at the point where lectin combines with N-acetyl-D-glucosamine. The enzyme collagenase is secreted by the fungus which dissolves collagen protein of Nematode cuticle.

Endoparasitic Fungi:

The Endoparasitic fungi do not extensively produce mycelium external to Nematode body. The Endoparasitic fungi are species of *Cephalosporum, Meria, Verticillium, Catenaria, Meristacrum* etc. *Catenaria anguillulae* produces zoospores which track down Nematodes by swarming, eventually encyst near Nematode body orifice, penetrate and colonize the prey. The encysted zoospores produce germ tube which penetrate Nematode through orifice or by dissolving cuticle. The infectious hyphae grow well inside Nematode body, digest content and lyse the prey. In *Meristacrum asterospermum* forms adhesive conidia which attach to the cuticle of Nematode. It germinate to form the hypha which swells and acts as infectious thallus. Adhesive spores are also produced by species of *Meria, Cephalosporium* and *Verticillium*. In *M. coniospora* an adhesive bud develops at the distal end of tear drop shaped spores.

Egg parasites:

There are a few saprophytic fungi which attack on nemtode eggs. When a fungal hypha comes in contact of an egg, a swollen structure at terminal portion develops at the point of contact. It gets attached to the egg, a swollen structure at terminal portion develops at the point of contact. It gets attached to the egg where from a marrow infectious tube develops that penetrates the shell of the egg. The hyphae swell up to form appressorium from which numerous irregularly branched absorption hyphae develops that consumes egg nutrients. Ex. *Dactyllele oviparasitica, Paecilomyces lilacinus* etc.

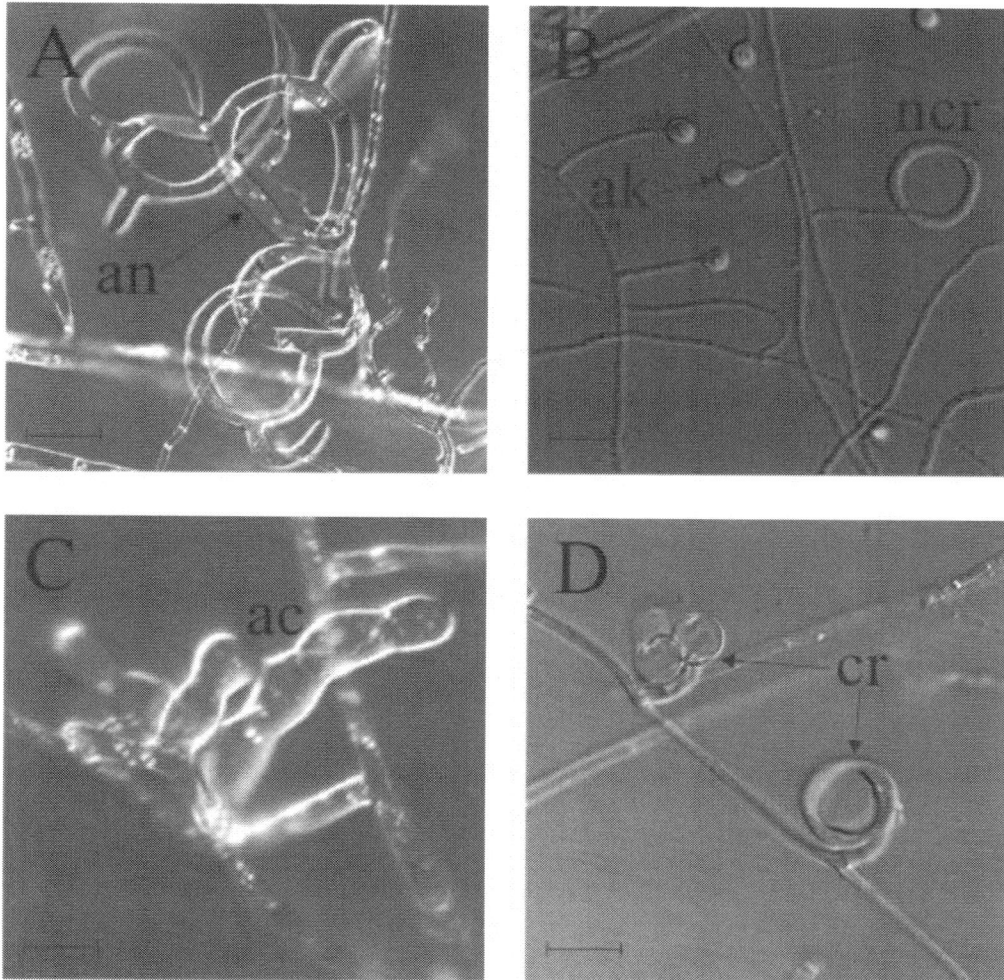

Fig.34.. Natural nematode-trapping devices. A, adhesive network (an), the most widely distributed trap. B, adhesive knob (ak) with nonconstricting rings ncr. C, adhesive column (ac) is a short erect branch consisting of a few swollen cells produced on a hypha. D, constricting ring (cr), the most sophisticated trapping device, captures prey actively; when a nematode enters a constricting ring, the three ring cells are triggered to swell rapidly inwards and firmly lasso the victim within 1 to 2 second. The ring at upper left in panel D has been triggered, that at bottom right is an unsprung trap. Scale bars = 10 μm. Modified from Yang *et al.*, 2007 using images kindly supplied by Prof. Xingzhong Liu and Dr Ence Yang, Institute of Microbiology, Beijing, China. Original images © (2007) National Academy of Sciences, USA.

****

35.

BIOFERTILIZER

A **biofertilizer** (also **bio-fertilizer**) is a substance which contains living micro-organisms which, when applied to seeds, plant surfaces, or soil, colonize the rhizosphere or the interior of the plant and promotes growth by increasing the supply or availability of primary nutrients to the host plant. Biofertilizers add nutrients through the natural processes of nitrogen fixation, solubilizing phosphorus, and stimulating plant growth through the synthesis of growth-promoting substances. The microorganisms in biofertilizers restore the soil's natural nutrient cycle and build soil organic matter. Through the use of biofertilizers, healthy plants can be grown, while enhancing the sustainability and the health of the soil. Biofertilizers can be expected to reduce the use of synthetic fertilizers and pesticides, but they are not yet able to replace their use. Since they play several roles, a preferred scientific term for such beneficial bacteria is "plant-growth promoting rhizobacteria"

Bio-fertilisers are living microorganisms of bacterial, fungal and algal origin. Their mode of action differs and can be applied alone or in combination.

Benefits of biofertilizers

- Biofertilizers fix atmospheric nitrogen in the soil and root nodules of legume crops and make it available to the plant.
- They solubilise the insoluble forms of phosphates like tricalcium, iron and aluminium phosphates into available forms.
- They scavenge phosphate from soil layers.
- They produce hormones and anti metabolites which promote root growth.
- They decompose organic matter and help in mineralization in soil.

- When applied to seed or soil, biofertilizers increase the availability of nutrients and improve the yield by 10 to 25% without adversely affecting the soil and environment.

Types and features of biofertilizers

Based on type of microorganism, the bio-fertilizer can also be classified as follows:

- **Bacterial Biofertilizers**: e.g. Rhizobium, Azospirilium, Azotobacter, Phosphobacteria.
- **Fungal Biofertilizers**: e.g. Mycorhiza
- **Algal Biofertilizers**: e.g. Blue Green Algae (BGA) and Azolla.
- **Actinimycetes Biofertilizer**: e.g. Frankia.

Bio-fertilizer are mostly cultured and multiplied it the laboratory. However, blue green algae and azolla can be mass-multiplied in the field.

Characteristics Features of common Biofertilizers

- **Rhizobium** : Rhizobium is relatively more effective and widely used biofertilizer. Rhizobium, in association wit legumes, fixes atmospheric N. The legumes and their symbiotic association with the rhizobium bacterium result in the formation of root nodules that fix atmospheric N. Successful nodulation of leguminous crop by rhizobium largely depends on the availability of a compatible stain for a particular legume. Rhizobium population in the soil is dependent on the presence of legumes crops in field. In the absence of legumes the population of rhizobium in the soil diminishes.
- **Azospirillum** : Azospirillum is known to have a close associative symbiosis with the higher plant system. These bacteria have association with cereals like; sorghum, maize, pearl millet, finger millet, foxtail millet and other minor millets and also fodder grasses.
- **Azotobacter** : It is a common soil bacterium. *A. chrococcum* is present widely in Indian soil. Soil organic matter is the important factor that decides the growth of this bacteria.
- **Blue Green Algae (BGA)** : Blue green algae are referred to as rice organisms because of their abundance in the rice field. Many species belonging to the genera, Tolypothrix, Nostic, Schizothrix, Calothrix, Anoboenosois and Plectonema are abundant in tropical conditions. Most of the nitrogen fixation BGA are filamenters, consisting of chain of vegetative

cell including specialized cells called heterocyst which function as a micronodule for synthesis and N fixing machinery.

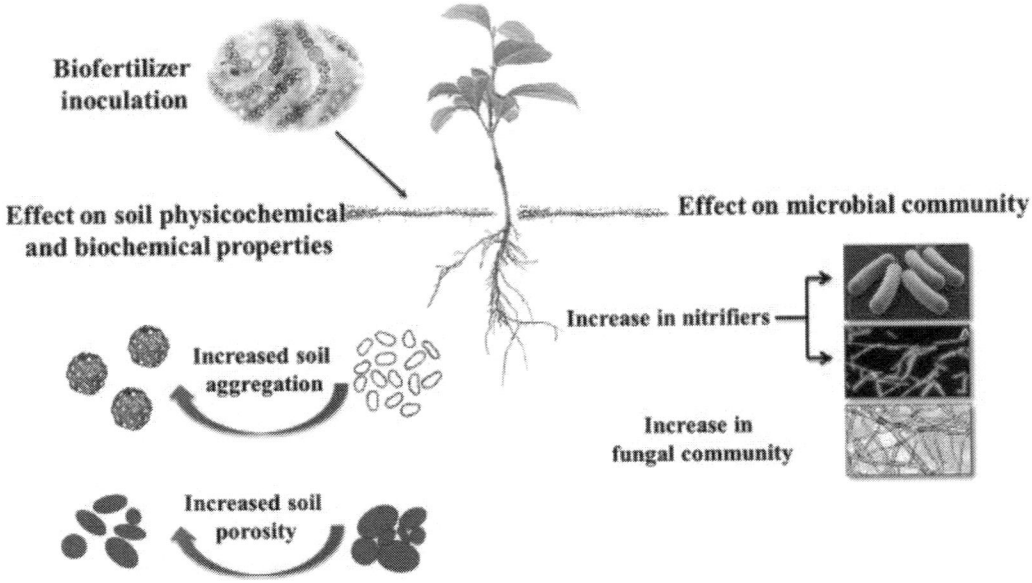

Fig.35. Role of Algae as a Biofertilizer

List of commonly produced bio-fertilizers in India

Name	Crops suited	Benefits usually seen	Remarks
Rhizobium strains	Legumes like pulses, groundnut, soybean	10-35% yield increase, 50-200 kg N/ha.	Fodders give better results. Leaves residual N in the soil.
Azotobacter	Soil treatment for non- legume crops including dry land crops	10-15% yield increase- adds 20-25 kg N/ha	Also controls certain diseases.
Azospirillum	Non-legumes like maize, barley, oats, sorghum, millet, Sugarcane, rice etc	10-20% yield increase	Fodders give higher/enriches fodder response. Produces growth promoting substances. It can be applied to legumes as co-inoculant

Phosphate Solubilizers (there are 2 bacterial and 2 fungal species in this group)	Soil application for all crops	5-30% yield increase	Can be mixed with rock phosphate.
Blue-green algae and Azolla	Rice/wet lands	20-30 kg N/ha, Azolla can give biomass up to 40-50 tonnes and fix 30-100 kg N/ha	Reduces soil alkalinity, can be used for fishes as feed. They have growth promoting hormonal effects.
Microhizae (VAM)	Many trees, some crops, and some ornamental plants	30-50% yield increase, enhances uptake of P. Zn, S and Water.	Usually inoculated to seedlings.

Biofertilizers recommended for crops

- Rhizobium + Phosphotika at 200 gm each per 10 kg of seed as seed treatment are recommended for pulses such as pigeonpea, green gram, black gram, cowpea etc, groundnut and soybean.
- Azotobacter + Phosphotika at 200 gm each per 10 kg of seed as seed treatment are useful for wheat, sorghum, maize, cotton, mustard etc.
- For transplanted rice, the recommendation is to dip the roots of seedlings for 8 to 10 hours in a solution of Azospirillum + Phosphotika at 5 kg each per ha.

Application of biofertilizers to crops

Seed treatment

Each packet (200g) of inoculant is mixed with 200 ml of rice gruel or jaggery solution. The seeds required for one hectre are mixed in the slurry so as to have uniform coating of the inoculants over the seeds and then shade dried for 30 minutes. The treated seeds should be used within 24 hous. One packet of inoculant is sufficient to treat to 10 kg seeds. Rhizobium, Azospirillum, Azotobacter and Phosphobacteria are applied as seed treatment.

Seedling root dip

This method is used for transplanted crops. Five packets (1.0 kg) of the inoculants are required for one ha and mixed with 40 litres of water. The root portion of the seedlings is dipped in the solutions for 5 to 10 minutes and then transplanted. Azospirillum is used for seedling root dip particularly for rice.

Soil treatment

4 kg each of the recommended biofertilizers are mixed in 200 kg of compost and kept overnight. This mixture is incorporated in the soil at the time of sowing or planting.

Use of VAM Biofertilizer

- The inoculum should be applied 2-3 cm below the soil at the time of sowing.
- The seeds are sown or cuttings planted just above the VAM inoculums so that the roots may come in contact with the inoculums and cause infection.
- Bulk inoculums of 100gm is sufficient for one meter square area.
- Seedlings raised in the polythene bags need 5-10 g of bulk inoculums for each bag.
- At the time of planting of saplings, VAM inoculums is to be applied at the rate of 20g /seedling in each spot.
- In the existing tree, inoculums of 200g is required for each tree.

Use of Blue Green Algae (BGA)

Fig. 35.1. Algae-based biofertilizer from runoff water

- Algal culture is applied as dried flakes at 10 kg/ha over the standing water in field rice.

- This is done two days after transplanting in loamy soils and six days after planting in clayey soils.
- The field is kept water logged for few days immediately after algal application.
- The biofertilizer is to be applied for 3-4 consecutive seasons in the same field.

Use of Azolla

- **Green manure:** Azolla is applied @ 0.6-1.0 kg/m^2 (6.25-10.0 t/ha) and incorporated before transplanting of rice.
- **Dual crop:** Azolla is applied @ of 100 g/m^2 (1.25t/ha), one to three days after transplanting of rice and allowed to multiply for 25-30 days. Azolla fronds can be incorporated into the soil at the time of first weeding.

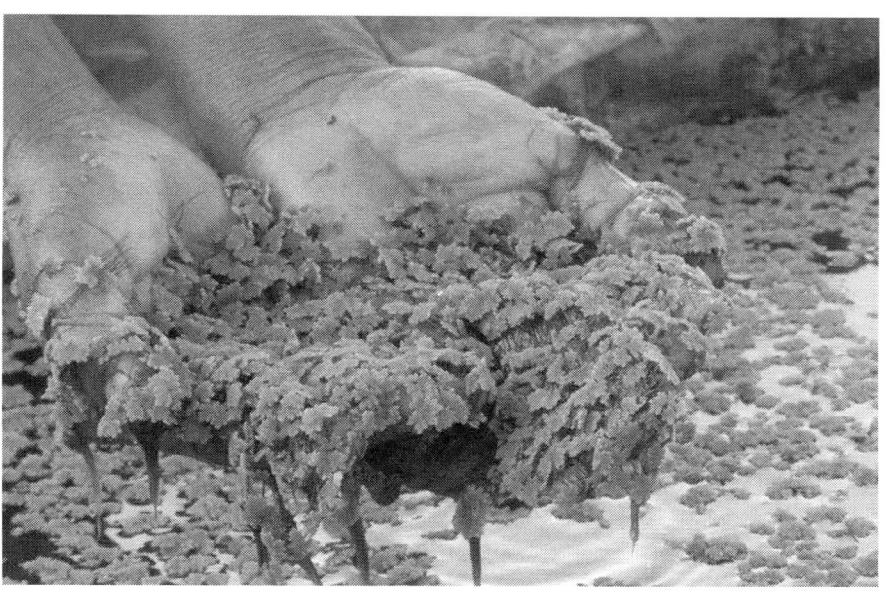

Fig. 35.2. Azola: An Eco-friendly Bio-Fertilizer

Tips to get good response to biofertilizer application

- Biofertilizer product must contain good effective strain in appropriate population and should be free from contaminating microorganisms.
- Select right combination of biofertilizers and use before expiry date.
- Use suggested method of application and apply at appropriate time as per the information provided on the label.

- For seed treatment adequate adhesive should be used for better results.
- For problematic soils use corrective methods like lime or gypsum pelleting of seeds or correction of soil pH by use of lime.
- Ensure the supply of phosphorus and other nutrients.

Precautions to take while using biofertilizers

- Biofertilizer packets need to be stored in cool and dry place away from direct sunlight and heat.
- Right combinations of biofertilizers have to be used.
- As Rhizobium is crop specific, one should use for the specified crop only.
- Other chemicals should not be mixed with the biofertilizers.
- While purchasing one should ensure that each packet is provided with necessary information like name of the product, name of the crop for which intended, name and address of the manufacturer, date of manufacture, date of expiry, batch number and instructions for use.
- The packet has to be used before its expiry, only for the specified crop and by the recommended method of application.
- Biofertilizers are live product and require care in the storage
- Both nitrogenous and phosphatic biofertilizers are to be used to get the best results.
- It is important to use biofertilizers along with chemical fertilizers and organic manures. Biofertilizers are not replacement of fertilizers but can supplement plant nutrient requirements.

****

36.

MYCORRHIZA

"Myco" – "rhiza" literally means "fungus" – "root" and describes the mutually beneficial relationship between the plant and root fungus. These specialized fungi colonize plant roots in a symbiotic manner and extend far into the soil. Mycorrhizal fungal filaments in the soil are truly extensions of root systems and are more effective in nutrient and water absorption than the roots themselves. More than 95 percent of terrestrial plant species form a symbiotic relationship with beneficial mycorrhizal fungi, and have evolved this symbiotic relationship over the past several hundred million years. These fungi predate the evolution of terrestrial plants, and it was the partnership with mycorrhizal fungi that allowed plants to begin to colonize dry land and create life on Earth as we know it. The mycorrhizal symbiotic relationship centres on the plant's ability to produce carbohydrates through photosynthesis and share some of these sugars with the fungus in return for otherwise unavailable water and nutrients that are sourced from the soil or growing media by the extensive network of mycelial hyphae produced by the fungus. It's a two-way relationship of sharing resources between two species, thus a classic symbiotic mutualism. The endomycorrhizal fungi rely on the plant, and the plant's performance and survival are enhanced by the fungus.

Fig. 36. Mycorrhizae Help Feed Your Plants https://www.finegardening.com

Mycorrhizal fungi can colonize plants from three main sources of inoculum: spores, colonized root fragments, and vegetative hyphae. Collectively, these inoculants are called "propagules," and this is the standard unit of measure that is listed on most commercially available mycorrhizal products.

To colonize plant roots, these propagules must be present in the substrate and in close proximity to actively growing roots of a compatible plant. The growing root tips emit root exudates as they push through the substrate which signal the fungi to colonize the roots and establish the symbiosis. Once the roots are colonized, then the process is self-sustaining as the mycelia continue to grow with the plant's root system and additional spores and hyphae are produced.

AMF propagules can be incorporated into the substrate prior to or during planting or they can be top-dressed on the surface and watered into a porous substrate. They can also be applied as a dip or slurry at the time of sticking a cutting, seeding, or at the time of transplanting. The propagules can also be applied as a drench to the soil and watered-in, applied to the outer surface of the root ball before transplanting, or used in transplant hole and backfill soil.

Different Types of Mycorrhizal Fungi

2 Major Types of Mycorrhizal Fungi

Endomycorrhizal Fungi	Ectomycorrhizal Fungi
• Form symbiotic relationships with approximately 85% of plant families.	• Form symbiotic relationships with about 10% of plant families.
• Pair with most commercially produced plants, including green, leafy, and fruiting or flowering plants.	• Mainly pair with conifers and many American hardwoods.
• Penetrate into the root cortex and form nutrient exchange structures within the root cells (arbuscules, vessicles, etc).	• Do not penetrate into the root cell walls, but form a sheath around the root, and nutrient exchange structures known as a "Hartig net."

Other(s)
- Brassica Family is non-mycorrhizal
- Ericaceae and Orchids have specific species of mycorrhizal fungi (less commercially available)

Source https://mycorrhizae.com/how-it-works/

Examples of Endomycorrhizal Fungi:

- *Glomus intraradices* (AKA *Rhizophagus irregularis*)
- *Glomus mosseae*
- *Glomus aggregatum*
- *Glomus etunicatum*
- Glomus deserticola
- *Glomus clarum*
- *Glomus monosporum*
- *Paraglomus brasilianum*
- *Gigaspora margarita*

Examples of Ectomycorrhizal Fungi:

- *Rhizopogon villosulus*
- *Rhizopogon luteolus*
- *Rhizopogon amylopogon*
- *Rhizopogon fulvigleba*
- *Pisolithus tinctorius*
- *Suillus granulatus*
- *Laccaria bicolor*
- *Laccaria laccata*
- *Scleroderma cepa*
- *Scleroderma citrinum*

Examples of Mycorrhiza

Orchid Mycorrhiza

As mentioned above, some orchids cannot photosynthesize prior to the seedling stage. Other orchids are entirely non-photosynthetic. All orchids, however, depend on the sugars provided by their fungal partner for at least some part of their lives. Orchid seeds require fungal invasion in order to germinate because, independently, the seedlings cannot acquire enough nutrients to grow. In this relationship, the orchid parasitizes the fungus that invades its roots. Once the seed coat ruptures and roots begin to emerge, the hyphae of orchidaceous mycorrhiza penetrate the root's cells and create hyphal coils, or pelotons, which are sites of nutrient exchange.

Arbuscular Mycorrhiza

Arbuscular mycorrhizae are the most widespread of the micorrhizae species and are well known for their notably high affinity for phosphorus and ability for nutrient uptake. They form arbuscules, which are the sites of exchange for nutrients such as phosphorus, carbon, and water. The fungi involved in this mycorrhizal association are members of the zygomycota family and appear to be obligate symbionts. In other words, the fungi cannot grow in the absence of their plant host.

Fig.36.1. Wheat is arbuscular mycorrhizal

Ericaceous Mycorrhiza

Ericaceous mycorrhizae is generally found on plants of the order Ericales and in inhospitable, acidic environments. While they do penetrate and invaginate the root cells, ericoid mycorrhiza do not create arbuscules. They do, however, help regulate the plant's acquisition of minerals including iron, manganese, and aluminum. Additionally, mycorrhizal fungi form hyphal coils outside of the root cells, significantly increasing root volume.

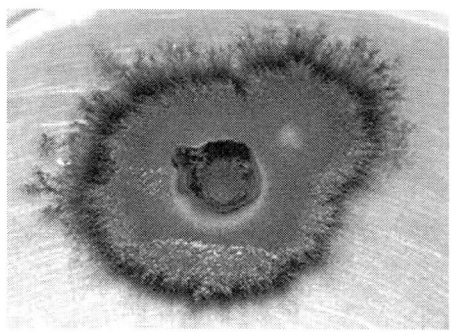

Fig. 36.2. An ericoid mycorrhizal fungus isolated from *Woollsia pungens*

Arbutoid Mycorrhiza

Arbutoid mycorrhiza are a type of endomycorrhizal fungi that look similar to ectomycorrhizal fungi. They form a fungal sheath that encompasses the roots of the plant; however, the hyphae of the arbutoid mycorrhiza penetrate the cortical cells of plant roots, differentiating it from ectomycorrhizal fungi.

Ectotrophic Mycorrhiza

The fungi involved in this mycorrhizal association are from the Ascomyota and Basidiomyota families. They are found in many trees in cooler environments. Unlike their wood-rotting family members, these fungi are not adapted to degrade cellulose and other plant materials; instead, they derive their nutrients and sugars from the roots of their living plant host.

Plant Benefits from Mycorrhizae

Mycorrhiza associations are particularly beneficial in areas where the soil does not contain sufficient nitrogen and phosphorus, as well as in areas where water is not easily accessible. Because the mycorrhizal mycelia are much finer and smaller in diameter than roots and root hairs, they vastly increase the surface area for absorption of water, phosphorus, amino acids, and nitrogen—almost like a second set of roots! As these nutrients are essential for plant growth, plants with mycorrhizal associations have a leg-up on their non-mycorrhizal associated counterparts that rely solely on roots for the uptake of materials. Without mycorrhiza, plants can be out-competed, possibly leading to a change in the plant composition of the area.

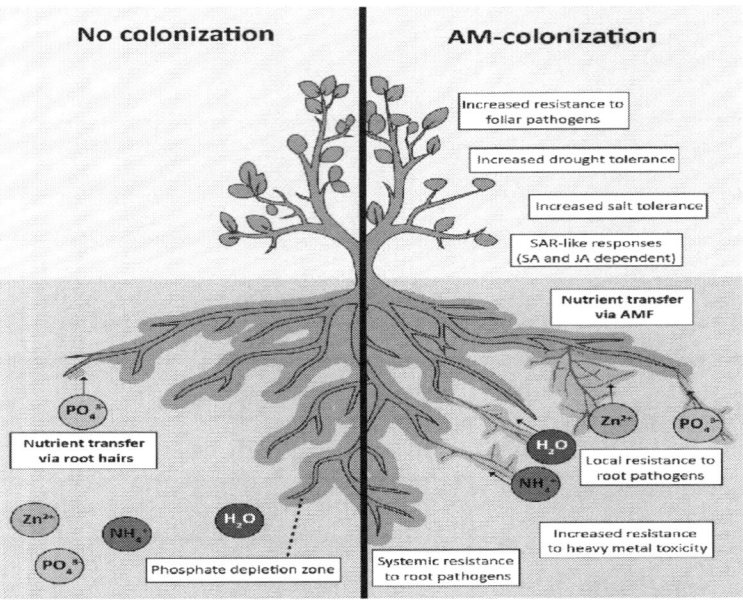

Fig. 36.3. Positive effects of arbuscular mycorrhizal (AM) colonization

Additionally, studies have found that plants with mycorrhizal associations are more resistant to certain soil-borne diseases. In fact, mycorrhizal fungi can be an effective method of disease control. In the case of sheathing mycorrhiza, they create a physical barrier between pathogens and plant roots. Mycorrhiza also thicken the root's cell walls through lignifications and the production of other carbohydrates; compete with pathogens for the uptake of essential nutrients; stimulate plant production of metabolites that increases resistance to disease; stimulate flavonolic wall infusions that prevent lesion formation and invasion by pathogens; and increase plant root concentrations of orthodihydorxy phenol and other allochemicals to deter pathogenic activity. In addition to disease resistance, mycorrhizal fungi can also impart to its host plant resistance to toxicity and resistance to insects, ultimately improving plant fitness and vigor.

In more complex relationships, mycorrhizal fungi can connect individual plants within a mycorrhizal network. This network functions to transport materials such as water, carbon, and other nutrients from plant to plant, and even provides some type of defense communication via chemicals signifying an attack on an individual within the network. Not only can plants use these signals to start producing natural insect repellants, they can also use them to start producing an attractant to bring in natural predators of the plant's pests!

In some cases, mycorrhizal fungi allow plants to bypass the need for soil uptake, such as trees in dystrophic forests. Here, phosphates and other nutrients are taken directly from the leaf litter via mycorrhizal hyphae.

Mycorrhizal fungi are also able to interact with and change the environment in the favor of the host plants—namely, by improving soil structure and quality. The filaments of mycorrhizal fungi create humic compounds, polysaccharides, and glycoproteins that bind soils, increase soil porosity, and promote aeration and water movement into the soil. In environments that have highly compacted or sandy soils, improved soil structure can be more important for plant survival than nutrient uptake.

Some ectomycorrhizal associations create structures that host nitrogen-fixing bacteria, which would largely contribute to the amount of nitrogen taken up by plants in nutrient-poor environments, and would play a large part in the nitrogen cycle. The mycorrhizal fungi, however, do not fix nitrogen themselves.

****

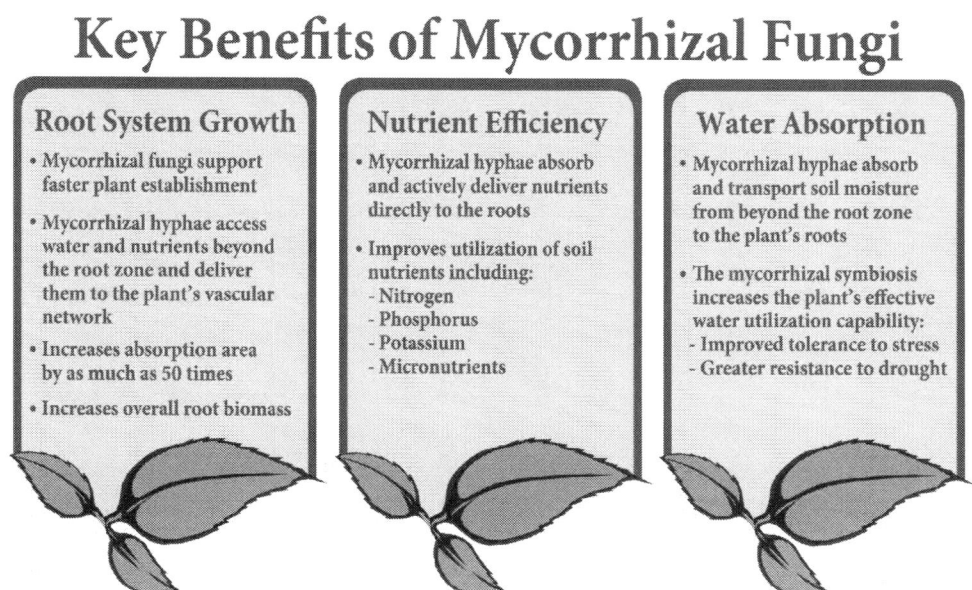

Other Benefits of Mycorrhizae

The symbiotic relationship with mycorrhizal fungi provides many additional benefits to plants and their environments, along with the top-three listed above. These additional benefits include Improved Soil Structure, Greater Transplant Success, Increased Stress Tolerance, Reduced Nutrient Runoff, and many more. Please contact our staff if you have any questions about how mycorrhizal inoculation can benefit your operation, plants/projects, and bottom line.

Fungi Benefits from Plants

When the plant is provided with enough water and nutrients, it is able to photosynthesis and produce glucose and sucrose—some of which is made directly accessible to the mycorrhizal fungi. The fungi are also provided with photosynthetically fixed carbon from the host, which functions as a trigger for nitrogen uptake and transport by the fungi. All of this is necessary for fungal growth and reproduction.

****

37. THE IMMUNOLOGY AND VACCINE

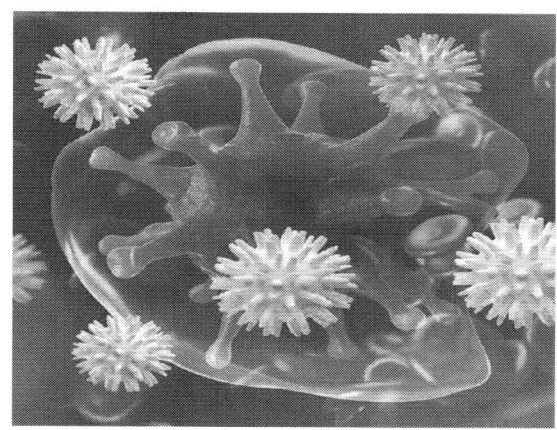

In the process of evolution, the body of organisms has developed the property of distinguishing "self" and "nonself", the "self" is accepted and "nonself" is rejected or degraded. For the rejection or degradation of "nonself", organisms' body has evolved a remarkably versatile defence system called immune system that operates via various mechanisms, which are collectively grouped under the name immunity or resistance (defence). The immunity or resistance (defence) is therefore the insusceptibility of the body to the effect of "nonself" factors like pathogenic microorganisms and their toxins and other kinds of foreign substances. The latin term immunis, meaning "exempt" is the source of the term immunity, the English world. The branch of biology that deals with immunity or resistance is called immunology. The immunity operation by the immune system of organism's body against the "nonself" is called immune response, which can be considered to perform two interrelated functions: recognition and response. Recognition is highly specific and distinguishes one foreign pathogen from the other.

Furthermore, it discriminates between foreign molecules and the body's own cells and molecules. Response, however, is of two types: effector response and memory response. When a foreign pathogen is recognised, the immune system of the body recruits a variety of cells and molecules to neutralize and eliminate the invader.

This response is called effector response. In memory response, the immune system retains a memory of its encounter with an invader so that a second encounter with the same invader sometime later evokes a faster and more intense immune response.

Timeline of immunology:

- 1549 – The earliest account of inoculation of smallpox (variolation) occurs in Wan Quan's (1499–1582) *Douzhen Xinfa*
- 1718 – Smallpox inoculation in Ottoman Empire realized by West. Lady Mary Wortley Montagu, the wife of the British ambassador to Constantinople, observed the positive effects of variolation on the native population and had the technique performed on her own children.
- 1796 – First demonstration of smallpox vaccination (Edward Jenner)
- 1837 – Description of the role of microbes in putrefaction and fermentation (Theodore Schwann)
- 1838 – Confirmation of the role of yeast in fermentation of sugar to alcohol (Charles Cagniard-Latour)
- 1840 – Proposal of the germ theory of disease (Jakob Henle)
- 1850 – Demonstration of the contagious nature of puerperal fever (childbed fever) (Ignaz Semmelweis)
- 1857–1870 – Confirmation of the role of microbes in fermentation (Louis Pasteur)
- 1862 – Phagocytosis (Ernst Haeckel)
- 1867 – Aseptic practice in surgery using carbolic acid (Joseph Lister)
- 1876 – Demonstration that microbes can cause disease-anthrax (Robert Koch)
- 1877 – Mast cells (Paul Ehrlich)
- 1878 – Confirmation and popularization of the germ theory of disease (Louis Pasteur)
- 1880 – 1881 -Theory that bacterial virulence could be attenuated by culture in vitro and used as vaccines. Proposed that live attenuated microbes produced immunity by depleting host of vital trace nutrients. Used to make chicken cholera and anthrax "vaccines" (Louis Pasteur)
- 1883 – 1905 – Cellular theory of immunity via phagocytosis by macrophages and microphages (polymorhonuclear leukocytes) (Elie Metchnikoff)
- 1885 – Introduction of concept of a "therapeutic vaccination". Report of a live "attenuated" vaccine for rabies (Louis Pasteur and Pierre Paul Émile Roux).
- 1888 – Identification of bacterial toxins (diphtheria bacillus) (Pierre Roux and Alexandre Yersin)
- 1888 – Bactericidal action of blood (George Nuttall)

- 1890 – Demonstration of antibody activity against diphtheria and tetanus toxins. Beginning of humoral theory of immunity. (Emil von Behring) and (Kitasato Shibasaburō)
- 1891 – Demonstration of cutaneous (delayed type) hypersensitivity (Robert Koch)
- 1893 – Use of live bacteria and bacterial lysates to treat tumors- "Coley's Toxins" (William B. Coley)
- 1894 – Bacteriolysis (Richard Pfeiffer)
- 1896 – An antibacterial, heat-labile serum component (complement) is described (Jules Bordet)
- 1900 – Antibody formation theory (Paul Ehrlich)
- 1901 – Blood groups (Karl Landsteiner)
- 1902 – Immediate hypersensitivity anaphylaxis (Paul Portier) and (Charles Richet)
- 1903 – Intermediate hypersensitivity, the "Arthus reaction" (Maurice Arthus)
- 1903 – Opsonization[2]
- 1905 – "Serum sickness" allergy (Clemens von Pirquet and (Bela Schick)
- 1909 – Paul Ehrlich proposes "immune surveillance" hypothesis of tumor recognition and eradication
- 1911 – 2nd demonstration of filterable agent that caused tumors (Peyton Rous)
- 1917 – Hapten (Karl Landsteiner)
- 1921 – Cutaneous allergic reactions (Otto Prausnitz and Heinz Küstner)
- 1924 – Reticuloendothelial system
- 1938 – Antigen-Antibody binding hypothesis (John Marrack)
- 1940 – Identification of the Rh antigens (Karl Landsteiner and Alexander Weiner)
- 1942 – Anaphylaxis (Karl Landsteiner and Merill Chase)
- 1942 – Adjuvants (Jules Freund and Katherine McDermott)
- 1944 – hypothesis of allograft rejection
- 1945 – Coombs test a.k.a. antiglobulin test (AGT)
- 1946 – Identification of mouse MHC (H2) by George Snell and Peter A. Gorer
- 1948 – Antibody production in plasma B cells

- 1949 – Growth of polio virus in tissue culture, neutralization with immune sera, and demonstration of attenuation of neurovirulence with repetitive passage (John Enders) and (Thomas Weller) and (Frederick Robbins)
- 1951 – vaccine against yellow fever
- 1953 – Graft-versus-host disease
- 1953 – Validation of immunological tolerance hypothesis
- 1957 – Clonal selection theory (Frank Macfarlane Burnet)
- 1957 – Discovery of interferon by Alick Isaacs and Jean Lindenmann[3]
- 1958–1962 – Discovery of human leukocyte antigens (Jean Dausset and others)
- 1959–1962 – Discovery of antibody structure (independently elucidated by Gerald Edelman and Rodney Porter)
- 1959 – Discovery of lymphocyte circulation (James Gowans)
- 1960 – Discovery of lymphocyte "blastogenic transformation" and proliferation in response to mitogenic lectins-phytohemagglutinin (PHA) (Peter Nowell)
- 1961–1962 Discovery of thymus involvement in cellular immunity (Jacques Miller)
- 1960 – Radioimmunoassay – (Rosalyn Sussman Yalow)
- 1961 – Demonstration that glucocorticoids inhibit PHA-induced lymphocyte proliferation (Peter Nowell)
- 1963 – Development of the plaque assay for the enumeration of antibody-forming cells in vitro by Niels Jerne and Albert Nordin
- 1963 – Gell and Coombs classification of hypersensitivity
- 1964–1968 – T and B cell cooperation in immune response
- 1965 – Discovery of lymphocyte mitogenic activity, "blastogenic factor" (Shinpei Kamakura) and (Louis Lowenstein) (J. Gordon) and (L.D. MacLean)
- 1965 – Discovery of "immune interferon" (gamma interferon) (E.F. Wheelock)
- 1965 – Secretory immunoglobulins
- 1967 – Identification of IgE as the reaginic antibody (Kimishige Ishizaka)
- 1968 – Passenger leukocytes identified as significant immunogens in allograft rejection (William L. Elkins and Ronald D. Guttmann)
- 1969 – The lymphocyte cytolysis Cr51 release assay (Theodore Brunner) and (Jean-Charles Cerottini)

- 1971 – Peter Perlmann and Eva Engvall at Stockholm University invented ELISA
- 1972 – Structure of the antibody molecule
- 1973 – Dendritic Cells first described by Ralph M. Steinman
- 1974 – Immune Network Hypothesis (Niels Jerne)
- 1974 – T-cell restriction to MHC (Rolf Zinkernagel and (Peter C. Doherty)
- 1975 – Generation of monoclonal antibodies (Georges Köhler) and (César Milstein)[4]
- 1975 – Discovery of Natural Killer cells (Rolf Kiessling, Eva Klein, Hans Wigzell)
- 1976 – Identification of somatic recombination of immunoglobulin genes (Susumu Tonegawa)
- 1980–1983 – Discovery and characterization of interleukins, 1 and 2 IL-1 IL-2 (Robert Gallo, Kendall A. Smith, Tadatsugu Taniguchi)
- 1983 – Discovery of the T cell antigen receptor TCR (Ellis Reinherz) (Philippa Marrack) and (John Kappler)[5] (James Allison)
- 1983 – Discovery of HIV (Luc Montagnier) (Françoise Barré-Sinoussi) (Robert Gallo)
- 1985–1987 – Identification of genes for the T cell receptor
- 1986 – Hepatitis B vaccine produced by genetic engineering
- 1986 – Th1 vs Th2 model of T helper cell function (Timothy Mosmann)
- 1988 – Discovery of biochemical initiators of T-cell activation: CD4- and CD8-p56lck complexes (Christopher E. Rudd)
- 1990 – Gene therapy for SCID
- 1991 – Role of peptide for MHC Class II structure (Scheherazade Sadegh-Nasseri & Ronald N. Germain)(http://doi.org/10.1038/353167a0)
- 1992 – Discovery of transitional B cells (David Allman & Michael Cancro)[6][7]
- 1994 – 'Danger' model of immunological tolerance (Polly Matzinger)
- 1995 – James P. Allison describes the function of CTLA-4
- 1995 – Regulatory T cells (Shimon Sakaguchi)
- 1995 – First Dendritic cell vaccine trial reported by Mukherji et al.
- 1996 – 1998 – Identification of Toll-like receptors
- 1997 – Discovery of the autoimmune regulator and the AIRE gene.
- 2000 – Characterization of M1 and M2 macrophage subsets by Charles Mills[8]

- 2001 – Discovery of FOXP3 – the gene directing regulatory T cell development
- 2005 – Development of human papillomavirus vaccine (Ian Frazer)
- 2006 – Antigen-specific NK cell memory first reported by Ulrich von Andrian's group after discovery by Mahmoud Goodarzi
- 2008 - Françoise Barré-Sinoussi and Luc Montagnier win the Nobel Prize in Physiology or Medicine for their discovery of human immunodeficiency virus
- 2010 – The first autologous cell-based cancer vaccine, PROVENGE, is approved by the FDA for the treatment of metastatic, asymptomatic stage IV prostate cancer. The treatment is marketed at a cost of $93,000 and imparts, on average, only an extra four months of life expectancy. The manufacturer, Dendreon Inc, declares bankruptcy in 2014.
- 2010 – First immune checkpoint inhibitor, ipilimumab (anti-CTLA-4), is approved by the FDA for treatment of stage IV melanoma
- 2011 – Carl H. June reports first successful use of CAR T-cells expressing the 4-1BB costimulatory signaling domain for the treatment of CD19+ malignancies
- 2014 – A second class of immune checkpoint inhibitor (anti-PD-1) is approved by the FDA for the treatment of melanoma. Two different drugs, pembrolizumab and nivolumab are approved within months of each other.
- 2016 – Halpert and Konduri first characterize the role of dendritic cell CTLA-4 in Th-1 immunity
- 2016 – A third class of immune checkpoint inhibitor, anti-PD-L1 (atezolizumab), is approved for the treatment of bladder cancer
- 2017 – The first autologous CAR T-cell therapy tisagenlecleucel also known as Kymriah is approved for the treatment of pediatric B-ALL. Marketed at a cost of $475,000, the treatment provides an 83% rate of durable remission among poor prognosis patients for whom a bad outcome would otherwise be expected.
- 2017 – The Indian government approves Apceden, the first potentially curative dendritic cell vaccine for the treatment of prostate, ovarian, colon, and lung cancers. It is developed by an Indian Biotechnology company, APAC Biotech.
- 2018 - James_P._Allison and Tasuku_Honjo won the Nobel Prize in Physiology or Medicine for their discovery of cancer therapy by inhibition of negative immune regulation

Two fundamentally different types of immunity

(i) Innate (nonspecific or natural) immunity and

(ii) Acquired (specific or adaptive) immunity.

The innate immunity includes general mechanisms inherited as part of the innate structure and function of each vertebrate, and acts as a first line of defence.

This immunity lacks immunological memory, i.e., it operates to the same extent each time a microbial pathogen or foreign substance is encountered by the vertebrate host. Contrary to it, the acquired immunity is a more highly evolved system of specific responses that recognize and selectively eliminate specific microbial pathogens and foreign substances.

Moreover, the acquired immunity retains memory of its first encounter with foreign invaders and improves on repeated exposure to them. Substances that are recognised by the immune system as 'foreign' and provoke immune responses are referred to as antigens.

The antigens cause specific cells to generate special type of proteins called antibodies (immunoglobulins), which bind to and inactivate the antigen. In addition to the specific cells that generate antibodies, there are other cells of immune system that destroy the foreign invaders.

The above mentioned two immunities (innate and acquired) usually work together to provide a high degree of protection for vertebrates. In some circumstances, however, the immune system of the body fails in providing protection because of some deficiency in its components.

At other times, it becomes an aggressor and turns its awesome powers against its own host resulting in immune disorders (malfunctions). Hypersensitivities (allergies), autoimmune diseases, immunodeficiencies, and transplantation (tissue) rejection are such immune disorders.

Antigen is a substances usually protein in nature and sometimes polysaccharide, that generates a specific immune response and induces the formation of a specific antibody or specially sensitized T cells or both.

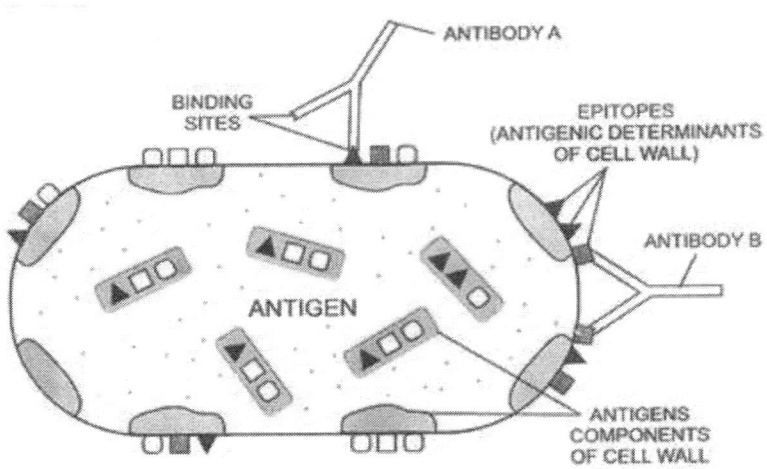

Diagram showing an antigen with epitopes (antigenic determinants). Two attached antibodies are also shown.

Fig.37.

Although all antigens are recognized by specific lymphocytes or by antibodies, only some antigens are capable of activating lymphocytes. Molecules that stimulate immune responses are called **Immunogens.**

Epitope is immunologically active regions of an immunogen (or antigen) that binds to antigen-specific membrane receptors on lymphocytes or to secreted antibodies. It is also called **antigenic determinants.**

Autoantigens, for example, are a person's own self antigens. Examples: Thyroglobulin, DNA, Corneal tissue, etc.

Alloantigens are antigens found in different members of the same species (the red blood cell antigens A and B are examples).

Heterophile antigens are identical antigens found in the cells of different species. Examples: Forrssman antigen, Cross-reacting microbial antigens, etc.

Adjuvants are substances that are non-immunogenic alone but enhance the immunogenicity of any added immunogen.

Chemical Nature of Antigens (Immunogens)

A. Proteins
The vast majority of immunogens are proteins. These may be pure proteins or they may be glycoproteins or lipoproteins. In general, proteins are usually very good immunogens.

B. Polysaccharides
Pure polysaccharides and lipopolysaccharides are good immunogens.

C. Nucleic Acids
Nucleic acids are usually poorly immunogenic. However, they may become immunogenic when single stranded or when complexed with proteins.

D. Lipids
In general lipids are non-immunogenic, although they may be haptens.

Types of Antigen On the basis of order of their class (Origin)

1. Exogenous antigens

- These antigens enters the body or system and start circulating in the body fluids and trapped by the APCs (Antigen processing cells such as macrophages, dendritic cells, etc.)
- The uptakes of these exogenous antigens by APCs are mainly mediated by the phagocytosis
- Examples: bacteria, viruses, fungi etc
- Some antigens start out as exogenontigens, and later become endogenous (for example, intracellular viruses)

2. Endogenous antigens

- These are body's own cells or sub fragments or compounds or the antigenic products that are produced.
- The endogenous antigens are processed by the macrophages which are later accepted by the cytotoxic T – cells.
- Endogenous antigens include xenogenic (heterologous), autologous and idiotypic or allogenic (homologous) antigens.
- Examples: Blood group antigens, HLA (Histocompatibility Leukocyte antigens), etc.

3. Autoantigens

- An autoantigen is usually a normal protein or complex of proteins (and sometimes DNA or RNA) that is recognized by the immune system of patients suffering from a specific autoimmune disease
- These antigens should not be, under normal conditions, the target of the immune system, but, due mainly to genetic and environmental factors, the normal immunological tolerance for such an antigen has been lost in these patients.
- Examples: Nucleoproteins, Nucleic acids, etc.

On the basis of immune response

1. Complete Antigen or Immunogen

- Posses antigenic properties denovo, i.e. ther are able to generate an immune response by themselves.
- High molecular weight (more than 10,000)
- May be proteins or polysaccharides

2. Incomplete Antigen or Hapten

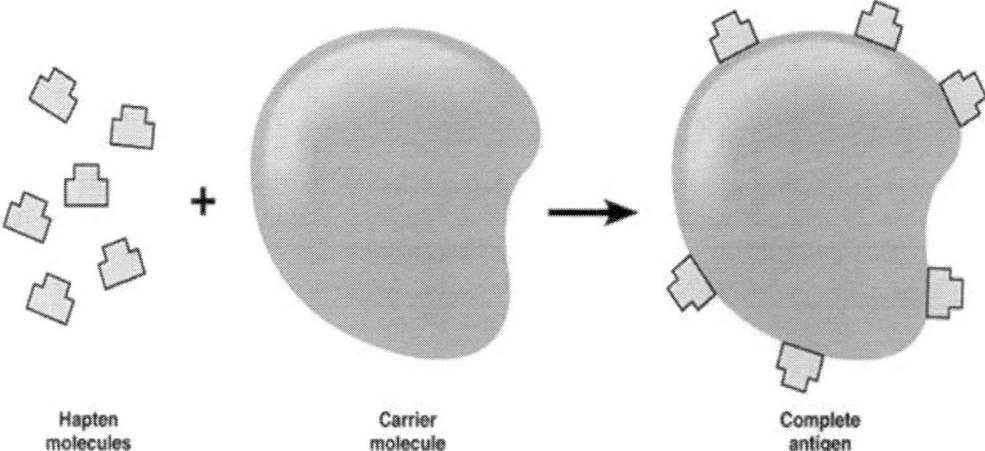

Fig. 37.1. Complete Antigen or Immunogen

- These are the foreign substance, usually non-protein substances

- Unable to induce an immune response by itself, they require carrier molecule to act as a complete antigen.
- The carrier molecule is a non-antigenic component and helps in provoking the immune response. Example: Serum Protein such as Albumin or Globulin.
- Low Molecular Weight (Less than 10,000)
- Haptens can react specifically with its corresponding antibody.
- Examples: Capsular polysaccharide of pneumococcus, polysaccharide "C" of beta haemolytic streptococci, cardiolipin antigens, etc.

Antibody, also called **immunoglobulin**, a protective protein produced by the immune system in response to the presence of a foreign substance, called an antigen. Antibodies recognize and latch onto antigens in order to remove them from the body. A wide range of substances are regarded by the body as antigens, including disease-causing organisms and toxic materials such as insect venom. When an alien substance enters the body, the immune system is able to recognize it as foreign because molecules on the surface of the antigen differ from those found in the body. To eliminate the invader, the immune system calls on a number of mechanisms, including one of the most important—antibody production. Antibodies are produced by specialized white blood cells called B lymphocytes (or B cells). When an antigen binds to the B-cell surface, it stimulates the B cell to divide and mature into a group of identical cells called a clone. The mature B cells, called plasma cells, secrete millions of antibodies into the bloodstream and lymphatic system. As antibodies circulate, they attack and neutralize antigens that are identical to the one that triggered the immune response. Antibodies attack antigens by binding to them. The binding of an antibody to a toxin, for example, can neutralize the poison simply by changing its chemical composition; such antibodies are called antitoxins. By attaching themselves to some invading microbes, other antibodies can render such microorganisms immobile or prevent them from penetrating body cells. In other cases the antibody-coated antigen is subject to a chemical chain reaction with complement, which is a series of proteins found in the blood. The complement reaction either can trigger the lysis (bursting) of the invading microbe or can attract microbe-killing scavenger cells that ingest, or phagocytose, the invader. Once begun, antibody production continues for several days until all antigen molecules are removed. Antibodies remain in circulation for several months, providing extended immunity against that particular antigen.

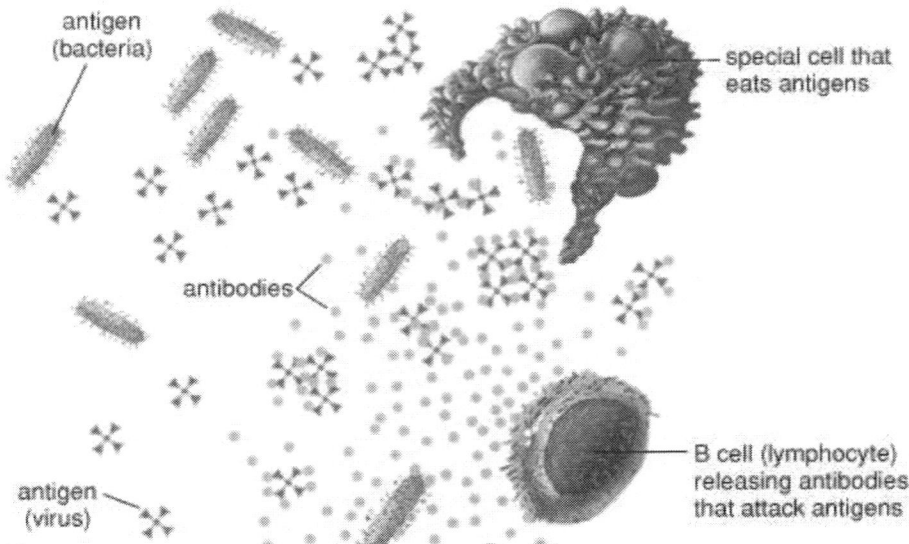

Fig. 37.2. Antigen; antibody; lymphocyte Phagocytic cells destroy viral and bacterial antigens by eating them, while B cells produce antibodies that bind to and inactivate antigens.*Encyclopædia Britannica, Inc.*

B cells and antibodies together provide one of the most important functions of immunity, which is to recognize an invading antigen and to produce a tremendous number of protective proteins that scour the body to remove all traces of that antigen. Collectively B cells recognize an almost limitless number of antigens; however, individually each B cell can bind to only one type of antigen. B cells distinguish antigens through proteins, called antigen receptors, found on their surfaces. An antigen receptor is basically an antibody protein that is not secreted but is anchored to the B-cell membrane. All antigen receptors found on a particular B cell are identical, but receptors located on other B cells differ. Although their general structure is similar, the variation lies in the area that interacts with the antigen—the antigen-binding, or antibody-combining, site. This structural variation among antigen-binding sites allows different B cells to recognize different antigens. The antigen receptor does not actually recognize the entire antigen; instead it binds to only a portion of the antigen's surface, an area called the antigenic determinant or epitope. Binding between the receptor and epitope occurs only if their structures are complementary. If they are, epitope and receptor fit together like two pieces of a puzzle, an event that is necessary to activate B-cell production of antibodies.

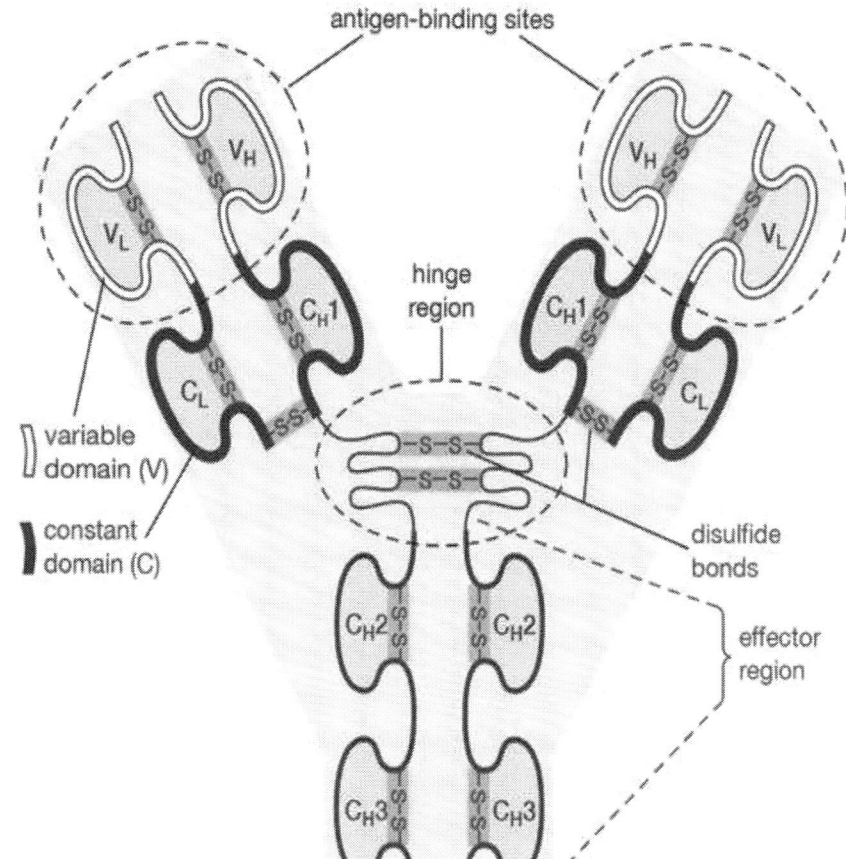

Fig.37.3. riable and constant domains of an antibody Variable (V) and constant (C) domains within the light (L) and heavy (H) chains of an antibody, or immunoglobulin, molecule. The folded shapes of the domains are maintained by disulfide bonds (—S—S—).*Encyclopædia Britannica, Inc.*

Isotypes

Antibodies can come in different varieties known as isotypes or classes. In placental mammals there are five antibody isotypes known as IgA, IgD, IgE, IgG, and IgM. They are each named with an "Ig" prefix that stands for immunoglobulin (a name sometimes used interchangeably with antibody) and differ in their biological properties, functional locations and ability to deal with different antigens, as depicted in the table. The different suffixes of the antibody isotypes denote the different types of heavy chains the antibody contains, with each heavy chain class named alphabetically: α (alpha), γ (gamma), δ (delta), ε (epsilon), and μ (mu). This gives rise to IgA, IgG, IgD, IgE, and IgM, respectively.

Antibody isotypes of mammals		
Class	**Subclasses**	**Description**
IgA	2	Found in mucosal areas, such as the gut, respiratory tract and urogenital tract, and prevents colonization by pathogens. Also found in saliva, tears, and breast milk.
IgD	1	Functions mainly as an antigen receptor on B cells that have not been exposed to antigens. It has been shown to activate basophils and mast cells to produce antimicrobial factors.
IgE	1	Binds to allergens and triggers histamine release from mast cells and basophils, and is involved in allergy. Also protects against parasitic worms
IgG	4	In its four forms, provides the majority of antibody-based immunity against invading pathogens. The only antibody capable of crossing the placenta to give passive immunity to the fetus.
IgM	1	Expressed on the surface of B cells (monomer) and in a secreted form (pentamer) with very high avidity. Eliminates pathogens in the early stages of B cell-mediated (humoral) immunity before there is sufficient IgG.

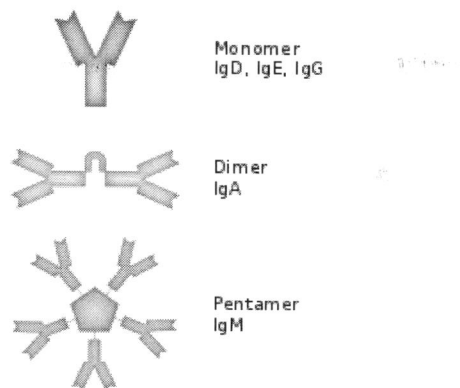

Fig.37.4 Antibody complexes

The antibody isotype of a B cell changes during cell development and activation. Immature B cells, which have never been exposed to an antigen, express only the IgM isotype in a cell surface bound form. The B lymphocyte, in this ready-to-respond form, is known as a "naive B lymphocyte." The naive B lymphocyte expresses both surface IgM and IgD. The co-expression of both of these immunoglobulin isotypes renders the B cell ready to respond to antigen. B cell activation follows engagement of the cell-bound antibody molecule with an antigen, causing the cell to divide and differentiate into an antibody-producing cell called a plasma cell. In this activated form, the B cell starts to produce antibody in a secreted form rather than a membrane-bound form. Some daughter cells of the activated B cells undergo isotype switching, a mechanism that causes the production of antibodies to change from IgM or IgD to the other antibody isotypes, IgE, IgA, or IgG, that have defined roles in the immune system.

Antibody isotypes not found in mammals		
Class	Types	Description
IgY		Found in birds and reptiles; related to mammalian IgG
IgW		Found in sharks and skates; related to mammalian IgD.

VACCINE

A **vaccine** is a biological preparation that provides active acquired immunity to a particular infectious disease. A vaccine typically contains an agent that resembles a disease-causing microorganism and is often made from weakened or killed forms of the microbe, its toxins, or one of its surface proteins. The agent stimulates the body's immune system to recognize the agent as a threat, destroy it, and to further recognize and destroy any of the microorganisms associated with that agent that it may encounter in the future. Vaccines can be prophylactic (to prevent or ameliorate the effects of a future infection by a natural or "wild" pathogen), or therapeutic (to fight a disease that has already occurred, such as cancer).

The administration of vaccines is called vaccination. Vaccination is the most effective method of preventing infectious diseases; widespread immunity due to vaccination is largely responsible for the worldwide eradication of smallpox and the restriction of diseases such as polio, measles, and tetanus from much of the world. The effectiveness of vaccination has been widely studied and verified; for example, vaccines that have proven effective include the influenza vaccine, the HPV vaccine, and the chicken pox vaccine. The World Health Organization (WHO) reports that licensed vaccines are currently available for twenty-five different preventable infections.

The terms *vaccine* and *vaccination* are derived from *Variolae vaccinae* (smallpox of the cow), the term devised by Edward Jenner to denote cowpox. He used it in 1798 in the long title of his *Inquiry into the Variolae vaccinae Known as the Cow Pox*, in which he described the protective effect of cowpox against smallpox. In 1881, to honor Jenner, Louis Pasteur proposed that the terms should be extended to cover the new protective inoculations then being developed

Vaccine Types

There are several different types of vaccines. Each type is designed to teach your immune system how to fight off certain kinds of germs — and the serious diseases they cause.

When scientists create vaccines, they consider:

- How your immune system responds to the germ
- Who needs to be vaccinated against the germ
- The best technology or approach to create the vaccine

Based on a number of these factors, scientists decide which type of vaccine they will make. There are 4 main types of vaccines:

- Live-attenuated vaccines
- Inactivated vaccines
- Subunit, recombinant, polysaccharide, and conjugate vaccines
- Toxoid vaccines

Live-attenuated vaccines

Live vaccines use a weakened (or attenuated) form of the germ that causes a disease.

Because these vaccines are so similar to the natural infection that they help prevent, they create a strong and long-lasting immune response. Just 1 or 2 doses of most live vaccines can give you a lifetime of protection against a germ and the disease it causes.

But live vaccines also have some limitations. For example:

- Because they contain a small amount of the weakened live virus, some people should talk to their health care provider before receiving them, such as people with weakened immune systems, long-term health problems, or people who've had an organ transplant.
- They need to be kept cool, so they don't travel well. That means they can't be used in countries with limited access to refrigerators.

Live vaccines are used to protect against:

- Measles, mumps, rubella (MMR combined vaccine)
- Rotavirus
- Smallpox
- Chickenpox
- Yellow fever

Inactivated vaccines

Inactivated vaccines use the killed version of the germ that causes a disease.

Inactivated vaccines usually don't provide immunity (protection) that's as strong as live vaccines. So you may need several doses over time (booster shots) in order to get ongoing immunity against diseases.

Inactivated vaccines are used to protect against:

- Hepatitis A
- Flu (shot only)
- Polio (shot only)
- Rabies

Subunit, recombinant, polysaccharide, and conjugate vaccines

Subunit, recombinant, polysaccharide, and conjugate vaccines use specific pieces of the germ — like its protein, sugar, or capsid (a casing around the germ).

Because these vaccines use only specific pieces of the germ, they give a very strong immune response that's targeted to key parts of the germ. They can also be used on almost everyone who needs them, including people with weakened immune systems and long-term health problems.

One limitation of these vaccines is that you may need booster shots to get ongoing protection against diseases.

These vaccines are used to protect against:

- Hib (*Haemophilus influenzae* type b) disease
- Hepatitis B
- HPV (Human papillomavirus)
- Whooping cough (part of the DTaP combined vaccine)
- Pneumococcal disease
- Meningococcal disease
- Shingles

Toxoid vaccines

Toxoid vaccines use a toxin (harmful product) made by the germ that causes a disease. They create immunity to the parts of the germ that cause a disease instead of the germ itself. That means the immune response is targeted to the toxin instead of the whole germ.

Like some other types of vaccines, you may need booster shots to get ongoing protection against diseases.

Toxoid vaccines are used to protect against:

- Diphtheria
- Tetanus

The future of vaccines

Did you know that scientists are still working to create new types of vaccines? Here are 2 exciting examples:

- **DNA vaccines** are easy and inexpensive to make — and they produce strong, long-term immunity.
- **Recombinant vector vaccines (platform-based vaccines)** act like a natural infection, so they're especially good at teaching the immune system how to fight germs.

****

38.

MICROBIOLOGY OF AIR

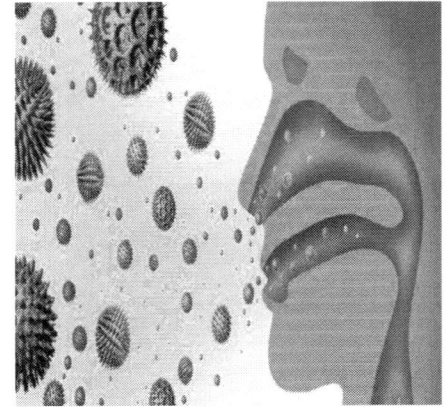

The earth's atmosphere is teeming with airborne microorganisms. These organisms are thought to exhibit correlations with air pollution and weather. Most airborne bacteria originate from natural sources such as the soil, lakes, oceans, animals, and humans. Many 'unnatural' origins are also known, such as sewage treatment, animal rendering, fermentation processes, and agricultural activities which disturb the soil. Viable airborne microorganisms are not air pollutants, but should be considered as a factor affecting air quality. Air is an unfavourable environment for microorganisms, in which they cannot grow or divide. It is merely a place which they temporarily occupy and use for movement.

There are 3 elementary limiting factors in the air

- A lack of adequate nutrients
- Frequent deficit of water (desiccation)
- Solar radiation

The atmosphere can be occupied for the longest time by those forms which, due to their chemical composition or structure, are resistant to desiccation and solar radiation. They can be subdivided into the following groups:

- Bacterial resting forms,
- Bacterial vegetative forms which produce carotenoidal dyes or special protective layers (capsules, special structure of cell wall),
- Spores of fungi,
- Viruses with envelopes

Resting forms of Bacteria

Endospores are the best known resting forms. These structures evolve within cells and are covered by a thick multi-layer casing. Consequently, endospores are unusually resistant to most unfavourable environment conditions and are able to survive virtually endlessly in the conditions provided by the atmospheric air. They are only produced by some bacteria, mainly by Bacillus and Clostridium genera. Because each cell produces only one endospore, these spore forms cannot be used for reproduction.

Another type of resting form is produced by very common soil bacteria, the actinomycetes. Their special vertical, filiform cells, of the so-called air mycelium, undergo fragmentation producing numerous ball-shaped formations. Due to the fact that their production is similar to the formation of fungal, they are also called conidia. Contrary to endospores, the conidia are used for reproduction. There are also other bacterial resting forms, among others, the cysts produced by azotobacters - soil bacteria capable of molecular nitrogen assimilation.

Resistant Vegetative Cells of Bacteria

The production of carotenoidal dyes ensures cells with solar radiation protection. Carotenoids, due to the presence of numerous double bonds within a molecule (-C=C-), serve a purpose as antioxidants, because, as strong reducing agents, they are oxidizedby free radicals. Consequently, important biological macromolecules are being protected against oxidation (DNA, proteins etc.). Bacteria devoid of these dyes quickly perish due to the photodynamic effect of photooxidation. That explains why the colonies of bacteria, which settle upon open agar plates, are often colored. The ability to produce carotenoids is possessed especially by cocci and rod-shaped actinomycetes. Rod-shaped actinomycetes, e.g. *Mycobacterium tuberculosis*, besides being resistant to light, also demonstrate significant resistance to drying due to a high content of lipids within their cell wall. High survival rates in air are also a characteristic for the bacteria which possess a capsule, e.g. *Klebsiella* genus, that cause respiratory system illnesses.

Fungal Spores

Spores are special reproductive cells used for asexual reproduction. Fungi produce spores in astronomical quantities, for example the giant puffball (*Calvatia gigantea*) produces 20 billion spores, which get into the air and are dispersed over vast areas. A very common type of spores found in air is that of conidia.

Conidia are a type of spore formed by asexual reproduction. They form in the end-sections of vertical hyphae called conidiophores and are dispersed by wind. The spores of common mould fungi such as *Penicillium* and *Aspergillus* are examples of the above. Spore plants such as ferns, horsetails and lycopods also produce spores. Plant pollen is also a kind of spores.

Resistant Viruses

Besides cells, the air is also occupied by viruses. Among those that demonstrate the highest resistance are those with enveloped nucleocapsids, such as influenza viruses.

Among viruses without enveloped nucleocapsids, enteroviruses demonstrate a relatively high resistance. Of course, besides the previously mentioned resistant forms, the air is also occupied by more sensitive cells and viruses, but their survival is much shorter. It is believed, that among vegetative forms, gram-positive bacteria demonstrate greater resistance than Gram negative bacteria (especially for desiccation), mainly due to the thickness of their cell wall. Viruses are usually more resistant than bacteria.

Factors Affecting Growth of Microorganism in Air

There are several factors which influence the ability of a bioaerosol to survive in air:

- Particular resistance for a given microorganism (morphological characteristics)
- Meteorological conditions (inter alia, air humidity, solar radiation),
- Air pollution,
- The length of time in air.

Resistance of microorganisms

It is a species dependent feature, which relies on the microorganism's morphology and physiology.

Relative humidity

The content of water in air is one of the major factors determining the ability to survive. At a very low humidity and high temperature cells face dehydration, whereas high humidity may give cells protection against the solar radiation. Microorganisms react differently to humidity variations in air, but nevertheless most of them prefer high humidity. The morphology and biochemistry of cell-surrounding structures, which may change its conformation depending on the amount of water in air, are crucial. Actually, an exact mechanism of this is not known. Forms of resting spores with thick envelopes (e.g. bacterial endospores) are not particularly susceptible to humidity variations. Gram-negative bacteria and enveloped viruses (e.g. influenza virus, myxo) deal better with low air humidity which is contrary to gram-positive bacteria and non-enveloped viruses (e.g. enteroviruses) that have higher survival rates in high air humidity.

Temperature

Temperature can indirectly affect cells by changing the relative-air humidity (the higher the temperature, the lower the relative humidity) or a direct affect, causing, in some extreme situations, cell dehydration and protein denaturation (high temperatures) or crystallization of water contained within cells (temperatures below 0°C). Therefore, it can be concluded that low temperatures (but above 0°C) are optimal for the bioaerosol. According to some researchers the optimal temperatures are above 15°C.

Solar radiation

Solar radiation has a negative effect on the survival rate of the bioaerosol, both visible as well as ultraviolet (UV) and infrared radiation due to the following factors:

- Causes mutation,
- Leads to the formation of free radicals, which damage important macromolecules.
- Creates a danger of dehydration.

Visible light rays of about 400-700 nm wavelength, create the so-called photodynamic effect, which produces free radicals within cells, especially compounds such as peroxy and hydroxyl radicals. These radicals demonstrate strong oxidizing activities and may cause damage to crucial macromolecules, e.g. DNA or proteins.

UV radiation has a much larger affect on cells than visible light does, especially the rays of 230-275 nm wavelengths. The mechanism of this effect is based on changes to DNA, both directly (e.g. by creating thymine dimer and consequently causing mutation), as well as indirectly, by creating free radicals as in the case of the visible light.

In addition, infrared (IR) radiation may have a negative effect upon cells contained in air - heating up and consequently dehydration.

Biological aerosols

Microorganisms in air occur in a form of colloidal system or the so-called bioaerosol. Every colloid is a system where, inside its dispersion medium, particles of dispersed phase occur whose size is halfway between molecules and particles visible with the naked eye. In the case of biological aerosols, it's the air (or other gases) that has the function of the dispersion medium, whereas microorganisms are its dispersed phase. However, it is quite rare to have microbes independently occurring in air. Usually, they are bound with dust particles or liquid droplets (water, saliva etc.), thus the particles of the bioaerosol often exceed microorganisms in size and may occur in two phases:

- Dust phase (e.g. bacterial dust) or
- Droplet phase (e.g. formed as the result of water-vapour condensation or uring sneezing).

The dust particles are usually larger than the droplets and they settle faster. The difference in their ability to penetrate the respiratory tract is dependent on the size of the particles; particles of the droplet phase can reach the alveoli, but dust particles are usually retained in the upper respiratory tract. The number of microorganisms associated with one dust particle is greater than in the droplet phase. The average size of bioaerosols ranges from about 0.02 μm to 100 μm. The sizes of certain particles may change under the influence of outside factors (mainly humidity and temperature) or as a result of larger aggregates forming. By using size criterion, the biological aerosol can be subdivided into the following:

- Fine particles (less than 1μm) and
- Coarse particles (more than 1μm)

Fine particles are mainly viruses, endospores and cell fragments. They possess hygroscopic properties and make-up the so-called nucleus of condensation of water vapour. At high humidity water collects around these particles creating a droplet phase.

Then, the diameter of the particles increases. Coarse particles consist mainly of bacteria and fungi, usually associated with dust particles or with water droplets.

Biological aerosols as a human hazard source.

- What types of dangers are connected to the presence of microorganisms in air?
- Infectious diseases (viral, bacterial, fungal and protozoan),
- Allergic diseases,
- Poisoning (exotoxins, endotoxins, mycotoxins).

Bioaerosols may carry microorganisms other than those which evoke respiratory system diseases. The intestinal microorganisms contained in aerosols may, after settling down, get into the digestive system (e.g. by hands) causing various intestinal illnesses.

Infectious Airborne Diseases

The mucous membrane of the respiratory system is a specific type of a 'gateway' for most airborne pathogenic microorganisms. Susceptibility to infections is increased by dust and gaseous air-pollution, e.g. SO_2 reacts with water that is present in the respiratory system, creating H_2SO_4, which irritates the layer of mucous. Consequently, in areas of heavy air pollution, especially during smog, there is an increased rate of respiratory diseases. Bioaerosols may, among other things, carry microbes that penetrate organs via the respiratory system. After settling, microbes from the air may find their way onto the skin or, carried by hands, get into the digestive system (from there, carried by blood, to other systems, e.g. the nervous system). Fungi that cause skin infections, intestinal bacteria that cause digestive system diseases or nervous system attacking enteroviruses are all examples of the above.

Viral diseases

After penetrating the respiratory system with inhaled air, particles of viruses reproduce inside the cuticle cells of both the upper and lower respiratory system. After reproduction some of the viruses stay inside the respiratory system causing various ailments (runny nose, colds, bronchitis, pneumonia), whereas others leave the respiratory system to attack other organs (e.g. chickenpox viruses attack the skin). The most noteworthy viruses are:

Influenza (orthomyxoviruses) Influenza, measles, bronchitis, mumps and pneumonia among newborns (paramyxoviruses)

- German measles (similar to paramyxoviruses)
- Colds (rhinoviruses and koronaviruses)
- Cowpox and true pox (pox type viruses)
- Chickenpox (cold sore group of viruses)
- Foot-and-mouth disease (picorna type viruses)
- Meningitis, pleurodynia (enteroviruses)
- Sore throat, pneumonia (adenoviruses)

Bacterial diseases

Similarly to viruses, some bacteria that find their way to the respiratory system may also cause ailments of other systems. Especially staphylococcus infections assume various clinical forms (bone marrow inflammation, skin necrosis, intestinal inflammation, pneumonia). Often, a susceptible base for development of various bacterial diseases is first prepared by viral diseases, e.g. staphylococcus pneumonia is usually preceded by a flu or mumps. Bacterial airborne diseases include:

- Tuberculosis (*Mycobacterium tuberculosis*),
- Pneumonia (*Staphylococcus, Pneumococci, Streptococcus pneumoniae*, less frequently chromatobars of *Klebsiella pneumoniae*),
- Angina, scarlet fever, laryngitis (*Streptococcus*),
- Inflammation of upper and lower respiratory system and meningitis (*Haemophilus influenzae*),
- Whooping cough (chromatobars of *Bordetella pertussis*),
- Diphtheria (*Corynebacterium diphtheriae*),
- Legionnaires disease (chromatobars of *Legionella* genus, among others *L. pneumophila*),
- Nocardiosis (oxygen actinomycetes of *Nocardia* genus).

Fungal diseases

Many potentially pathogenic airborne fungi or the so-called saprophytes live in soil. They usually have an ability to break down keratin (keratinolysis) - difficult to decompose proteins found in horny skin formations, e.g. human or animal hair, feathers, claws. Some of the keratinolytic fungi, the so-called dermatophytes, cause mycosis of the outer skin (dermatosis), as the breakdown of keratin enables them to penetrate the epidermis. Other fungi, after penetrating the respiratory system, cause deep mycosis (organ), e.g. attacking lungs. The following are examples of airborne fungi diseases:

- Mycosis (*Microsporum racemosum*),
- Deep mycosis: aspergillosis (*Aspergillus fumigatus*), cryptococcus (*Cryptococcus neoformans*).

Protozoan diseases

Some protozoa, which are able to produce cysts that are resistant to dehydration and solar radiation, may also infect humans by inhalation. The most common example of the above is *Pneumocystis carinii* which causes pneumonia. Dangers connected with pathogenic bioaerosols do not concern only human diseases. Other significant diseases are those that attack cultivated plants or farm animals. The following are examples of the above:

- Blight - grain disease caused by *Puccinia graminis*, and
- Aphthous fever - very infectious disease that attacks artiodactylous animals.

Basic Sources of Bioaerosol Emission

There are two basic sources of bioaerosol:

Natural sources: These are mainly soil and water, from which microorganisms are being lifted up by the movement of air, and from organisms such as fungi, that produce gigantic amounts of spores that are dispersed by the wind. Therefore, there are always a given number of microorganisms in the air, as a natural background. It is estimated, that the air is considered to be clean, if the concentration of bacteria and fungi cells does not exceed $1000/m^3$ and $3000/m^3$ respectively. This latter statement is only true when the concentration of microorganisms consists of saprophytic organisms, not pathogenic organisms. If the concentration of microorganisms in the air exceeds the above values, or contains microorganisms dangerous to humans, then such air is considered to be microbiologically polluted.

Human activities: From the hygienic point of view, living sources of bioaerosols related to human activity, are more important than the natural sources. The emissions from these sources are dangerous due to the following two reasons

- They may distribute pathogenic microorganisms,
- They often cause a high increase of microorganisms in the air, significantly exceeding the natural background.

The emission sources of biological aerosols can have a localized character (e.g. aeration tank) or a surface character (e.g. sewage-irrigated field).

The most important sources of bioaerosol emission are:

- Agriculture and farming-food industry,
- Sewage treatment plants,
- Waste management.

Microbiology of Inside Air

Bacteria are microscopic organisms found in indoor environments typically come from human sources (skin and respiration) or from the outdoors. Like mold, most of the bacteria found in the air in buildings are saprobes meaning they grow on dead organic matter. As far as building envelopes are concerned the primary concern is about bacteria colonies that may grow in damp areas. Most of the bacteria are shed from human skin surfaces. It is not surprising to find hundreds of thousands of bacteria per gram of dust in carpets. As long as the bacterial types are a mixture of those listed below, there is generally no cause for concern. Bacteria may also enter with outdoor air or floodwater and grow in indoor environmental reservoirs. Common indoor reservoirs are water systems, air handling unit and wet organic material. Inadequately maintained air handling system is an important source for bacterial exposure that may lead to allergic type disease. Air handling system must be check for the contaminated water where chest tightness, cough, and fever are associated with a particular indoor environment.

The most abundant bacteria present include

Micrococcus sp

Micrococcus species are human shed bacteria and are caused by the normal shed of skin. It is found in areas of higher occupant density and/or inadequate ventilation. Micrococcus species are generally regarded as being harmless bacteria. Normally, these bacteria are removed through ventilation systems or cleaning procedures such as mopping or vacuuming.

Bacillus sp

Bacillus sp mainly associated with soil and dust. Appropriate temperature and moisture with deposited dust on hard surfaces allow for ideal growing conditions. Most are not serious pathogens.

Staphylococcus sp

Staphylococcus sp is an inhabitant and shed from of the skin surfaces. Among the *Staphylococcus* species that are commonly found indoors is *Staphylococcus aureus*, which is an important pathogen in hospital environments. It shouldn't be a matter of concern unless it is the predominating colony found on air or surface samples in indoor environment.

Gram positive rod

Gram positive rod bacteria mainly associated with soil and dust. Appropriate temperature and moisture allow for ideal growing conditions on carpet, wall, furniture's etc. Most are not serious pathogens. These bacteria can be removed by good house keeping practice and adequate ventilation systems.

Gram negative rod

These organisms are rare in indoor environments, if they found in higher concentration may be related to the bio aerosol of contaminated water or other contamination of wet/moist surfaces or materials, or possibly air handling units systems in which they are proliferating. Some Gram negative bacteria (or endotoxin extracted from their walls) have been shown to provoke symptoms of fever. Occasionally, growth in air handling units has been great enough for aerosols to be generated which contained sufficient allergenic cells to have caused the acute pneumonia like symptoms. If there has been a sewage spill or flood, then Gram negative bacteria are to be expected and such environments should be thoroughly cleaned with disinfectant.

Identification of bacteria by cultural analysis is based on morphology (e.g., spherical, rod-shaped, etc.), by staining reactions (e.g. Gram positive or negative) and by the pattern of results from a series of biochemical tests.

****

39. WATER MICROBIOLOGY

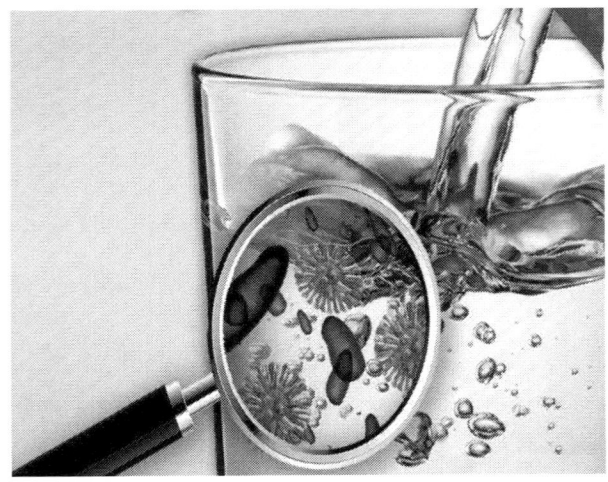

Water microbiology is concerned with the **microorganisms** that live in **water**, or can be transported from one **habitat** to another by water. Water can support the growth of many types of microorganisms. This can be advantageous. For example, the chemical activities of certain strains of yeasts provide us with beer and bread. As well, the growth of some **bacteria** in contaminated water can help digest the poisons from the water.

However, the presence of other **disease** causing microbes in water is unhealthy and even life threatening. For example, bacteria that live in the intestinal tracts of humans and other warm blooded animals, such as *Escherichia coli*, *Salmonella*, *Shigella*, and *Vibrio*, can contaminate water if feces enters the water. **Contamination** of drinking water with a type of *Escherichia coli* known as O157:H7 can be fatal. The contamination of the municipal water supply of Walkerton, Ontario, Canada in the summer of 2000 by strain O157:H7 sickened 2,000 people and killed seven people.

The intestinal tract of warm-blooded animals also contains viruses that can contaminate water and cause disease. Examples include rotavirus, enteroviruses, and coxsackievirus.

Another group of microbes of concern in water microbiology are **protozoa**. The two protozoa of the most concern are *Giardia* and *Cryptosporidium*. They live normally in the intestinal tract of animals such as beaver and **deer**. *Giardia* and *Cryptosporidium* form dormant and hardy forms called cysts during their life cycles. The cyst forms are resistant to **chlorine**, which is the most popular form of drinking water disinfection, and can pass through the filters used in many **water treatment** plants. If ingested in drinking water they can cause debilitating and prolonged diarrhea in humans, and can be life threatening to those people with impaired immune systems. *Cryptosporidium* contamination of the drinking water of Milwaukee, Wisconsin with in 1993 sickened more than 400,000 people and killed 47 people.

Many microorganisms are found naturally in fresh and **saltwater.** These include bacteria, cyanobacteria, protozoa, **algae**, and tiny animals such as rotifers. These can be important in the food chain that forms the basis of life in the water. For example, the microbes called cyanobacteria can convert the **energy** of the **sun** into the energy it needs to live. The plentiful numbers of these organisms in turn are used as food for other life. The algae that thrive in water is also an important food source for other forms of life.

A variety of microorganisms live in fresh water. The region of a water body near the shoreline (the littoral zone) is well lighted, shallow, and warmer than other regions of the water. Photosynthetic algae and bacteria that use **light** as energy thrive in this zone. Further away from the shore is the limnitic zone. Photosynthetic microbes also live here. As the water deepens, temperatures become colder and the **oxygen** concentration and light in the water decrease. Now, microbes that require oxygen do not thrive. Instead, purple and green **sulfur** bacteria, which can grow without oxygen, dominate. Finally, at the bottom of fresh waters (the benthic zone), few microbes survive. Bacteria that can survive in the absence of oxygen and sunlight, such as methane producing bacteria, thrive.

Saltwater presents a different environment to microorganisms. The higher **salt** concentration, higher **pH**, and lower **nutrients**, relative to **freshwater,** are lethal to many microorganisms. But, salt loving (halophilic) bacteria abound near the surface, and some bacteria that also live in freshwater are plentiful (i.e., *Pseudomonas* and *Vibrio*). Also, in 2001, researchers demonstrated that the ancient form of microbial life known as **archaebacteria** is one of the dominant forms of life in the **ocean**. The role of archaebacteria in the ocean food chain is not yet known, but must be of vital importance.

Another microorganism found in saltwater are a type of algae known as dinoflagellelates. The rapid growth and multiplication of dinoflagellates can turn the water red. This "red tide" depletes the water of nutrients and oxygen, which can cause many **fish** to die. As well, humans can become ill by eating contaminated fish.

Water can also be an ideal means of transporting microorganisms from one place to another. For example, the water that is carried in the hulls of ships to stabilize the vessels during their ocean voyages is now known to be a means of transporting microorganisms around the globe. One of these organisms, a bacterium called *Vibrio cholerae*, causes life threatening diarrhea in humans.

Drinking water is usually treated to minimize the risk of microbial contamination. The importance of drinking water treatment has been known for centuries. For example, in pre-Christian times the storage of drinking water in jugs made of **metal** was practiced. Now, the anti-bacterial effect of some metals is known. Similarly, the boiling of drinking water, as a means of protection of water has long been known.Chemicals such as chlorine or chlorine derivatives has been a popular means of killing bacteria such as *Escherichia coli* in water since the early decades of the twentieth century. Other bacteria-killing treatments that are increasingly becoming popular include the use of a gas called **ozone** and the disabling of the microbe's genetic material by the use of ultraviolet light. Microbes can also be physically excluded form the water by passing the water through a filter. Modern filters have holes in them that are so tiny that even particles as miniscule as viruses can be trapped.An important aspect of water microbiology, particularly for drinking water, is the testing of the water to ensure that it is safe to drink. Water quality testing can de done in several ways. One popular test measures the turbidity of the water. Turbidity gives an indication of the amount of suspended material in the water. Typically, if material such as **soil** is present in the water then microorganisms will also be present. The presence of particles even as small as bacteria and viruses can decrease the clarity of the water. Turbidity is a quick way of indicating if water quality is deteriorating, and so if action should be taken to correct the water problem.

In many countries, water microbiology is also the subject of legislation. Regulations specify how often water sources are sampled, how the sampling is done, how the analysis will be performed, what microbes are detected, and the acceptable limits for the target microorganisms in the water sample. Testing for microbes that cause disease (i.e., *Salmonella typhymurium* and *Vibrio cholerae*) can be expensive and, if the bacteria are present in low numbers, they may escape detection. Instead, other more numerous bacteria provide an indication of fecal **pollution** of the water. *Escherichia coli* has been used as an indicator of fecal pollution for decades. The bacterium is present in the intestinal tract in huge numbers, and is more numerous than the disease-causing bacteria and viruses. The chances of detecting *Escherichia coli* is better than detecting the actual disease causing microorganisms. *Escherichia coli* also had the advantage of not being capable of growing and reproducing in the water (except in the warm and food-laden waters of tropical countries). Thus, the presence of the bacterium in water is indicative of recent fecal pollution. Finally, *Escherichia coli* can be detected easily and inexpensively.

****

40.

SOIL MICROORGANISMS

Microorganisms are very small forms of life that can sometimes live as single cells, although many also form colonies of cells. A microscope is usually needed to see individual cells of these organisms. Many more microorganisms exist in topsoil, where food sources are plentiful, than in subsoil. They are especially abundant in the area immediately next to plant roots (called the rhizosphere), where sloughed-off cells and chemicals released by roots provide ready food sources. These organisms are primary decomposers of organic matter, but they do other things, such as provide nitrogen through fixation to help growing plants, detoxify harmful chemicals (toxins), suppress disease organisms, and produce products that might stimulate plant growth. Soil microorganisms have had another direct importance for humans—they are the source of most of the antibiotic medicines we use to fight diseases.

Bacteria

Bacteria live in almost any habitat. They are found inside the digestive system of animals, in the ocean and fresh water, in compost piles (even at temperatures over 130°F), and in soils. Although some kinds of bacteria live in flooded soils without oxygen, most require well-aerated soils. In general, bacteria tend to do better in neutral pH soils than in acid soils.

In addition to being among the first organisms to begin decomposing residues in the soil, bacteria benefit plants by increasing nutrient availability. For example, many bacteria dissolve phosphorus, making it more available for plants to use. Bacteria are also very helpful in providing nitrogen to plants, which they need in large amounts but is often deficient in agricultural soils. You may wonder how soils can be deficient in nitrogen when we are surrounded by it—78% of the air we breathe is composed of nitrogen gas. Yet plants as well as animals face a dilemma similar to that of the Ancient Mariner, who was adrift at sea without fresh water: "Water, water, everywhere nor any drop to drink." Unfortunately, neither animals nor plants can use nitrogen gas (N_2) for their nutrition. However, some types of bacteria are able to take nitrogen gas from the atmosphere and convert it into a form that plants can use to make amino acids and proteins. This conversion process is known as *nitrogen fixation*.

Some nitrogen-fixing bacteria form mutually beneficial associations with plants. One such symbiotic relationship that is very important to agriculture involves the nitrogen-fixing rhizobia group of bacteria that live inside nodules formed on the roots of legumes. These bacteria provide nitrogen in a form that leguminous plants can use, while the legume provides the bacteria with sugars for energy.

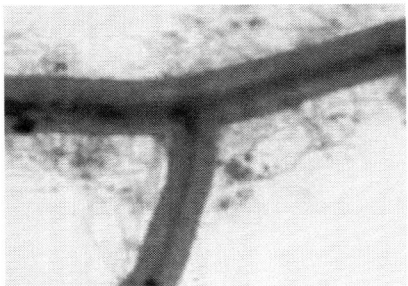

Fig.40. Root heavily infected with mycorrhizal fungi (note round spores at the end of some hyphae). Photo by Sara Wright.

People eat some legumes or their products, such as peas, dry beans, and tofu made from soybeans. Soybeans, alfalfa, and clover are used for animal feed. Clovers and hairy vetch are grown as cover crops to enrich the soil with organic matter, as well as nitrogen, for the following crop. In an alfalfa field, the bacteria may fix hundreds of pounds of nitrogen per acre each year. With peas, the amount of nitrogen fixed is much lower, around 30 to 50 pounds per acre.

The actinomycetes, another group of bacteria, break large lignin molecules into smaller sizes. Lignin is a large and complex molecule found in plant tissue, especially stems, that is difficult for most organisms to break down. Lignin also frequently protects other molecules like cellulose from decomposition. Actinomycetes have some characteristics similar to those of fungi, but they are sometimes grouped by themselves and given equal billing with bacteria and fungi.

Fungi

Fungi are another type of soil microorganism. Yeast is a fungus used in baking and in the production of alcohol. Other fungi produce a number of antibiotics. We have all probably let a loaf of bread sit around too long only to find fungus growing on it. We have seen or eaten mushrooms, the fruiting structures of some fungi. Farmers know that fungi cause many plant diseases, such as downy mildew, damping-off, various types of root rot, and apple scab. Fungi also initiate the decomposition of fresh organic residues. They help get things going by softening organic debris and making it easier for other organisms to join in the decomposition process. Fungi are also the main decomposers of lignin and are less sensitive to acid soil conditions than bacteria. None are able to function without oxygen. Low soil disturbance resulting from reduced tillage systems tends to promote organic residue accumulation at and near the surface. This tends to promote fungal growth, as happens in many natural undisturbed ecosystems.

RELATIVE AMOUNTS OF BACTERIA AND FUNGI

All soils contain both bacteria and fungi, but they may have different relative amounts depending on soil conditions. The general ways in which you manage your soil—the amount of disturbance, the degree of acidity permitted, and the types of residues added—will determine the relative abundance of these two major groups of soil organisms. Soils that are disturbed regularly by intensive tillage tend to have higher levels of bacteria than fungi. So do flooded rice soils, because fungi can't live without oxygen, while many species of bacteria can. Soils that are not tilled tend to have more of their fresh organic matter at the surface and to have higher levels of fungi than bacteria. Because fungi are less sensitive to acidity, higher levels of fungi than bacteria may occur in very acid soils. Despite many claims, little is known about the agricultural significance of bacteria versus fungal-dominated soil microbial communities, except that bacteria-prevalent soils are more characteristic of more intensively tilled soils that tend to also have high nutrient availability and enhanced nutrient levels as a result of more rapid organic matter decomposition.

Many plants develop a beneficial relationship with fungi that increases the contact of roots with the soil. Fungi infect the roots and send out root-like structures called *hyphae* (see figure 4.2). The hyphae of these *mycorrhizal* fungi take up water and nutrients that can then feed the plant. The hyphae are very thin, about 1/60 the diameter of a plant root, and are able to exploit the water and nutrients in small spaces in the soil that might be inaccessible to roots. This is especially important for phosphorus nutrition of plants in low-phosphorus soils. The hyphae help the plant absorb water and nutrients, and in return the fungi receive energy in the form of sugars, which the plant produces in its leaves and sends down to the roots. This symbiotic interdependency between fungi and roots is called a mycorrhizal relationship. All things considered, it's a pretty good deal for both the plant and the fungus. The hyphae of these fungi help develop and stabilize larger soil aggregates by secreting a sticky gel that glues mineral and organic particles together

MYCORRHIZAL FUNGI

Mycorrhizal fungi help plants take up water and nutrients, improve nitrogen fixation by legumes, and help to form and stabilize soil aggregates. Crop rotations select for more types of and better performing fungi than does mono cropping. Some studies indicate that using cover crops, especially legumes, between main crops helps maintain high levels of spores and promotes good mycorrhizal development in the next crop. Roots that have lots of mycorrhizae are better able to resist fungal diseases, parasitic nematodes, drought, salinity, and aluminum toxicity. Mycorrhizal associations have been shown to stimulate the free-living nitrogen-fixing bacteria azotobacter, which in turn also produce plant growth–stimulating chemicals.

Algae

Algae, like crop plants, convert sunlight into complex molecules like sugars, which they can use for energy and to help build other molecules they need. Algae are found in abundance in the flooded soils of swamps and rice paddies, and they can be found on the surface of poorly drained soils and in wet depressions. Algae may also occur in relatively dry soils, and they form mutually beneficial relationships with other organisms. Lichens found on rocks are an association between a fungus and an alga.

Protozoa

Protozoa are single-celled animals that use a variety of means to move about in the soil. Like bacteria and many fungi, they can be seen only with the help of a microscope. They are mainly secondary consumers of organic materials, feeding on bacteria, fungi, other protozoa, and organic molecules dissolved in the soil water. Protozoa—through their grazing on nitrogen-rich organisms and excreting wastes—are believed to be responsible for mineralizing (releasing nutrients from organic molecules) much of the nitrogen in agricultural soils.

****

41. ENVIRONMENTAL MICROBIOLOGY

Environmental microbiology is the study of the composition and physiology of microbial communities in the environment. The environment in this case means the soil, water, air and sediments covering the planet and can also include the animals and plants that inhabit these areas. Environmental microbiology also includes the study of microorganisms that exist in artificial environments such as bioreactors.

Molecular biology has revolutionized the study of microorganisms in the environment and improved our understanding of the composition, phylogeny, and physiology of microbial communities. The current molecular toolbox encompasses a range of DNA-based technologies and new methods for the study of RNA and proteins extracted from environmental samples. Currently there is a major emphasis on the application of "omics" approaches to determine the identities and functions of microbes inhabiting different environments.

Microbial life is amazingly diverse and microorganisms literally cover the planet. It is estimated that we know fewer than 1% of the microbial species on Earth. Microorganisms can survive in some of the most extreme environments on the planet and some can survive high temperatures, often above 100°C, as found in geysers, black smokers, and oil wells. Some are found in very cold habitats and others in highly salt|saline, acidic, or alkaline water.

An average gram of soil contains approximately one billion (1,000,000,000) microbes representing probably several thousand species. Microorganisms have special impact on the whole biosphere. They are the backbone of ecosystems of the zones where light cannot approach. In such zones, chemosynthetic bacteria are present which provide energy and carbon to the other organisms there. Some microbes are decomposers which have ability to recycle the nutrients. So, microbes have a special role in biogeochemical cycles. Microbes, especially bacteria, are of great importance in the sense that their symbiotic relationship (either positive or negative) have special effects on the ecosystem.

Microorganisms are cost effective agents for in-situ remediation of domestic, agricultural and industrial wastes and subsurface pollution in soils, sediments and marine environments. The ability of each microorganism to degrade toxic waste depends on the nature of each contaminant. Since most sites are typically comprised of multiple pollutant types, the most effective approach to microbial biodegradation is to use a mixture of bacterial species/strains, each specific to the degradation of one or more types of contaminants. It is vital to monitor the composition of the indigenous and added bacterial consortium in order to evaluate the activity level of the bacteria, and to permit modifications of the nutrients and other conditions for optimizing the bioremediation process.

Oil Biodegradation

Petroleum oil is toxic and pollution of the environment by oil causes major ecological concern. Oil spills of coastal regions and the open sea are poorly containable and mitigation is difficult, however much of the oil can be eliminated by the hydrocarbon-degrading activities of microbial communities, in particular the hydrocarbonoclastic bacteria (HCB). These organisms can help remediate the ecological damage caused by oil pollution of marine habitats. HCB also have potential biotechnological applications in the areas of bioplastics and biocatalysis.

Degradation of Aromatic Compounds by Acinetobacter

Acinetobacter strains isolated from the environment are capable of the degradation of a wide range of aromatic compounds. The predominant route for the final stages of assimilation to central metabolites is through catechol or protocatechuate (3,4-dihydroxybenzoate) and the beta-ketoadipate pathway, and the diversity within the genus lies in the channelling of growth substrates, most of which are natural products of plant origin, into this pathway.

Analysis of Waste Biotreatment

Biotreatment, the processing of wastes using living organisms, is an environmentally friendly alternative to other options. Bioreactors have been designed to overcome the various limiting factors of biotreatment processes in highly controlled systems. This versatility in the design of bioreactors allows the treatment of a wide range of wastes under optimized conditions. It is vital to consider various microorganisms and a great number of analyses are often required.

Environmental Genomics of Cyanobacteria

The application of molecular biology and genomics to environmental microbiology has led to the discovery of a huge complexity in natural communities of microbes. Diversity surveying, community fingerprinting, and functional interrogation of natural populations have become common, enabled by a battery of molecular and bioinformatics techniques. Recent studies on the ecology of Cyanobacteria have covered many habitats and have demonstrated that cyanobacterial communities tend to be habitat-specific and that much genetic diversity is concealed among morphologically simple types. Molecular, bioinformatics, physiological, and geochemical techniques have combined in the study of natural communities of these bacteria.

****

42. FOOD MICROBIOLOGY

Food microbiology encompasses the study of microorganisms which have both beneficial and deleterious effects on the quality and safety of foods. It focuses on the general biology of the microorganisms that are found in foods including: their growth characteristics, identification, and pathogenesis. Specifically, areas of interest which concern food microbiology are: food poisoning, food spoilage, and food legislation. Pathogens in product, or harmful microorganisms, result in major public health problems worldwide and are the leading causes of illnesses and death. In the United States alone, food borne illness has been estimated to cause 5,000 deaths and 76 million illnesses per year.

Factors affecting microbial growth in food

- Intrinsic Factors
- Environmental Factors
- Implicit Factors
- Processing Factors

Intrinsic Factors

- Nutrients
- pH and buffering capacity
- Redox potential
- Water activity
- Antimicrobial constituents

- Antimicrobial structures

Nutrient content

- The concentration of key nutrients can, to some extent, determine the rate of microbial growth.

- The relationship between the two, known as Monod equation, is mathematically identical to the Michaelis-Menten equation of enzyme kinetics.

- It reflects the dependence of microbial growth on rate-limiting enzyme reaction.

pH and buffering capacity

- As measured with the glass electrode, pH is equal to the negative logarithm of the hydrogenion activity.

- For aqueous solutions pH 7 corresponds to neutrality, pH values below 7 are acidic and those above 7 indicate an alkaline environment.

- In general bacteria grow fastest in the pH range 6.0-8.0, yeasts 4.5–6.0 and filamentous fungi 3.5-10.0.

Redox Potential

Redox potential (also known as reduction / oxidation potential, 'ROP', pe, ε,) is a measure of the tendency of a chemical species to acquire electrons from or lose electrons to an electrode and thereby be reduced or oxidised respectively. Redox potential is measured in volts (V), or millivolts (mV). Each species has its own intrinsic redox potential; for example, the more positive the reduction potential (reduction potential is more often used due to general formalism in electrochemistry), the greater the species' affinity for electrons and tendency to be reduced. ORP can reflect the antimicrobial potential of the water.

Water activity

The two most important chemical composition factors that affect how a food is preserved are water content and acidity. Water content includes moisture level, but an even more important measurement is water activity. Water activity (aw) refers to the energy status of water in the food, which affects whether or not chemical reactions occur and/or microorganisms will grow. The content of the food—such as sugar, salt, protein, or starch—"binds" the water, making it less available. Foods with lower water activities are less prone to spoilage by microorganisms and have fewer undesirable chemical changes occur during storage.

Antimicrobial Barriers and Constituents

Physical barrierto infection: skin. Shell, husk or rind of the product. Usually composed from macromoleculesrelatively resistant to degradationand provides unhospitable environment for microorganisms by having a low water activity, a shortage of readily available nutrients and often antimicrobial compoundssuch as short chain fatty acids(on animal skin) or essential oils(on plant surfaces).

Effect of antimicrobial substances

Some substances present inenvironment display negative effect on microorganisms, based on their specific composition (antimicrobial).

Microbistatic
-compoundsstop division of microorganism
Microbicidal
–compounds killing microorganisms

Effect of concentration (stimulatory effect)

Types of antimicrobial effects

Compounds demaging structure of cell or its function (cell wall, cytoplasmatic membrane, ribosomes)

Compounds affecting microbial enzymes (oxidative agents, chelating agents, heavy metals, antimetabolites)

Compounds reacting with DNA (chemical mutagenes–alkylating or deaminating agents, cytostatics)

Presence of antibacterial compounds (biocides) in food

•Some foods contain natural antimicrobial compounds (spices,mineraloils,garlic,mustard,honey)

•Raw cow milk contain lactoferrin,lactoperoxidase system,lysozym,kasein

•Eggs contain lysozym, conalbumin, ovotransferrin, avidin

Environmental factors

- Relative humidity
- Temperature
- Gaseous atmosphere

Relative humidity

•Relative humidity and water activity are interrelated, thus relative humidity is essential a measure of the water activity of the gas phase.

•When food comodity having a low water activity are stored in an atmosphere of high relative humidity -water will transfer from the gas phase to the food.

Temperature Is one of the most important environmental factors, controlling the rate of cell division (multiplication) of microorganisms

•We recognize 3 basic temperatures

- minimal temperature
- optimal temperature
- maximal temperature

Microorganisms are divided to groups according their demands on optimal temperature for division and metabolisms

1. Psychrophilic bacteria (12-15 °C)
2. Psychrotrophic bacteria (25-30 °C)
3. Mezophilic bacteria (30-40 °C)
4. Thermophilic bacteria (50-70 °C)

Gaseous atmosphere

•Oxygen forms 21% of the earth atmosphere
•Effect of carbon dioxide is not uniform.
•Anaerobic microorganisms
•Aerobic microorganisms
•Facultatively anaerobic microorganism
•Microaerobic microorganisms

Primary source of microorganisms found in food

The foods of plants and animal origin carry several microorganisms associated with their natural habitat. Plants carry typical micro-flora on their surface and also get contaminated from outside sources. Animals carry microorganisms on their surface and intestine, and also contain contaminants from surrounding environment. Through their excretions and secretions animals release microorganisms in to surrounding environment. Besides, both plants and animals carry pathogenic microorganisms capable of causing human illness. The food associated microorganisms are influenced by the availability of specific nutritional requirements and the environmental parameters. The primary sources of entry of microorganisms in to foods are from the soil, water, air, during handling, processing transportation and storage of foods.

Soil

Soil being the rich source of several kinds of microorganisms immediately contaminates the plants and edible plant parts, and the surface of animals with the soil associated microorganisms. As the soil particles are carried in to aquatic environment through wind, rain and other means contamination of water takes place with several soil micro-flora. Therefore, it is not uncommon to find several microorganisms both in soil and water environment. These soil derived microorganisms form part of the the microbial flora involved in spoilage of foods of plant and animal source. Thus, there is a need to reduce the load of soil microorganisms in foods which can be achieved by removing the soil by washing the surface of foods with good quality water, and by avoiding contact with soil/ dust.

Water

Natural waters not only contain several microorganisms native to the aquatic environment but also from soil, raw/treated sewage and pollutants entering the water body. The microbial numbers and types vary in different water bodies depending on the nutrient status. Thus, all kinds of microorganisms found in water are likely to be associated with the aquatic organisms as surface flora. Use of such water for food processing will add microorganisms from water to food. Sewage waters containing human pathogenic microorganisms contaminate foods when such waters are used without proper treatment. The water used in food processing should meet agreeable chemical and bacteriological characteristics.

Air

Air contains several microorganisms which may get deposited on the food being processed and handled. Though the air does not contain natural flora of microorganisms, whatever microorganisms encountered are those associated with the suspended solid material and water droplets. The sources of microorganisms to air are from dust, dry soil, and water spray from natural surface waters, droplets of moisture from coughing, sneezing and talking by food handlers, from sporulating moulds growing on walls, ceilings, floor, foods and food ingredients. Thus, it is likely that the microorganisms persisting in air get deposited on the food being processed and contribute for microbial load and subsequent spoilage of food. The number of microorganisms present in air depends on factors such as extent of movement of air, sunshine, humidity, location and amount of suspended dust in air. Quiet air allows settling of microorganisms but the moving air brings in microorganisms and keeps them suspended. Thus, the number of microorganisms in air is increased by air currents caused by movement of people, by ventilation and by breeze. The rain or snow removes microorganisms from the air.

Foodborne illness

Foodborne illness (also **foodborne disease** and colloquially referred to as **food poisoning**) is any illness resulting from the spoilage of contaminated food, pathogenic bacteria, viruses, or parasites that contaminate food, as well as toxins such as poisonous mushrooms and various species of beans that have not been boiled for at least 10 minutes.

Symptoms vary depending on the cause, and are described below in this article. A few broad generalizations can be made. For contaminants requiring an incubation period, symptoms may not manifest for hours to days, depending on the cause and on quantity of consumption. Longer incubation periods tend to cause sufferers to not associate the symptoms with the item consumed, so they may misattribute the symptoms to gastroenteritis, for example.

Symptoms often include vomiting, fever, and aches, and may include diarrhea. Bouts of vomiting can be repeated with an extended delay in between, because even if infected food was eliminated from the stomach in the first bout, microbes, like bacteria (if applicable), can pass through the stomach into the intestine and begin to multiply. Some types of microbes stay in the intestine, some produce a toxin that is absorbed into the bloodstream, and some can directly invade deeper body tissues.

Common Foodborne Pathogens

Bacteria

Bacteria are the largest group of problematic foodborne pathogens by far. They are small, one-celled microbes that come in many shapes and are capable of reproducing themselves. Typical cell shapes include spherical (cocci), rod-shaped (bacilli), and curved or comma-shaped (spirillar). These shapes can be seen under the microscope when the bacteria are stained in the laboratory with a Gram stain or dye. Whether or not bacterial cells stain Gram-positive (retaining a crystal violet color) or Gram-negative (those losing the color) also aids in identifying what bacteria are present and what treatments to administer. An important substructure of bacteria is the flagella, a hair-like tail that is responsible for bacterial movement. Bacteria are also classified and identified on the basis of their flagella. Much of modern foodborne microbiology is devoted to keeping pathogenic bacteria out of food products and preventing their growth if they are present. *Salmonella*, *E. coli* O157:H7, *Listeria*, and *Shigella* are well known species of foodborne bacteria.

Viruses

Viruses are thought to be the leading cause of foodborne illness in the United States based on the percentage of people ill, even though there are only a few viruses that are important foodborne pathogens. Viruses are much smaller than bacteria and cannot live outside a host, such as an animal or the human body. They are not cells but look more like particles (they have a protein coat, not a cell wall); reproducing only when they invade living cells. Although they do not multiply in food products, it can take only a few viral particles to make a person sick. Viruses are easily transferred from one food product to another, from contaminated water to foods, and from infected food handlers to foods. The two most well-known foodborne viruses are Hepatitis A and Norovirus (also known as Norwalk virus). Antibiotic drugs will not help in treatment because antibiotics fight against bacteria not viruses.

Parasites

There are about 20 different species of that are known to cause illness in humans from contaminated food or water. They range in size from microscopic single-celled organisms known as *protozoa* to visible worms known as *helminthes*. But, what they all have in common is that they derive their nourishment from other living organisms known as host organisms. When the parasites live and reproduce in the tissues and organs of animal and human hosts they can then be excreted in feces and go on to infect other individuals. There is a hard shell covering to some varieties of protozoa that permit them to survive for lengthy periods of time in water waiting to infect another host. Examples of protozoan parasites include *Cyclospora*, *Giardia*, and *Cryptosporidium*. A well-known foodborne *helminth* is *Trichinella*, an intestinal roundworm.

Fungal and Marine Pathogens

There are several types of molds (fungi) that are foodborne pathogens, and algae found in plankton can cause paralytic shellfish poisoning. Several other types of toxins found in seafood can also cause illness. Mad Cow Disease, also known as Bovine Spongiform Encephalopathy (BSE), is a degenerative brain disease of cattle caused by prion particles that can be passed to humans who consume beef contaminated by the brain, spinal cord, or nervous tissue of diseased animals. Heavy-metal contamination and synthetic plastics such as melamine have also been found in recent years to cause human illness and is the subject of ongoing research.

Foodborne Pathogens

Some microorganisms can cause big health problems when consumed in contaminated foods or beverages. The world of foodborne microbes contains a mix of approximately 250 different types of bacteria, viruses, parasites, molds, and algae that are known to cause disease in humans and are therefore called foodborne pathogens. What they all have in common is that they are most often too small to be seen without a microscope, they have simpler structures and functions than higher plants and animals, and they are able to be cultured in laboratory settings with prescribed methods that aid in their identification.

The term foodborne pathogen loosely describes the microbes that are found in animals (in farm/zoo animals and pets) and in the environment (soil, water and air) that make people sick regardless of how they became infected. Usually, infection happens by direct ingestion of a contaminated product, but it can also happen by contact with other individuals or contact with an animal or pet. Some foodborne microbes make people ill by forming toxins in foods that affect the gut or the neurological system. When an illness is caused by a ingesting a toxin and causes an intoxication it will generally make people sick faster than other foodborne pathogens which cause an infection.

Food Spoilage

Spoilage organisms alter food which results in changes in texture, appearance and organoleptic qualities of the food, making it unsuitable for human consumption. Spoilage is often the result of a succession. One organism creates an environment conducive to the growth of another. Common microbial food spoilage are:

I. Putrefaction- Protein + proteolytic microorganism's — amino acids + amines + ammonia + H2S

II. Fermentation- Carbohydrates + fermenting microrganism's —acids + alcohols + gases

III. Rancidity- Fatty foods + lypolytic microorganism's — fatty acids + glycerol

Fermented Foods

Fermented foods are foods and beverages that have undergone controlled microbial growth and fermentation [1]. Fermentation is an anaerobic process in which microorganisms like yeast and bacteria break down food components (e.g. sugars such as glucose) into other products (e.g. organic acids, gases or alcohol). This gives fermented foods their unique and desirable taste, aroma, texture and appearance.

There are thousands of different types of fermented foods, including:

- Cultured milk and yoghurt
- Wine
- Beer
- Cider
- Tempeh
- Miso
- Kimchi
- Sauerkraut
- Fermented sausage.

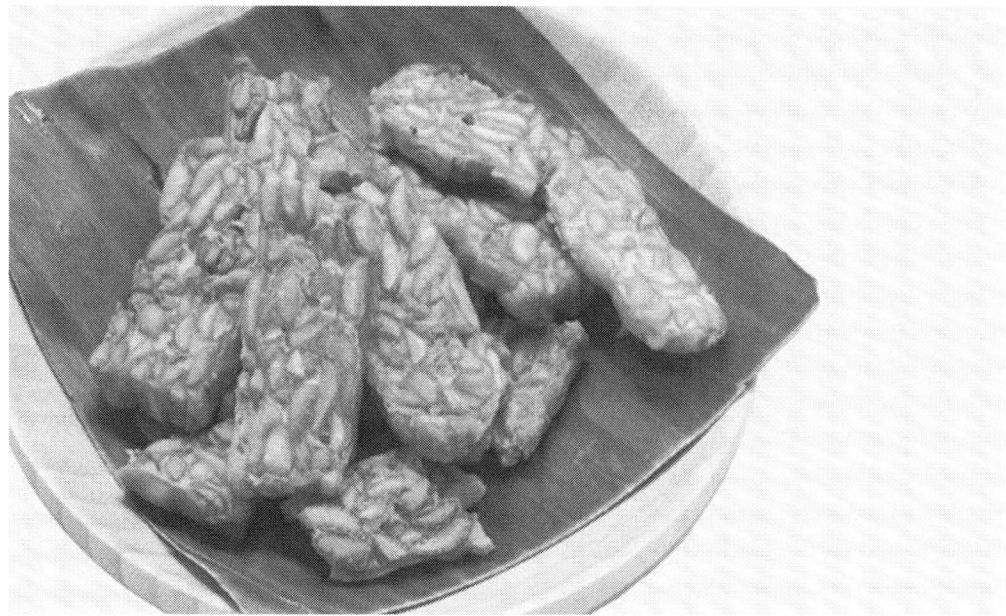

Fig.42.Tempeh is made from fermented soy beans.

Most foods can be fermented from whole foods like vegetables, fruits, cereals, dairy, meat, fish, eggs, legumes, nuts and seeds. While these foods are nutritious in their original form, through fermentation, they have the potential to carry additional health benefits – especially when they contain probiotics and prebiotics.

Probiotics

Many people know probiotics as 'good' or 'friendly' bacteria for the gut, with Lactobacillus and Bifidobacterium being the most well-known. Probiotics are live microorganisms or bacteria that provide a health benefit to the human body.

Experts believe that most strains from commonly studied species, such as Lactobacillus and Bifidobacterium, benefit the gut by creating a more favourable gut environment . They also agreed that probiotics support a healthy immune system, however, some strains may be more effective than others.

Several other benefits such as supporting organ health (e.g. lungs, reproductive, skin) and mood are promising, but there is not enough evidence to say that all probiotics have these effects.

Many fermented foods contain probiotics because they are added or they naturally occur in the food. For example, Lactobacilli is a probiotic strain that is commonly found in yoghurt and naturally lives on the surface of some foods such as vegetables and fruit. This means that not all fermented foods contain probiotics, especially many commercially produced foods that are pasteurized, which kills any bacteria (along with their associated health benefits).

Prebiotics are food ingredients that the microorganisms in your body (e.g. gut bacteria) use or 'feed' on to grow and live, leading to health benefits [6]. The most reported and researched prebiotics to have documented health benefits in humans are the non-digestible oligosaccharides fructans and galactans.

Good sources of these include:

- asparagus
- garlic
- onions
- wheat
- chicory
- Jerusalem artichokes
- tomato
- barley
- honey
- rye
- milk (human and cow's milk).

However, most fruits and vegetables, and legumes contain some type of prebiotic. As with probiotics, prebiotics have primarily been associated with improving the gut environment.

Benefits of fermented foods

Fermented foods have historically been valued for their improved shelf life and unique taste, aroma, texture and appearance. They also allow us to consume otherwise inedible foods. For example, table olives must be fermented in order to remove their bitter-tasting phenolic compounds.

Many health benefits have been associated with fermented foods, including reduced risk of cardiovascular disease, high blood pressure, diabetes, obesity and inflammation. They have also been linked to better weight management, better mood and brain activity, increased bone health and better recovery after exercise. When looking at heart health, probiotics may help to decrease total and low-density lipoprotein (LDL) cholesterol however the evidence for this is still very limited.

One explanation for all of these effects is the production of bioactive peptides, vitamins and other compounds produced by the microorganisms involved in fermentation and have key roles in the body, such as blood health, nerve function and immunity.

It's important to remember that these health benefits are likely dependent on the type of fermented food and microorganisms involved. For example, yoghurt consumption has been associated with reduced risk of type 2 diabetes [9-11], while fermented milk that contains Lactobacillus helveticus has been associated with reduced muscle soreness.

Food preservation

Food preservation is known "as the science which deals with the process of prevention of decay or spoilage of food thus allowing it to be stored in a fit condition for future use". Preservation ensures that the quality, edibility and the nutritive value of the food remains intact. Preservation involves preventing the growth of bacteria, fungi and other microorganisms as well as retarding oxidation of fats to reduce rancidity. The process also ensures that there is no discolouration or aging. Preservation also involves sealing to prevent re-entry of microbes. Basically food preservation ensures that food remains in a state where it is

- not contaminated by pathogenic organisms or chemicals
- does not lose optimum qualities of colour, texture, flavor and nutritive value

Drying is the oldest method of food preservation. This method reduces water activity which prevents bacterial growth. Drying reduces weight so foods can be carried easily. Sun and wind are both used for drying as well as modern applications like Bed dryers, Fluidized bed dryers, Freeze Drying, Shelf dryers, Spray drying and commercial food dehydrators and Household oven. Meat and fruits like apples, apricots and grapes are some examples of drying with this method.

Freezing is keeping prepared food stuffs in cold storages. Potatoes can be stored in dark rooms but potato preparations need to be frozen.

Smoking is the process that cooks, flavours and preserves food exposing it to the smoke from burning wood. Smoke is antimicrobial and antioxidant and most often meats and fish are smoked. Various methods of smoking are used like hot smoking, Cold smoking, Smoke roasting and Smoke baking. Smoking as a preservative enhances the risk of cancer.

Vacuum packing creates a vacuum by making bags and bottles airtight. Since there is no oxygen in the created vacuum bacteria die. Usually used for dry fruit.

Salting and Pickling: Salting also known as curing removes moisture from foods like meat. Pickling means preserving food in brine (salt solution) or marinating in vinegar (acetic acid) and in Asia, oil is used to preserve foods. Salt kills and inhibits growth of microorganisms at 20% of concentration. There are various methods of pickling like chemical pickling and fermentation pickling. In commercial pickles sodium benzoate or EDTA is added to increase shelf life.

Sugar is used in syrup form to preserve fruits or in crystallized form if the material to be preserved is cooked in the sugar till crystallization takes place like candied peel and ginger. Another use is for glazed fruit that gets superficial coating of sugar syrup. Sugar is also used with alcohol to preserve luxury foods like fruit in brandy.

Lye also known as Sodium hydroxide turns food alkaline and prevents bacterial growth.

Canning and bottling means sealing cooked food in sterile bottles and cans. The container is boiled and this kills or weakens bacteria. Foods are cooked for various lengths or time. Once the can or bottle is opened the food is again at risk of spoilage.

Jellying is preserving food by cooking in a material that solidifies to form a gel. Fruits are generally preserved as jelly, marmalade or fruit preserves and the jellying agent is pectin that is naturally found in fruit. Sugar is also added.

Potting is a traditional British way of preserving meat by placing it in a pot and sealing it with a layer of fat.

Jugging is preserving meat by stewing it in an earthenware jug or casserole. Brine or wine is used to stew meat in and sometimes the animal's blood.

Burial in the ground preserves food as there is lack of light and oxygen and it has cool temperatures, pH level, or desiccants in the soil. Used to preserve cabbages and root vegetables.

Pulsed Electric Field Processing is a new method of preservation that uses brief pulses as strong electric field to process cells. This is still at an experimental stage.

Modified atmosphere preserves food by operating on the atmosphere around it. Salad crops that are difficult to preserve are packaged in sealed bags with an atmosphere modified to reduce the oxygen concentration and increase the carbon dioxide concentration.

Controlled use of organism is used on cheese, wine and beer as they are preserved for a longer time. This method uses benign organisms to preserve food by introducing them to food where they make an environment which is not suitable for harmful pathogens to grow.

High pressure food preservation is a method that presses foods inside a vessel by exerting 70,000 pounds per square inch or more of pressure. This disables microorganisms and prevents spoilage but food retains its appearance, texture and flavour.

Modified Atmosphere Packaging extends the shelf life of fresh food products. The atmospheric air inside a package is substituted with a protective gas mix which ensures that the product will stay fresh for as long as possible.

Besides these there is **Pasteurisation and Irradiation** are also used.

****

43. INDUSTRIAL MICROBIOLOGY

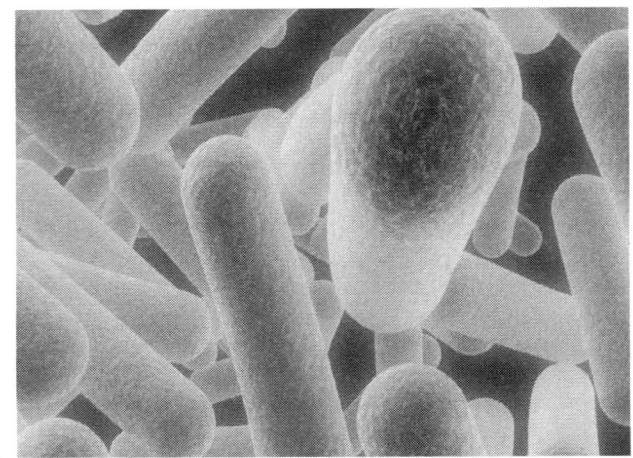

Industrial microbiology is a branch of biotechnology that applies microbial sciences to create industrial products in mass quantities, often using microbial cell factories. There are multiple ways to manipulate a microorganism in order to increase maximum product yields. Introduction of mutations into an organism may be accomplished by introducing them to mutagens. Another way to increase production is by gene amplification, this is done by the use of plasmids, and vectors. The plasmids and/ or vectors are used to incorporate multiple copies of a specific gene that would allow more enzymes to be produced that eventually cause more product yield. The manipulation of organisms in order to yield a specific product has many applications to the real world like the production of some antibiotics, vitamins, enzymes, amino acids, solvents, alcohol and daily products. Microorganisms play a big role in the industry, with multiple ways to be used. Medicinally, microbes can be used for creating antibiotics in order to treat antibiotics. Microbes can also be used for the food industry as well. Microbes are very useful in creating some of the mass produced products that are consumed by people. The chemical industry also uses microorganisms in order to synthesis amino acids and organic solvents. Microbes can also be used in an agricultural application for use as a biopesticide instead of using dangerous chemicals and or inoculants to help plant proliferation.

Industrial fermentation is the intentional use of fermentation by microorganisms such as bacteria and fungi as well as eukaryotic cells like CHO cells and insect cells, to make products useful to humans. Fermented products have applications as food as well as in general industry. Some commodity chemicals, such as acetic acid, citric acid, and ethanol are made by fermentation. The rate of fermentation depends on the concentration of microorganisms, cells, cellular components, and enzymes as well as temperature, pH and for aerobic fermentation oxygen. Product recovery frequently involves the concentration of the dilute solution. Nearly all commercially produced enzymes, such as lipase, invertase and rennet, are made by fermentation with genetically modified microbes. In some cases, production of biomass itself is the objective, as in the case of baker's yeast and lactic acid bacteria starter cultures for cheese making.

In general, fermentations can be divided into four types:

- Production of biomass (viable cellular material)
- Production of extracellular metabolites (chemical compounds)
- Production of intracellular components (enzymes and other proteins)
- Transformation of substrate (in which the transformed substrate is itself the product)

These types are not necessarily disjoint from each other, but provide a framework for understanding the differences in approach. The organisms used may be bacteria, yeasts, molds, algae, animal cells, or plant cells. Special considerations are required for the specific organisms used in the fermentation, such as the dissolved oxygen level, nutrient levels, and temperature.

Major products of industrial microbiology

In **Industrial Microbiology**, **microbes** are used to synthesize a number of products valuable to human beings. This industry has provided products that have deeply changed our lives and life spans. There are various industrial products that are derived from microbes such as **beverages, food additives, products for human and animal health, and biofuels**.

1. Beverages

Microbes especially yeast have been used from time immemorial for the production of beverages like **wine, beer, whiskey, brandy or rum**. For this purpose, the yeast *Saccharomyces cerevisiae* (commonly called **Brewer's Yeast**) is used for fermenting **malted cereals** and **fruit juices** to produce **ethanol**.

Fig. 43. Microbes are important agents in the Beverages Industry.

Among these beverages, **Wine** and **Beer** are produced **without distillation** whereas **whiskey, brandy, and rum** are **distilled beverages**.

2. Antibiotics

Antibiotics produced by microbes are regarded was one of the **most significant discoveries of the twentieth century** and have made major contributions towards the welfare of human society.

Many antibiotics are produced by **microorganisms**, predominantly by **Actinomycetes** in the genus **Streptomycin** (e.g. **Tetracycline, Streptomycin, Actinomycin D**) and by **filamentous fungi** (e.g. **Penicillin, Cephalosporin**)

3. Organic acids

Microbes are also used for the commercial and industrial production of certain **organic acids**. These compounds can be produced directly from **glucose** (e.g. **gluconic acid**) or formed **as end products** from **pyruvate or ethanol**. Examples of acids producing microorganisms are *Aspergillus Niger* (**a fungus**) of **Citric acid**, *Acetobacter acute* (**a bacterium**) of **Acetic Acid**, *Lactobacillus* (**a bacterium**) of lactic acid and many others.

4. Amino Acids

Amino acids such as **Lysine and Glutamic acid** are used in the food industry as nutritional supplements in bread products and as flavor enhancing compounds such as **Monosodium Glutamate (MSG)**.

Fig. 43.1. Crystals of the food additive monosodium glutamate (MSG).

Amino acids are generally synthesized as **primary metabolites by microbes**. However, when the rate and amount of synthesis of some amino acids **exceed the cell's need** for protein synthesis, then **cell excrete them** into the surrounding medium.

5. Enzymes

Many microbes **synthesize and excrete large quantities of enzymes** into the **surrounding medium**. Using this feature of these tiny organisms, many enzymes have been produced commercially. These include **Amylase, Cellulase, Protease, Lipase, Pectinase, Streptokinase**, and many others.

Enzymes are extensively used in **food processing** and **preservation**, **washing powders**, **leather industry, paper industry** and in scientific research.

6. Vitamins

Vitamins are some organic compounds which are capable of performing many **life-sustaining functions** inside our body. These compounds cannot be synthesized by humans, and therefore they have to be supplied in **small amounts in the diet**.

Microbes are capable of synthesizing the vitamins and hence they can be successfully used for the commercial production of many of the **vitamins** e.g. **thiamine, riboflavin, pyridoxine, folic acid, pantothenic acid, biotin, vitamin b12, ascorbic acid, beta-carotene (pro-vitamin A), ergosterol** (provitamin **D**).

7. Biofuels

Organic solvents such as **ethanol, acetone, butanol,** and **glycerol** are some very important chemicals that are widely used in **petrochemical industries**. These chemicals can be commercially produced by using **microbes** and **low-cost raw materials** (e.g. wood, cellulose, starch).

"**Brazil was the first country to produce ethanol in large scale by yeast fermentation, utilizing sugarcane and cassava.**"

Yeast (**Saccharomyces cerevisiae**) is used for commercial production of ethanol. This alcohol is used as motor fuel and is often referred to as **green petrol**.

8. Single Cell Protein (SCP)

Single Cell Protein (SCP) can serve as **an alternate source of energy** when a larger portion of the world is suffering from hunger and malnutrition. Single cell proteins are **microbial cells** that are **rich in protein content** and can be **used as protein supplements for humans and animals**.

Microbes like *Spirulina* can be grown easily on materials like **waste water from potato processing plants (containing starch), straw, molasses, animal manure,** and **even sewage**, to produce large quantities and can serve as food rich in **protein, minerals, fats, carbohydrate, and vitamins**.

9. Steroids

These are a very important group of chemicals, which are used **as anti-inflammatory drugs**, and **as hormones such as estrogens and progesterone**, which are used in **oral contraceptives**.

Steroids are widely distributed in animals, plants, and fungi like yeasts. But, producing **steroids** from **animal sources** or **chemically synthesizing** them is **difficult**, but microorganisms can **synthesize steroids from sterols** or from related, easily obtained compounds.

10. Vaccines

Vaccines are also a product of **industrial microbiology**. Many **antiviral vaccines** are **mass-produced** in chicken eggs or cell cultures.

The production of vaccines against **bacterial diseases** usually requires the growth of large amounts of the bacteria. **Recombinant DNA technology** is increasingly important in the development and production of **subunit vaccines**.

11. Pharmaceutical Drugs

Many **pharmaceutical drugs** are also produced by **microbes** e.g. **Cyclosporin A**, that is used as **an immunosuppressive agent** in **organ-transplant patients**, is produced by the fungus *Trichoderma polysporum*.

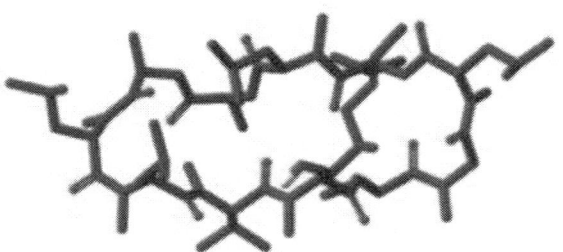

Fig.43.2. Neutron structure of the immunosuppressant cyclosporin A.

Statins produced by the yeast *Monascus purpureus* have been commercialized as **blood-cholesterol lowering agents**. It acts by competitively inhibiting the enzyme responsible for the synthesis of cholesterol.

****

44. PETROLEUM MICROBIOLOGY

Petroleum microbiology is a branch of microbiology that deals with the study of microorganisms that can metabolize or alter crude or refined petroleum products. These microorganisms, also called hydrocarbonoclastic microorganisms, can degrade hydrocarbons and, include a wide distribution of bacteria, methanogenic archaea, and some fungi. Not all hydrocarbonoclasic microbes depend on hydrocarbons to survive, but instead may use petroleum products as alternative carbon and energy sources. Interest in this field is growing due to the increasing role of bioremediation in oil spill cleanup

Petroleum is a natural product which is a mixture of aliphatic and aromatic hydrocarbons and various heterocyclics. It includes oxygen, nitrogen and sulphur containing compounds. The aliphatic ones include methane, ethane, propane and butane besides longer chain non-gaseous ones. Petroleum hydrocarbon is a natural resource that has been of immense benefit to man in a lot of ways. It is a major source of power at homes, in automobiles and industries. They are feed stocks of the petrochemical and allied industries. Products from these industries include: cosmetics, paints, inks, drugs, fertilizers, electronic casings, among others.

Microbes in Oil Prospecting

The presence of hydrocarbon-utilizing microorganisms has been proposed for prospecting for petroleum. The basis for this approach is the detection of microseepage of petroleum or some of its constituents, especially the volatile components, in the ground overlying a deposit using the presence of the hydrocarbon-utilizing microorganisms as indicators. It involves enriching the soil, sediment and water samples from a suspected seepage area for microbes that can metabolize gaseous hydrocarbons and demonstrating hydrocarbon consumption. The enrichment for these sets of microorganisms will be composed of mineral salts solution with added volatile hydrocarbon (ethane, propane, butane, isobutane).

Methane-oxidizing bacteria are poor indicators in petroleum prospecting because CH_4 can occur in the absence of petroleum deposits and also some methane-oxidizing bacteria are unable to oxidize other aliphatic hydrocarbons. Presumptive evidence for a hydrocarbon seepage and petroleum reservoir can be detected by the presence of bacteria that can oxidize ethane and longer chain hydrocarbons.

Factors affecting oil biodegradation

Abiotic factors

- **Temperature**

Rates of oil biodegradation generally decrease at lower temperatures. This is believed to be a result primarily of decrease in enzymatic activity, however, increased solubility and bioavailability of less soluble hydrophobic substances also play an important role. Higher temperatures enhance the rates of hydrocarbon metabolism with general optimum in the range of 30°C to 40°C, above which increased toxicity of oil might inhibit its biodegradation. Temperature also influence physical nature and composition of the oil: at low temperatures, the viscosity of the oil increases, the volatilization of toxic short-chain alkanes is reduced, and their water solubility increases, inhibiting the onset of biodegradation. Cold-adapted, *psychrophilic* and *psychrotrophic* microorganisms (for example, *Rhodococcus sp.*) are able to grow at temperatures around 0°C. They are widely distributed in nature because a large part of the Earth's biosphere is at temperatures below 5°C.

- **Oxygen**

The initial steps in the catabolism of aliphatic, cyclic, and aromatic hydrocarbons by bacteria and fungi involve the oxidation of the substrate by oxygenases for which molecular oxygen is required. Although anaerobic biodegradation has been shown to occur its ecological significance has been generally considered to be minor. Conditions of oxygen limitation normally exist in aquatic sediments and soils. Oxygen depletion can occur in the presence of easily utilizable substrates that increase microbial oxygen consumption.

- **Nitrogen and phosphorus**

Along with carbon, five other elements - hydrogen, nitrogen, oxygen, phosphorus, and sulfur - play a major role in life on Earth. The release of hydrocarbons into environments often produces excess of carbon over nitrogen and phosphorus which quickly become exhausted. It is well established that deprivation of nitrogen and phosphorus inhibits microbial oil degradation in such ecosystems as estuaries, seawater and marine sediments, freshwater lakes, groundwater, and soils.

- **Salinity**

Biodegradation rates of oil in fresh and marine water are comparable and depend on tolerance/adaptation of resident oil-degrading microorganisms to the specific salinity. Because abundance of microorganisms in highly halophilic conditions is greatly reduced, so is oil biodegradation. It was shown that rates of hydrocarbon utilization start to decrease noticeably in the salinity range 3.3 to 28.4%. Nonetheless, there are several reports about microorganisms able to oxidize petroleum hydrocarbons even in the presence of 30% w/v NaCl. Among such microorganisms are crude oil-degrading *Streptomyces albiaxialis*, and an *n*-alkane (C_{10}-C_{30})-degrading member of the *Halobacterium* group.

- **Pressure**

The importance of pressure as a variable in the biodegradation of hydrocarbons is most probably confined to the deep-sea environment where temperature factor is also at play. *Barophiles* (*piezophiles*) are microorganisms that require high pressure for growth, or grow better at pressures higher than atmospheric pressure. Little is known about ability of deep-sea microorganisms to cope with hydrocarbons. In general, oil which reaches the deep-ocean environment is degraded very slowly by resident microorganisms and some fractions persist for decades.

- **Water availability**

Water availability or water activity (also *water potential*) ranges from 0.0 to 0.99. Hydrocarbon biodegradation in terrestrial ecosystems may be limited by the available water. Optimal rates of biodegradation in sludge are observed at 30% to 90% of water saturation.

- **pH**

In contrast to most aquatic biosystems, soil pH can be highly variable ranging from 2.5 in mine refuse to 11.0 in alkaline deserts. Most heterotrophic bacteria and fungi favor a near neutral pH, with fungi being more tolerant to acidic conditions. It is common practice to add lime to bioremediate acid soils containing harmful organic compounds.

- **Wave and wind energy**

Usually, lowest rates of biodegradation of spilled oil occur in quiescent, low energy environments such as beaches, harbors, small lakes or ponds, in which the oil is relatively protected from dispersion by wind and wave action. The highest rates of biodegradation are observed in the areas of greatest wave and wind energy.

Biotic factors

Hydrocarbons in the environment are biodegraded primarily by bacteria and fungi. Algae and protozoa are important members of the microbial community in both aquatic and terrestrial ecosystems, but the extent of their involvement in hydrocarbon biodegradation is largely unknown and most likely is minor. There are three mechanisms for adaptation of microbial communities to chemical contaminants: (1) induction and depression of enzymes, (2) genetic changes (mutations, horizontal gene transfer), and (3) selective enrichment.

The third mechanisms has been most documented. Prior exposure of microbial community to hydrocarbons, either from anthropogenic sources (spills, oil disposals, etc.) or from natural sources (seeps and plant-derived hydrocarbons) is important in determining how rapidly subsequent hydrocarbon inputs can be biodegraded. Generally, turnover rates of common oil components can be from 10 to 400 times slower than in chronically contaminated environment.

Environmental Pollution by Petroleum Hydrocarbon

Pollution has been defined as an undesirable change in physical, chemical and biological characteristics of air, water and land. The sudden input of large amount of hydrocarbon (HCO(s)) associated with spillages stresses the environment in a way not imposed by natural hydrocarbons. Oil spillage as it is referred to have deleterious impact on flora, fauna and microbiota of the ecosystem. The economic life of the populace in the affected area is disrupted and the fragile ecobalance is usually disturbed. Farm lands, navigational activities, availability of clean potable water and fishing resorts are badly affected. Spill incidence remains amongst the various mean by which petroleum HCO pollute the environment. Crude oil spillage can occur at different stages of production and transportation either for export or refining processes. The spillages can be categorized depending on the barrels of oil spilled. It was categorized into (i) Minor (ii) Medium (iii) Major according to the NNPC Inspectorate division.

Minor Impact: < 25 bbl in inland water or 250 bbl on land that does not pose a threat to public health. Medium Impact: Discharge of oil 25 – 250 bbl on inland water or 250 – 2,500 bbl on land offshore and coastal waters.

Major Impact: Any discharge over > 250 bbl in inland water or > 2,500 on land offshore and coastal water. Any uncontrolled well blow out, pipe rupture or storage tank failure.

Effect of Oil pollution

Marine ecosystem is the ultimate recipient of surface run-off from soil and rivers which finally go into lagoons. Devastating oil spills occur more often in the oceans. e.g Gulf of Mexico. The

response of microbes to spillages in various ecosystems varies but usually there is increase in number of hydrocarbon degraders. Algal diversity and consumption will shift as a result of pollution. The species diversity in the coral reefs is badly affected. Invertebrates such as zooplanktons, crustaceans as well as vertebrates are adversely affected. The larval stages of shrimps are very susceptible to oil spill. Studies on the toxicity of oil to mangroves reveals that *Avicenia sp* will die within a week while it is mild on *Rhizophora* sp.

Freshwater ecosystem

It has been reported that there are drastic changes in the levels of microbial populations and the ability of indigenous microorganisms to degrade the hydrocarbon spilled/polluting the environment. Nitrogen fixation is reduced to about 8% after exposure to HCO. It has been reported, that some hydrocarbons not originally present in the crude were found present, after exposure to the estuarine community in the water body.

Soil ecosystem

Contamination of the terrestrial ecosystem affects not only the microbiota of the soil but also the resident microcommunity. It has negative effects on the pH community both by contact toxicity and indirect deleterious effects. Among the various effects are oxygen deprivation of roots and generation of phytotoxic compounds such as hydrogen sulphides. Herbaceous vegetation is quickly killed.

Major Pathways of Petroleum hydrocarbon degradation

Susceptibility to degradation varies with the type and size of the hydrocarbon molecule. N alkane of intermediate chain length (C10 – C24) is degraded most rapidly. Short chains are toxic to many microorganisms, but generally evaporate from oil slick, rapidly. Very long chain alkanes become increasingly resistant to biodegradation. As the chain length increases and the alkanes exceed a molecular weight of 500, the alkanes do not serve as carbon sources. Branching, in general reduces the rate of biodegradation. Aromatic compounds, especially of the condensed polynuclear type are degraded more slowly than Alkanes. Alicyclic compounds are frequently unable to serve as the sole carbon source for microbial growth unless they have a sufficiently long aliphatic side chain, but they can be degraded via cometabolism by two or more co-operating microbial strains withcomplementary metabolic capabilities.

List of some hydrocarbon-utilizing microorganisms

Pseudomonas aeruginosa
P. fluoresens
P. putida
Alcanivorax borkumensis
Corynebacterium
Micrococcus
Serratia marscense
Norcadia sp
Mycobacterium
Rhodococcus erythropolis

Alkanes

The preliminary attack on alkanes occurs by enzymes that have a strict requirement formolecular oxygen, that is monooxygenases or dioxygenases. One atom of O_2 is incorporated intothe alkane, yielding a primary alcohol.

Microbial Desulfurization

Sulfur is usually the third most abundant element in crude oil, normally accounting for 0.05 to 5%, but up to 14% in heavier oils. Most of the sulphur in crude oil is organically bound, mainly in the form of condensed thiophenes, and refiners use expensive physicochemical methods, including hydrodesulfurization to remove sulphur from crude oil. These high costs are driving the search for more efficient desulfurization methods, including biodesulphurization. In developing a lower cost biologically based desulfurization alternative, promoting selective metabolism of the sulphur component (attacking the C-S bonds) without simultaneously degrading the nonsulphur (CC bonds) fuel components in organic sulphur will be the most important consideration.

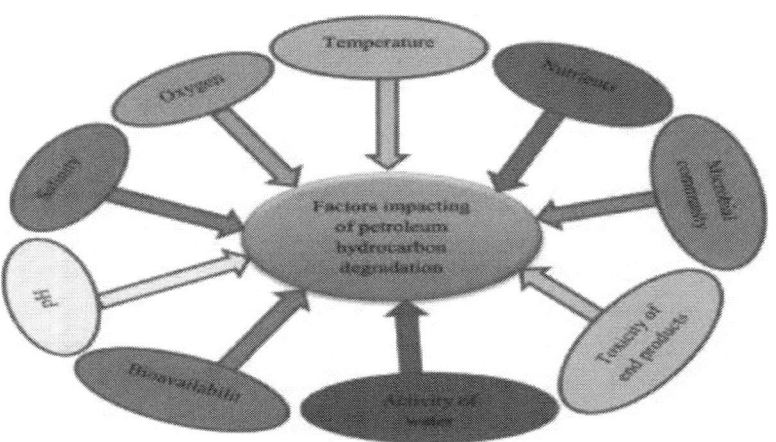

Fig.44. Principles of microbial degradation of petroleum hydrocarbons

Aerobically grown strains, such as *Rhodococcus erythropolis* and related species, remove the sulfur from compounds such as dibenzothiophene (DBT) without degrading the carbon ring

structure. These strains can use sulfur from DBT as a sole source of sulphur, which facilitates astrategy for isolation of desulfurizing organisms. Other aerobic selective desulfurizing microbesinclude *Nocardia* spp., *Agrobacterium* sp. Strain MC501, *Mycobacterium* spp. *Gordona* sp.Strain CYKS1, *Klebsiella* spp, *Xanthomonas* spp, and the thermophile *Paenibacillus .Rhodococcus* sp. strain was isolated from a mixed culture obtained from a sulfur-limitedcontinous-culture system capable of using organically bound sulphur. Strain IGTS8 converts DBTto dibenzothiophene-5-oxide (DBTO), then to dibenzene-5,5-dioxide (DBTO2), then to 2-(2-hydroxybiphenyl)-benzene sulphinate (HPBS), and finally to 2-hydroxybiphenyl (HBP) to release inorganic sulphur in a pathway involving two monooxygenases and a desulphinase. This enzyme system also transforms alkyl- and arylsubstituted DBT. Since the HBP product partitions into theoil phase, its fuel value is not lost. The flammability and explosive risks from the above oxygen requiring process have led to consideration of cloning the desulfurization genes into anaerobic hosts, which would hyperproduce the enzymes for addition to the crude oil. (http://www.unaab.edu.ng)

****

45. SCOPE AND APPLICATIONS OF MICROBIOLOGY

- Microorganisms are present everywhere on earth which includes humans, animals, plants and other living creatures, soil, water and atmosphere.
- Microbes can multiply in all three habitats except in the atmosphere. Together their numbers far exceed all other living cells on this planet.
- Microorganisms are relevant to all of us in a multitude of ways. The influence of microorganism in human life is both beneficial as well as detrimental also.
- For example microorganisms are required for the production of bread, cheese, yogurt, alcohol, wine, beer, antibiotics (e.g. penicillin, streptomycin, chloromycetin), vaccines, vitamins, enzymes and many more important products.
- Microorganisms are indispensable components of our ecosystem. Microorganisms play an important role in the recycling of organic and inorganic material through their roles in the C, N and S cycles, thus playing an important part in the maintenance of the stability of the biosphere.
- They are also the source of nutrients at the base of all ectotropical food chains and webs. In many ways all other forms of life depend on the microorganisms.

- Microorganisms also have harmed humans and disrupted societies over the millennia. Microbial diseases undoubtedly played a major role in historical events such as decline of the Roman empire and conquest of the new world.
- In addition to health threat from some microorganisms many microbes spoil food and deteriorate materials like iron pipes, glass lenses, computer chips, jet fuel, paints, concrete, metal, plastic, paper and wood pilings.
- There is vast scope in the field of microbiology due to the advancement in the field of science and technology.
- The scope in this field is immense due to the involvement of microbiology in many fields like medicine, pharmacy, diary, industry, clinical research, water industry, agriculture, chemical technology and nanotechnology.
- The study of microbiology contributes greatly to the understanding of life through enhancements and intervention of microorganisms. There is an increase in demand for microbiologists globally.
- **Genetics:** Mainly involves engineered microbes to make hormones, vaccine, antibiotics and many other useful products for human being.
- **Agriculture:** The influence of microbes on agriculture; the prevention of the diseases that mainly damage the useful crops.
- **Food science:** It involves the prevention of spoilage of food and food borne diseases and the uses of microbes to produce cheese, yoghurt, pickles and beer.
- **Immunology:** The study of immune system which protect the body from pathogens.
- **Medicine:** deals with the identification of plans and measures to cure diseases of human and animals which are infectious to them.
- **Industry:** it involves use of microbes to produce antibiotics, steroids, alcohol, vitamins and amino acids etc.
- **Agricultural microbiology** – try to combat plant diseases that attack important food crops, work on methods to increase soil fertility and crop yields etc. Currently there is a great interest in using bacterial or viral insect pathogens as substitute for chemical pesticides.
- **Microbial ecology** – biogeochemical cycles – bioremediation to reduce pollution effects

- **Food and dairy microbiology** – try to prevent microbial spoilage of food and transmission of food borne diseases such as botulism and salmonellolis. Use microorganisms to make foods such as cheese, yogurt, pickles and beers.
- **Industrial microbiology** – used to make products such as antibiotics, vaccines, steroids, alcohols and other solvents, vitamins, amino acids and enzymes.
- **Microbial physiology and Biochemistry** – study the synthesis of antibiotics and toxins, microbial energy production, microbial nitrogen fixation, effects of chemical and physical agents on microbial growth and survival etc.
- **Microbial genetics and Molecular biology** – nature of genetic information and how it regulated the development and function of cells and organisms. Development of new microbial strains that are more efficient in synthesizing useful products.
- **Genetic engineering** – arisen from work of microbial genetics and molecular biology. Engineered microorganisms are used to make hormones, antibiotics, vaccines and other products. New genes can be inserted into plants and animals.

Applications of Microbiology

Microbiology is one of the most applied branches of science. Its outstanding applications in the field of food microbiology, medical microbiology, industrial microbiology, soil microbiology, water and wastewater microbiology, microbial technology (biotechnology), extraction of metals and environmental microbiology including the use of microorganisms as biosensors is as given below.

1. It provides us with information about different types of microorganisms enabling us to understand their structure and functions; identifications and differentiations; their classifications; nomenclatures (naming), requirements regarding their nutrition; their isolation and purification; as plant and human pathogens; to derive phylogenetic relationships (relationships according to developmental stages in the evolution of an organism) and to understand the origin of life itself.

2. Microorganisms as food: Besides comestible fungi like mushrooms, microorganisms are also being used as single cell protein in the form of yeasts, bacteria, cyanobacteria, fungi as human food or animal feed. The production of the algal microbes as Chlorella (green alga and Spirulina (cyanobacterium) are being produced in Japan, Taiwan, Mexico, Israel, Thailand and America. Production of cellulose or lignocellulose utilizing microorganisms serves as human food as such or in the form of their products. Microbial products are also used as animal feed.

3. Microorganisms are used in production of a large number of, fermented foods such as leavened bread, sourdough bread, fermented milk products and flavours. The fermented milk products are yoghurt, cheese and several other products.

4. The important fermented vegetables are sauerkraut (from cabbage) and Kimchi (from other fermented vegetables in Korea).

5. Fermented meats and fermented fish are used in different parts of the world due to their increased retentivity, otherwise the meats and fish are highly perishable.

6. Beer, vinegar, tempeh, soya sauce, rice wine too are fermented products.

7. Microbiology has been very useful in preservation of food by heat processing, by pasteurization and appertization (commercially sterile food), by calculating thermal death values, prevention of spoilage of canned foods, aspectic packaging, irradiation, UV radiation, ionizing radiation, high pressure processing, i.e., pascalization, low temperature storage (chill storage and freezing), chemical preservatives (organic acids, esters, nitrite, and sulphur dioxide).

In food microbiology one learns about bacterial and nonbacterial agents of food borne illness. Among the helminthes and nematodes are: Platyhelminthus (i.e. liver flukes and tapeworms) and roundworms (e.g., *Trichinella spiralis*). The protozoa that cause food borne diseases are Giardia lamblia and Entamoeba histolytica.

8. Microbial diseases: Microorganisms are the causative agents of a large number of diseases which have been described under a separate chapter.

9. Industrial Microbiology: A large number of products of microbial metabolism after microbial processing of raw materials are produced on industrial scale. A separate chapter has been given on 'Industrial Microbiology'.

10. Energy from microbial sources: A number of substrates can be used as a source of energy as biogas from methanogenic microorganisms. The microbes like Methanobacterium and Methanococcus can utilize CO_2 as an electron acceptor finally producing methane. A new species of Methanobacterium, i.e., M. cadomensis strain 23 has been evolved in Japan for faster production of methane. Ethanol can also be used for the production of gasohol by mixing 80 per cent gasoline and 20 per cent ethanol.

11. Degradation of cellulose and lignin: Trichoderina reesei can be used to degrade cellulose since it produces extracellular cellulase. The white rot fungus *Sporotrichum pulverulentum* is a cellulase negative organism but a mutant of it has been prepared which can degrade kraft and wood lignocellulase actively. It has been possible to produce biological pulp without any chemical treatment for delignification.

12. Mining and extraction of metals: *Thiobacillus ferrooxidans* and combination of *Leptospirillum ferroxidans* and *Thiobacillus organoparpus* can be used to degrade pyrite (FeS_2) and chalcopyrite ($CuFeS_2$). The archaeal species *Sulfolobus acidocaldarius* and *S. brierlevi* are capable of oxidizing sulphur and iron for energy depending on CO_2 or other simple organic compounds for carbon. The pyrite and chalcopyrite are also degraded by these archaeobacterial species.

13. Recombinant DNA and genetic recombination: Recombinant DNA is a wonderful product of genetic engineering, i.e., manufacturing and manipulating genetic material in vitro. The process of joining DNA from different sources is genetic recombination. A large number of restriction enzymes/restriction endonucleases have been obtained from various microorganisms that can cut or cleave double stranded DNA leaving staggered ends.

14. Hybridoma and preparation of monoclonal antibodies: Hybridoma is a cell made by fusing an antibody-producing B-cell with a cancer cell. The resulting hybrid myeloma or hybridoma cells have properties of both parent cells immortality and the ability to secrete large amounts of a single specific type of antibody. This was discovered by Kohler.

15. Harvesting DNA biotechnology for public health engineering programmes: Such programmes include production of interferon which is an antiviral protein produced by certain animal cells in response to a viral infection, production of human insulin production of somatotropin a human growth hormone and production of a large number of other hormones and vaccines.

The vaccines for cholera, diphtheria, tetanus, pertussis, viral hepatitis type A, type B, influenza, mumps, measles (rubella) plague, poliomyelitis, rabies, rubbela, typhoid, typhus and yellow fever have been developed so far.

16. Microbial technology of nitrogen fixation exploiting symbiotic microorganisms in association with lower or higher plants and asymbiotic or nonsymbiotic (by nitrogen fixing microorganisms independently).

Detailed information is covered under a separate chapter on 'biofertilizers'. In nature, in legume root nodules a red pigment containing protein called leghaemoglobin is involved in the process of nitrogen fixation. The key enzyme responsible for biological conversion of molecular nitrogen to ammonia is nitrogenase.

17. Making faster and smarter computers:

The Archaeobacterium Halobacterium halobium grows in nature in solar evaporation ponds having high concentration of salts. Such salty ponds are found around San Francisco Bay located on the Western coast of USA.

It has been found that the plasma membrane of Halobacterium halobium fragments into two fractions, when the cell is broken down. These two fractions are red and purple. The purple fraction is important in making computer parts (chips). The purple colour is due to a protein which is 75% of purple membrane and has been referred to as bacteriorhodopsin.

Robert Birge at Syracuse University's Centre of Molecular Electronics has grown Halobacerium halobium in 5-litre batches and has extracted the protein bacteriorhodopsin from the cells and developed the computer chips which are made up of a thin layer of bacteriorhodopsin.

The chips so made from the bacterial source can store more information than the conventional silicon chips and process the information faster more like a human brain. The only drawback is that one needs to store the protein chips at -4°C. But Birge believes that this problem will be overcome soon.

MICROBIOLOGY MULTIPLE CHOICE QUESTIONS & ANSWERS

1. Which of the following is the most accurate method for microbial assay of antibiotics?
a) Physical assay
b) Chemical assay
c) Biological assay
d) Chemical and biological assay
View Answer

Answer: b
Explanation: Chemical-assay methods are generally more accurate and require less time than biological methods, but they are less sensitive, and caution must be used lest biologically inactive degradation products give misleading results.

2. The international unit of penicillin is defined by how much amount of International Standard?
a) 1 mg
b) 0.262 µg
c) 0.5988 µg
d) 0.5 mg
View Answer

Answer: c
Explanation: The international unit of penicillin is the amount of activity produced under defined conditions by 0.5988 µg of the International Standard, which is a sample of pure benzyl-penicillin.

3. Zone of inhibition is observed in tube-dilution method.
a) True
b) False
View Answer

Answer: b
Explanation: Small paper disks impregnated with known amounts of chemotherapeutic agents are placed upon the surface of an inoculated plate. So after incubation in disk-plate methods, a zone of inhibition around the disk indicates that the organism was inhibited by the drug.

4. Pigs respond dramatically to the addition of which of the following antibiotics to their diet?
a) aureomycin
b) terramycin
c) penicillin
d) oxytetracycline
View Answer

Answer: d
Explanation: It has been suggested that pigs respond dramatically to the addition of oxytetracycline to their diet because the antibiotic inhibits the growth of Clostridium perfringens in their intestines and prevents a chronic but subclinical toxemia.

5. Which of the following antibiotic have a sparing effect on the B12 in the diet?
a) Streptomycin
b) Tetracycline
c) Anthramycin
d) Chloramphenicol
View Answer

Answer: a
Explanation: Streptomycin may have a "sparing effect" on the B12 in the diet, making it available in greater quantities for utilization by the animals.

6. Antibiotics are not effective against plant pathogens.
a) True
b) False
View Answer

Answer: b
Explanation: Some antibiotics are effective against plant pathogens and are attractive for the treatment of plant diseases. The extent of this practice is limited mainly by economic factors, i.e, the cost of the antibiotic.

7. Nystatin is produced from the strain _____
a) Streptomyces noursei
b) Streptomyces aureofaciens
c) Streptomyces kanamyceticus
d) Streptomyces fradiae
View Answer

Answer: a
Explanation: Nystatin is an antifungal agent produced during fermentation by a strain of Streptomyces noursei.

8. Griseofulvin is useful in the therapy of nonsystemic fungal infections.
a) True
b) False
View Answer

Answer: b
Explanation: Griseofluvin is used in the treatment of many superficial fungal infections of the skin and body surfaces and is also effective in the treatment of systemic mycoses.

9. Acycloguanosine is a nucleoside analog which is active against _____

a) Influenza A virus
b) HIV virus
c) Herpes virus
d) Influenza B virus
View Answer

Answer: c
Explanation: Acycloguanosine is a nucleoside analog that is active against the herpes virus in animals. Its mode of action appears to be that of inhibition of nucleotide utilization.

10. Which of the following has its antiviral action attributed to the interference of protein synthesis?
a) Amantadine
b) Interferons
c) Acycloguanosine
d) 5'-iododeoxyuridine
View Answer

Answer: b
Explanation: The antiviral action of interferons is attributed to interference of protein synthesis. Interferons is among the most promising chemotherapeutic agents for treating viral diseases.

11. Amantadine is very effective against influenza A virus.
a) True
b) False
View Answer

12. Which of the following is effective in the control of tuberculosis in humans?
a) Nitrofurans
b) Nalidixic Acid

c) Sibromycin
d) Isoniazid
View Answer

Answer: d
Explanation: Isonicotinic acid Hydrazide or Isoniazid has proved to be very useful in the control of tuberculosis in humans and is more effective when given alternately with streptomycin.

13. Anthramycin is which of the following type of antibiotic?
a) Antiviral
b) Antitumor
c) Antifungal
d) Antibacterial
View Answer

Answer: b
Explanation: The anthramycin group (anthramycin, sibromycin, tomaymycin and neothramycin) is an example of potent antitumor agents. Their antitumor activity is directed towards DNA structure and function.

"Antibiotics and their Mode of Action".

1. Bacterial cells grown in a medium exposed to high osmotic pressure, changes shape from rod-shaped to _____ shaped.
a) spherical
b) rod shaped
c) irregular
d) elongated
View Answer

Answer: a
Explanation: The high osmotic pressure prevents the cells from bursting. Rod-shaped cells become spherical because they lack the cell structure which imparts shape.

2. Cell-wall biosynthesis is inhibited by antibiotics by inhibiting the biosynthesis of which of the following?

a) lipopolysaccharide
b) cellulose
c) peptidoglycan
d) proteins
View Answer

Answer: c
Explanation: Antibiotics exert their microbial effect by inhibiting biosynthesis of the peptidoglycan polymer, resulting in the inhibition of cell-wall formation. This results in the inability of the bacterium to survive because of the absence of a protective covering.

3. Structurally all penicillins have only beta-lactam present in them.
a) True
b) False
View Answer

Answer: b
Explanation: All penicillins have a common basic nucleus, a fused beta-lactam-thiazolidine ring with different side chains that give each its unique properties.

4. The crystalline sodium or potassium salts are slightly soluble in _____
a) ether
b) dioxane
c) water
d) chloroform
View Answer

Answer: d
Explanation: Natural penicillins can be prepared as salts of sodium, potassium, procaine, and other bases. The crystalline sodium or potassium salts are freely soluble in water, ester, ethers, ethyl alcohol, and dioxane but only slightly soluble in chloroform and benzene.

5. Which of the following does not affect the activity of penicillin?
a) bile
b) hydrochloric acid
c) cysteine
d) sodium hydroxide
View Answer

Answer: a
Explanation: The natural penicillins are inactivated by heat, cysteine, sodium

hydroxide, penicillinase, and hydrochloric acid. They are not affected by the action of saliva or bile.

6. Benzylpenicillin is the chemical name for which of the following penicillin?
a) Penicillin G
b) Penicillin V
c) Penicillin F
d) Phenethicilin
View Answer

Answer: a
Explanation: Peniciilin G is also known as Benzylpenicillin which is natural penicillin and has a basic core of 6-aminopenicillanic acid different side chain.

7. Ampicillin is a bactericidal antibiotic.
a) True
b) False
View Answer

Answer: a
Explanation: Ampicillin is another semisynthetic penicillin which is strongly bactericidal and lacks toxicity, but is not resistant to penicillinases.

8. Streptomyces orientalis produces which of the following antibiotics?
a) Cephalosporins
b) Cycloserine
c) Bacitracin
d) Vancomycin
View Answer

Answer: d
Explanation: Vancomycin is an antibiotic produced by Streptomyces orientalis. It is a complex chemical entity consisting of amino acids and sugars.

9. Which of the following interferes with the regeneration of the monophosphate form of bactoprenol from the pyrophosphate form?
a) Vancomycin
b) Ampicillin
c) Bacitracin
d) Cephalosporins
View Answer

Answer: c
Explanation: Bacitracin interferes with regeneration of the monophosphate form of bactoprenol from the pyrophosphate form. It is a polypeptide chemically.

10. Polymyxins inhibits the growth of the microbes by carrying out which of the following actions?
a) inhibition of cell-wall synthesis
b) damage to cytoplasmic membrane
c) inhibition of nucleic acid and protein synthesis
d) inhibition of specific enzyme systems
View Answer

Answer: b
Explanation: Polymyxins, gramicidins, and tyrocidines cause damage to cytoplasmic membrane. They adversely affect the normal permeability characters of the cell membrane.

11. Streptomycin is produced by which of the following organisms?
a) Stretomyces noursei
b) Streptomyces nodosus
c) Streptomyces fradiae
d) Streptomyces griseus
View Answer

Answer: d
Explanation: Streptomycin is produced by Streptomyces griseus, a soil organism isolated by Schatz, Bugie and Waksman, who reported on its antibiotic activities in 1944.

12. Antibiotic produced by Streptomyces rimosus is _____
a) chlortetracycline
b) oxytetracycline
c) tetracycline
d) doxycycline
View Answer

Answer: b
Explanation: Antibiotic produced by Streptomyces rimosus is oxytetracycline whereas Streptomyces aureofaciens produces chlortetracycline. They are broad spectrum antibiotics with similar antimicrobial spectra, and cross resistance of bacteria to them is common.

13. Which of the following inhibits protein synthesis by combining with the 50S subunit ribosome?

a) Streptomycin
b) Tetracycline
c) Chloramphenicol
d) Penicillin
View Answer

Answer: c
Explanation: Chloramphenicol inhibits protein synthesis by combining with the 50S subunit ribosome. The transpeptidation and translocation functions associated with this site are blocked.

14. Tyrocidines are more effective against _____
a) Gram-positive organisms
b) Gram-negative organisms
c) Mycoplasmas
d) Spirochetes
View Answer

Answer: a
Explanation: Tyrocidines and gramicidines are more effective against Gram-positive organisms whereas polymyxins are particularly effective against Gram-negative bacteria.

Microbiological Assay of Antibiotics, Antifungal, Antiviral and Antitumour Antibiotics".

1. Which of the following is the most accurate method for microbial assay of antibiotics?
a) Physical assay
b) Chemical assay
c) Biological assay
d) Chemical and biological assay
View Answer

Answer: b
Explanation: Chemical-assay methods are generally more accurate and require less time than biological methods, but they are less sensitive, and caution must be used lest biologically inactive degradation products give misleading results.

2. The international unit of penicillin is defined by how much amount of International Standard?
a) 1 mg
b) 0.262 µg
c) 0.5988 µg

d) 0.5 mg
View Answer

Answer: c
Explanation: The international unit of penicillin is the amount of activity produced under defined conditions by 0.5988 µg of the International Standard, which is a sample of pure benzyl-penicillin.

3. Zone of inhibition is observed in tube-dilution method.
a) True
b) False
View Answer

Answer: b
Explanation: Small paper disks impregnated with known amounts of chemotherapeutic agents are placed upon the surface of an inoculated plate. So after incubation in disk-plate methods, a zone of inhibition around the disk indicates that the organism was inhibited by the drug.

4. Pigs respond dramatically to the addition of which of the following antibiotics to their diet?
a) aureomycin
b) terramycin
c) penicillin
d) oxytetracycline
View Answer

Answer: d
Explanation: It has been suggested that pigs respond dramatically to the addition of oxytetracycline to their diet because the antibiotic inhibits the growth of Clostridium perfringens in their intestines and prevents a chronic but subclinical toxemia.

5. Which of the following antibiotic have a sparing effect on the B12 in the diet?
a) Streptomycin
b) Tetracycline
c) Anthramycin
d) Chloramphenicol
View Answer

Answer: a
Explanation: Streptomycin may have a "sparing effect" on the B12 in the diet, making it available in greater quantities for utilization by the animals.

6. Antibiotics are not effective against plant pathogens.
a) True
b) False
View Answer

Answer: b
Explanation: Some antibiotics are effective against plant pathogens and are attractive for the treatment of plant diseases. The extent of this practice is limited mainly by economic factors, i.e, the cost of the antibiotic.

7. Nystatin is produced from the strain _____
a) Streptomyces noursei
b) Streptomyces aureofaciens
c) Streptomyces kanamyceticus
d) Streptomyces fradiae
View Answer

Answer: a
Explanation: Nystatin is an antifungal agent produced during fermentation by a strain of Streptomyces noursei.

8. Griseofulvin is useful in the therapy of nonsystemic fungal infections.
a) True
b) False
View Answer

Answer: b
Explanation: Griseofluvin is used in the treatment of many superficial fungal infections of the skin and body surfaces and is also effective in the treatment of systemic mycoses.

9. Acycloguanosine is a nucleoside analog which is active against _____

a) Influenza A virus
b) HIV virus
c) Herpes virus
d) Influenza B virus
View Answer

Answer: c
Explanation: Acycloguanosine is a nucleoside analog that is active against the herpes virus in animals. Its mode of action appears to be that of inhibition of nucleotide utilization.

10. Which of the following has its antiviral action attributed to the interference of protein synthesis?
a) Amantadine
b) Interferons
c) Acycloguanosine
d) 5'-iododeoxyuridine
View Answer

Answer: b
Explanation: The antiviral action of interferons is attributed to interference of protein synthesis. Interferons is among the most promising chemotherapeutic agents for treating viral diseases.

11. Amantadine is very effective against influenza A virus.
a) True
b) False
View Answer

Answer: a
Explanation: Amantadine is a low-molecular weight compound which is very effective against influenza A virus; it is not effective against influenza B. The incidence of influenza A infection is greatly reduced by the use of this drug.

12. Which of the following is effective in the control of tuberculosis in humans?
a) Nitrofurans
b) Nalidixic Acid
c) Sibromycin
d) Isoniazid
View Answer

Answer: d
Explanation: Isonicotinic acid Hydrazide or Isoniazid has proved to be very useful in the control of tuberculosis in humans and is more effective when given alternately with streptomycin.

13. Anthramycin is which of the following type of antibiotic?
a) Antiviral
b) Antitumor
c) Antifungal
d) Antibacterial
View Answer

Answer: b
Explanation: The anthramycin group (anthramycin, sibromycin, tomaymycin and

neothramycin) is an example of potent antitumor agents. Their antitumor activity is directed towards DNA structure and function.

"Interactions among Soil Microorganisms".

1. The association which involves the exchange of nutrients between two species is referred to as _____
a) mutualism
b) syntrophism
c) commensalism
d) antagonism
View Answer

Answer: b
Explanation: Syntrophism is a mutualistic association that involves the exchange of nutrients between two species.

2. The degradation of complex molecules in soil by fungi for utilization by bacteria is an example of which type of association?
a) Neutralism
b) Mutualism
c) Commensalism
d) Antagonism
View Answer

3. Which of the following types of association is present among Staphylococcus aureus and Aspergillus terreous?
a) antagonism
b) mutualism
c) parasitism
d) commensalism
View Answer

Answer: a
Explanation: Staphylococcus aureus produces a diffusable antifungal material that causes distortions and hyphal swellings in Aspergillus terreous. This is an antagonistic relationship between these two organisms.

4. Lytic enzymes which destroy are secreted by which of the following microorganism?
a) fungi
b) algae
c) staphylococcus

d) myxobacteria
View Answer

Answer: d
Explanation: Myxobacteria (slime bacteria) and streptomycetes are antagonistic because they secrete potent lytic enzymes which destroy other cells by digesting their cell wall or other protective surface layers.

5. Fungi produces which of the following inhibitory toxic product?
a) cyanide
b) fatty acids
c) methane
d) sulphides
View Answer

Answer: a
Explanation: Cyanide is produced by certain fungi in concentrations toxic to other microorganisms.

6. Which of the following comes under the category of positive association?
a) neutralism
b) parasitism
c) commensalism
d) ammensalism
View Answer

Answer: c
Explanation: The phenomenon of commensalism is a positive type of association because it refers to a relationship between organisms in which one species of a pair benefits, the other is not affected.

7. Parasitism results from competition among organisms for essential nutrients.
a) True
b) False
View Answer

Answer: b
Explanation: Parasitism is defined as a relationship between organisms in which one organism lives in or on another organism referred to as a host.

8. The dominant mineral particles in most soils are compounds of _____
a) sodium
b) potassium
c) magnesium

d) iron
View Answer

Answer: d
Explanation: The dominant mineral particles in most soils are compounds of silicon, aluminum, iron and lesser amounts of other minerals, including calcium, magnesium, potassium, sodium, nitrogen etc.

9. Bacteria are likely to be more prevalent in soils of vineyards, orchards and apiaries.
a) True
b) False
View Answer

Answer: b
Explanation: Yeasts are likely to be more prevalent in soils of vineyards, orchards, and apiaries, where special conditions, particularly the presence of sugars favor their growth.

10. Which of the following organisms are known to grow on the surfaces of freshly exposed rocks?
a) green algae
b) diatoms
c) cyanobacteria
d) yeast
View Answer

Answer: c
Explanation: Cyanobacteria, the oxygenic photosynthetic bacteria, are known to grow on the surfaces of freshly exposed rocks where the accumulation of their cells results in the simultaneous deposition of organic matter.

"Soil Microbiology – Nitrogen Cycle".

1. Ammonia oxidizers and nitrite oxidizers are _____
a) Gram-negative chemolithotrophs
b) Gram-positive chemolithotrophs
c) Gram-negative photolithotrophs
d) Gram-positive photolithotrophs
View Answer

Answer: a
Explanation: Oxidation of ammonia to nitrite by ammonia-oxidizing bacteria and

oxidation of nitrite to nitrate by nitrite-oxidizing bacteria are both Gram-negative chemolithotrophs.

2. Alanine gives pyruvic acid on deamination.
a) True
b) False
View Answer

Answer: a
Explanation: The process of removal of the amino group from amino acids is known as deamination. Alanine in presence of alanine deaminase gives pyruvic acid and ammonia as its deamination reaction products.

3. How much time does nitrifying bacteria require to grow at an incubation of 250 to 300 C?
a) 1 day
b) 2-3 days
c) 15 days
d) 1 to 4 months
View Answer

Answer: d
Explanation: For the growth of nitrifying bacteria a relatively large inoculum is used, and incubation is in the dark at 250 to 300 C for a period of 1 to 4 months.

4. Nitrosococcus nitrosus is a nitrite-oxidizing bacteria.
a) True
b) False
View Answer

Answer: b
Explanation: Nitrosococcus nitrosus is a type of ammonia-oxidizing bacteria. They usually have an extensive membrane system within their cytoplasm and frequently form cysts and zooglea.

5. Which among the following is not an ammonia-oxidizing bacteria?
a) Nitrosomonas europaea
b) Nitrosovibrio tenuis
c) Nitrospina gracilis
d) Nitrosococcus oceanus
View Answer

Answer: c
Explanation: Nitrospina gracilis is a nitrite-oxidizing bacteria. They are involved in the oxidation of nitrite to nitrate.

6. Agrobacterium is involved in which of the following processes?
a) Ammonification
b) Nitrification
c) Reduction of nitrate to ammonia
d) Denitrification
View Answer

Answer: d
Explanation: The transformation of nitrates to gaseous nitrogen is known as denitrification. Species of several bacteria like Agrobacterium, Bacillus, Pseudomonas etc are involved in this process.

7. Which of the following conditions decreases the level of denitrification?
a) Abundance of organic matter
b) Acidic pH
c) Elevated temperatures
d) Availability of oxygen
View Answer

Answer: b
Explanation: The process of denitrification is enhanced in soils by the presence of abundance of organic matter, by elevated temperatures, by neutral or alkaline pH. The availability of oxygen has a dual effect on denitrification.

8. Which of the following are not the features of component II of nitrogenase enzyme complex?
a) component II is nitrogenase reductase
b) component II is known as the MoFe protein
c) contains sulfur
d) not active without component I
View Answer

Answer: b
Explanation: Component II of nitrogenase enzyme complex is nitrogenase reductase. It is a smaller molecule and is designated the Fe protein. Both molecules contain sulfur and neither is active without the other.

9. Which of the following is symbiotic nitrogen fixing bacteria?
a) Rhizobium trifolii
b) Clostridium pasteurianum

c) Azotobacter sp.
d) Escherichia coli
View Answer

Answer: a
Explanation: Symbiotic nitrogen fixation is accomplished by bacteria of the genus Rhizobium in association with legumes like Rhizobium trifolii.

his set of Microbiology test focuses on "World of Bacteria II – Endospore – Forming Gram – Positive Bacteria".

1. Which of the following are mesophilic saprophytes?
a) B.polymyxa
b) B.anthracis
c) B.subtilis
d) B.thuringiensis
View Answer

Answer: c
Explanation: Bacillus subtilis are mesophilic saprophytes and are widely distributed in nature. This means they can grow best within a temperature range of approximately 25 to 40 degrees Celsius.

2. Which of the following causes "milky disease" of Japanese beetle grubs?
a) B.thuringiensis
b) B.popilliae
c) B.sphaericus
d) B.anthracis
View Answer

Answer: b
Explanation: B.popilliae is a pathogenic species of Bacillus species that causes "milky disease" of Japanese beetle grubs.

3. Which of the following species is associated with spoilage of canned goods?
a) B.stearothermophiles
b) B.cereus
c) B.subtilis
d) B.sphaericus
View Answer

Answer: a
Explanation: B.stearothermophilus is a thermophilic species having a maximum of

65 to 75 degrees Celsius. The endospores are highly resistant to heat and, therefore, this species is one of those associated with spoilage of canned goods.

4. Which of the following genus of species play an active role in the decomposition of urea?
a) Bacillus
b) Sporosarcina
c) Clostridium
d) Desulfotomaculum
View Answer

Answer: b
Explanation: Sporosarcinae are widely distributed in fertile soil, where they play an active role in the decomposition of urea.

5. C.perfringens is the major causative agent of _____
a) botulism
b) tetanus
c) gas gangrene
d) anthrax
View Answer

Answer: c
Explanation: Clostridium perfringens is the major causative agent of the wound infection known as gas gangrene.

6. Which of the following Clostridium species has the ability to fix Nitrogen?
a) C.difficile
b) C.pasteurianum
c) C.tetani
d) C.thermosaccharolyticum
View Answer

Answer: b
Explanation: C.pasteurianum is a mesophilic soil clostridium that is particularly noted for its ability to fix Nitrogen.

7. The genus Desulfotomaculum obtain energy by anaerobic respiration.
a) True
b) False
View Answer

Answer: a
Explanation: The members of the genus Desulfotomaculum obtain energy by

anaerobic respiration, with sulphate serving as the terminal electron acceptor and organic substrates such as lactic or pyruvic acid serving as the electron donors.

8. Pseudomembranous colitis is a disease of _____
a) stomach
b) wounds
c) bowel
d) limbs
View Answer

Answer: c
Explanation: C.difficile causes pseudomembranous colitis, a severe disease of the bowel.

9. Bacillus species cannot fix Nitrogen.
a) True
b) False
View Answer

Answer: b
Explanation: Bacillus polymyxa has the ability to fix Nitrogen under anaerobic conditions.

This set of Microbiology Multiple Choice Questions & Answers (MCQs) focuses on "World of Bacteria I – Rickettsias and Chlamydias".

1. Which of the following infect arthropods only?
a) Rickettsieae
b) Ehrlichieae
c) Wolbachieae
d) Rochalimaea
View Answer

Answer: c
Explanation: The Wolbachieae are not pathogenic for vertebrates, they infect arthropods only.

2. The Rickettsias and Chlamydias are similar in all respects.
a) True
b) False
View Answer

Answer: b
Explanation: The Rickettsias synthesize ATP as a source of energy whereas Chlamydias cannot synthesize ATP.

3. Classical typhus fever is transmitted by which of the following arthropod?
a) ticks
b) mites
c) fleas
d) lice
View Answer

Answer: d
Explanation: Classical typhus fever is caused by Rickettsia species and the arthropod vector which transmits the disease is lice.

4."Growing epicellularly" means _____
a) on the surface of host cells
b) within the host cell
c) engulfing the host cell
d) damaging the cell membrane of host cell
View Answer

Answer: a
Explanation: Growing epicellularly means growing on the surface of host cells. Rochalimaea grow epicellularly rather than in the cytoplasm or nucleus.

5. Which of the following genus of Rickettsiaceae have a high resistance to heat?
a) Rickettsia
b) Rochalimaea
c) Coxiella
d) Wolinella
View Answer

Answer: c
Explanation: The organisms belonging to Coxiella have an unusually high resistance to heat [may survive a temperature of 62 degree C for 30 mins], probably due to the occurrence of endospore-like structures in the cells.

6. C.burnetti causes which of the following diseases?
a) Q fever
b) Trench fever
c) Rocky Mountain spotted fever
d) Typhoid
View Answer

Answer: a
Explanation: C.burnetti is the causative agent of the Q fever, a type of pneumonia. It belongs to the family Rickettsiaceae.

7. Biting flies occur in which of the following place?
a) Australia
b) North America
c) South America
d) Africa
View Answer

Answer: c
Explanation: Biting flies occur along the western slopes of the Andes mountains in South America. They transmit Oroya fever in humans.

8. Initial body in the reproduction of Chlamydias is larger than the elementary body.
a) True
b) False
View Answer

Answer: a
Explanation: Within the vacuole the elementary body is reorganized into a reticulate body known as the initial body, which is two or three times the size of the elementary body and contains a less dense arrangement of nucleoid material.

9. Which is the most prevalent sexually transmitted disease in the US?
a) Syphilis
b) AIDS
c) Lymphogranuloma venerum
d) Nongonococcal urethritis
View Answer

Answer: d
Explanation: Nongonococcal urethritis is the most prevalent sexually transmitted disease in the United States and is caused by the organisms belonging to the genus Chlamydia.

"Animals and Plants – Viruses and Vaccination".

1. From which of the following animal was the material isolated which was used for the vaccination for the first time?
a) cat
b) cow
c) pig

d) goat
View Answer

Answer: b
Explanation: Jenner used the material isolated from cows to be used as vaccination and it provided protection against natural smallpox infection.

2. Vaccination was invented by _____
a) Jenner
b) Pasteur
c) Watson
d) Crick
View Answer

Answer: a
Explanation: In 1796 Jenner first vaccinated an 8-year old boy with material removed from cow and it gave protection against the smallpox virus.

3. Causative agent of tobacco mosaic disease was filterable.
a) True
b) False
View Answer

Answer: a
Explanation: In 1892, Dmitrii Ivanowski discovered that the causative agent of tobacco mosaic disease was filterable. Viruses that can pass through porcelain filters are known as filterable viruses.

4. Yellow fever virus can be attenuated by serial passage on cultures of _____
a) embryonated eggs
b) tissue
c) chick embryo tissue
d) pig embryo tissue
View Answer

5. Effective poliomyelitis vaccines were developed by culturing the virus of poliomyelitis on the kidney cells of which animal?
a) cow
b) monkey
c) giraffe
d) pig
View Answer

Answer: b
Explanation: Enders, Robbins, and Weller laid the foundation for the development of effective poliomyelitis vaccines by culturing the virus of poliomyelitis on monkey kidney cells in 1949.

6. For which viral disease, vaccine has been recently developed through the use of tissue culture?
a) Measles
b) Mumps
c) Rabies
d) S mallpox
View Answer

Answer: a
Explanation: Among the virus diseases for which vaccines have been recently developed through the use of tissue culture is measles (rubeola).

7. Rubella vaccines contain viruses are isolated only in African green monkey cells.
a) True
b) False
View Answer

Answer: b
Explanation: Rubella vaccines contain viruses either isolated in African green monkey cells and attenuated by further cell passage (in primary duck embryo cells) or isolated and passed to diploid human embryo cells.

8. Translation of mRNA into proteins takes place in the _____
a) host cell nucleus
b) host cell cytoplasm
c) viral nucleus
d) viral cytoplasm
View Answer

Answer: b
Explanation: Translation of mRNA into proteins takes place in the host cell cytoplasm and uses ribosomes, transfer RNAs, and enzymes in the cytoplasm.

"Immunological Preparations – Classification of Vaccines – 1".

1. DPT is a combination vaccination.
a) True
b) False
View Answer

Answer: a
Explanation: DPT (also DTP and DTwP) is a class of combination vaccines against three infectious diseases in humans: diphtheria, pertussis (whooping cough), and tetanus. A combination vaccine is one which can provide immunogenicity to more than one type of disease. Some more examples of recombination vaccines are DTaP, Hep-B.

2. The only recombinant vaccine that proved highly effective and is currently approved for human use is the vaccine against _____
a) Rubella virus
b) Poxvirus
c) HPV infection
d) Dengue virus
View Answer

Answer: c
Explanation: The only recombinant vaccine that proved highly effective and is currently approved for human use is the vaccine against HBV infection. It includes the recombinant hepatitis B virus surface antigen (HBsAg), made by DNA-transfected yeast or mammalian cells.

3. Who developed the Hepatitis B vaccine?
a) Pasteur
b) Salk
c) Jenner
d) Pablo DT Valenzuela
View Answer

Answer: d
Explanation: Pablo DT Valenzuela, Research Director of Chiron Corporation: succeeded in making the antigen in yeast. Thus they invented the world's first recombinant vaccine. The blood-derived hepatitis B vaccine was withdrawn then from the marketplace.

4. Virus proteins have been expressed in _____
a) Bacterial and yeast cells
b) Yeast and mammalian cells
c) Mammalian cells
d) Mammalian, yeast and bacterial cells
View Answer

Answer: d
Explanation: Virus proteins have been expressed in bacteria, yeast, mammalian cells, and viruses. This helps in the generation of vaccines. Bacterial cells help in

rapid growth and thus the rapid yield of the vaccine. Yeast gives us a post-translational modification. Mammalian cells are hard to culture but some of the proteins can easily be manufactured in mammalian cell culture.

5. GARDASIL is an example of _____
a) DNA vaccines
b) Recombinant vector vaccines
c) Live vector vaccines
d) RNA vaccines
View Answer

Answer: b
Explanation: The Hepatitis B vaccine currently used in the United States called Gardasil is a recombinant vaccine. A recombinant vector vaccine is manufactured by the recombinant DNA process. It can be effective against many more than one form of the disease.

6. Gardasil protects against which of the HPV types _____
a) HPV- 6, 11, 16, 18
b) HPV – 5, 6, 16, 18
c) HPV – 1, 2, 5, 11
d) HPV – 1, 2, 16, 18
View Answer

Answer: a
Explanation: There are various strains of HPV and Gardasil protects against HPV types 6, 11, 16 and 18 in both males and females. These can then be purified and used in the vaccine. After their invention of GARDASIL, breast cancers at initial stages are completely curable.

7. The HPV protein-encoding genes are expressed in _____ vectors to create large amounts of protein.
a) Bacteria
b) Yeast
c) Virus
d) Protozoa
View Answer

Answer: b
Explanation: The HPV protein-encoding genes are expressed in the yeast vectors to create large amounts of protein, which are then purified and used in the vaccine.

8. Generally, the whole protein molecule is necessary for immunogenicity.
a) True

b) False
View Answer

Answer: b
Explanation: Generally, the whole protein molecule is not necessary for immunogenicity; the immunogenic property is usually confined to a small portion of the protein molecule. Thus, the whole of the organism is not needed to inject into the patient body. It may be harmful. Thus, it is better to have proteins to cause immunogenic action.

9. Which polypeptide of cholera enterotoxin is nontoxic but immunogenic?
a) A1
b) A2
c) B
d) B1
View Answer

Answer: c
Explanation: Cholera enterotoxin (produced by Vibrio cholerae) consists of three polypeptides: A1, A2 and B polypeptides. The A polypeptides are toxic, while the B polypeptide is nontoxic but immunogenic. Thus, we can use the B polypeptide for the manufacture of vaccines.

10. What are the disadvantages of recombinant protein or polypeptide vaccines?
a) High cost and storage
b) Storage and transportation
c) Transportation and cost
d) Cost, transportation, storage
View Answer

Answer: d.
(source-https://www.sanfoundry.com)

MICROBIOLOGY TERMINOLOGY

A	B
Aerobes (aerobic)	Microorganisms that require oxygen to live and grow
Anaerobes (anaerobic)	Microorganims that do not require the presence of oxygen to live and reproduce
Autotrophic	Self-nourishing. Pertains to green plants and bacteria which form protein/carbohydrates from inorganic salts/ carbon dioxide
Bacteria	Most common group of microbes. Essential to life; many reside naturally in the body. Cause disease when a person is compromised
Binary fission	The process by which bacteria reproduce
Endotoxin	Bacterial toxin confined within the body of a bacterium, freed only when the bacteria is broken down
Exotoxin	A toxin produced by a microorganism and excreted into its surrounding medium. Most potent toxins known
Fomite	Inanimate object which act as reservoirs for infection.(Clothes/utensils/contaminated equipment/surgical instruments
Fungi	Eucaryotic organisms including mushrooms, molds and yeasts
Gram negative bacteria	Bacteria which appear red/pink following a staining process
Gram positive bacteria	Bacteria which retain a purple color when subjected to a staining process
Heterotrophic	Organisms which require complex organic food to grow and develop

Host	The organism from which a parasite gets its nourishment
Infection	Term used when nonresident flora invade a susceptible area/host
Inflammation	The body's response to infection. Signs/symtoms: redness, heat, swelling (edema), pain, pus
Microorganisms	Living organisms to small to be seen with the naked eye
Nosocomial infection	Infection that develops while a patient is in the hospital or is produced by microbes during hospitalization
Opportunist	A microbe with potential to cause disease given the opportune time or place
Pathogenic microorganism	A disease causing/producing microorganism
Prinary invader	First, initial microorganism causing infection
Protozoa	Eucaryotic, usually single-celled, animal like microorganism
Pyogenic organism	Pus producing microorganism
Septicemia	"Blood poisoning". Systemic disease associated with pathogenic microbes/toxins in blood
Spores	Formed by some bacteria as a means of survial
Vector	Animate carrier of microorganisms, many times an arthropod
Virulence	The degree of pathogenicity os a microorganism
Virus	Small, simple structured "particle" reliant of its host for survival. Causes hepatitis and HIV
Bacteremia	The presence of bacteria in blood

Acute infection	An infection of sudden onset and short duration
Algae	Group of microbes abundantly found, many beneficial. Example: seaweed
Carrier	Person who harbors a microorganism but is asymptomatic, may spread the disease
Chronic infection	Infection of slow progression, long duration
Infectious agent	Microorganism which produces a specific disease
Latent infection	Disease which reaches a stage where the patient is not symptomatic but still harbors the microorganism
Local infection	An infection confined within an area
Parasite	Organism that lives within, upon ar at the expense of a host
Primary disease	First disease or infection
Prion	An infectious protein affecting the central nervous system. Ex: mad cow disease Cruetsfeld-Jakob disease)
Pus	The liquid producto of inflammation, usually yellow in color
Resident microorganism	"Normal flora" Microorganism which are normally found in an area
Rickettsiae	Group of microorganisms transmitted by vectors. (Example: Rocky Mountain Spotted Fever)
Saprophyte	Microorganism whose main source of foof is dead and decaying organic matter
Secondary disase	A second illness encoutered due to a weakened resistance from an original disease
Shedder	Individuals who carry and shed staphylococcus at a higher volume/rate

Superinfection	Second infection due to the use of antibiotics or disruption of the normal flora
Suppuration	the process of pus formation
Systemic infection	An infection which occurs throughout the body
Toxin	Poisons released by microorganisms
Transient microorganism	Temporary flora, not part of the normal flora

REFERENCES

- Marco, M.L., et al., Health benefits of fermented foods: microbiota and beyond. Current opinion in biotechnology, 2017. 44:94-102.
- Bell, V., J. Ferrão, and T. Fernandes, Nutritional Guidelines and Fermented Food Frameworks. Foods, 2017. 6(8):65.
- Şanlier, N., B.B. GÖkcen, and A.C. Sezgin, Health benefits of fermented foods. Critical reviews in food science and nutrition, 2017:1-22.
- Hill, C., et al., Expert consensus document: The International Scientific Association for Probiotics and Prebiotics consensus statement on the scope and appropriate use of the term probiotic. Nature Reviews Gastroenterology and Hepatology, 2014. 11(8):506.
- Hotel, A.C.P. and A. Cordoba, Health and nutritional properties of probiotics in food including powder milk with live lactic acid bacteria. Prevention, 2001. 5(1):1-34.
- Gibson, G.R., et al., Expert consensus document: The International Scientific Association for Probiotics and Prebiotics (ISAPP) consensus statement on the definition and scope of prebiotics. Nature Reviews Gastroenterology and Hepatology, 2017. 14(8):491.
- Al-Sheraji, S.H., et al., Prebiotics as functional foods: a review. Journal of Functional Foods, 2013. 5(4):1542-1553.
- Cho, Y.A., Effect of Probiotics on Blood Lipis Concentrations: A meta-analysis of randomized controlled trials. Medicine, 2015. 94(43):e1714
- Chen, M., et al., Dairy consumption and risk of type 2 diabetes: 3 cohorts of US adults and an updated meta-analysis. BMC medicine, 2014. 12(1):215.
- Eussen, S.J., et al., Consumption of dairy foods in relation to impaired glucose metabolism and type 2 diabetes mellitus: the Maastricht Study. British Journal of Nutrition, 2016. 115(8):1453-1461.
- Soedamah-Muthu, S.S., et al., Consumption of dairy products and associations with incident diabetes, CHD and mortality in the Whitehall II study. British Journal of Nutrition, 2013. 109(4):718-726.
- Iwasa, M., et al., Fermented milk improves glucose metabolism in exercise-induced muscle damage in young healthy men. Nutrition journal, 2013. 12(1):83.

- Prescott, Lansing M, John P Harley, and Donald A Klein. *Microbiology*. Dubuque, IA: McGraw-Hill Higher Education, 2005. Print.

- Slonczewski, Joan, and John Watkins Foster. *Microbiology*. New York: W.W. Norton & Co., 2009. Print.
- Pelczar, Michael J, E. C. S Chan, and Noel R Krieg. *Microbiology*. New York: McGraw-Hill, 1993. Print.
- Chapelle, F.H. *Ground Water Microbiology and Geochemistry*. New York: John Wiley & Sons, 2000.
- Madigan, M.M., J. Martinko, and J. Parker. *Brock Biology of Microorganisms*. 8th ed Upper Saddle River, NJ: Prentice-Hall, 2000.
- Margesin R, Schinner F. Biodegradation and bioremediation of hydrocarbons in extreme environments. *Appl Microbiol Biotechnol. 2001 Sep;56(5-6):650-63.*
- Prince RC. Petroleum spill bioremediation in marine environments. *Crit Rev Microbiol. 1993;19(4):217-42.*
- Leahy JG, Colwell RR. Microbial degradation of hydrocarbons in the environment. *Microbiol Rev. 1990 Sep;54(3):305-15.*
- Karner, M.B., E.F. DeLong, and D.M. Karl. "Archae Dominance in the Mesopelagic Zone of the Pacific Ocean." *Nature* 409 (January 2001): 507–510.
- Ruiz, G.M., T.K. Rawlings, F.C. Dobbs, et al. "Global Spread of Microorganisms by Ships." *Nature* 406 (November 2000): 49.
- Needham, Joseph. (2000). Science and Civilization in China: Volume 6, Biology and Biological Technology, *Part 6, Medicine*. Cambridge: Cambridge University Press. Page 134.
- Wright, A.E.; Stewart, R.D. (1 September 1903). "An experimental investigation on the role of the blood fluids in connection with phagocytosis". Proceedings of the Royal Society of London. 72 (477–486): 357–370. doi:*10.1098/rspl.1903.0062.*
- Kolata, Gina (22 January 2015). "Jean Lindenmann, Who Made Interferon His Life's Work, Is Dead at 90". New York Times. *Retrieved 12 February 2015.*
- Köhler, G; Milstein, C (7 August 1975). "Continuous cultures of fused cells secreting antibody of predefined specificity". Nature. 256 (5517): 495–7. Bibcode:1975Natur.256..495K. doi:10.1038/256495a0. PMID 1172191.
- Haskins, K; Kubo, R; White, J; Pigeon, M; Kappler, J; Marrack, P (1 April 1983). "The major histocompatibility complex-restricted antigen receptor on T cells. I. Isolation with a monoclonal antibody". The Journal of Experimental Medicine. 157 (4): 1149–69. doi:10.1084/jem.157.4.1149. PMC *2186983*. PMID 6601175.

- Allman, DM; Ferguson, SE; Cancro, MP (15 October 1992). "Peripheral B cell maturation. I. Immature peripheral B cells in adults are heat-stable antigenhi and exhibit unique signaling characteristics". Journal of Immunology. 149 (8): 2533–40. PMID 1383316.
- Allman, DM; Ferguson, SE; Lentz, VM; Cancro, MP (1 November 1993). "Peripheral B cell maturation. II. Heat-stable antigen(hi) splenic B cells are an immature developmental intermediate in the production of long-lived marrow-derived B cells". Journal of Immunology. 151 (9): 4431–44. PMID 8409411.
- Mills CD; et al. (2000). "M-1/M-2 Macrophages and the Th1/Th2 Paradigm". J Immunol. 164 (12): 6166–6173. doi:*10.4049/jimmunol.164.12.6166*. PMID 10843666.
- Fritz, JM; Lenardo, MJ (2019). "Development of immune checkpoint therapy for cancer". J Exp Med. 216 (6): 1244-1254. doi:*10.1084/jem.20182395*.

Article Source
- http://ecoursesonline.iasri.res.in/mod/page/view.php?
- https://www.caister.com/highveld/microbiology/environmental
- https://biodifferences.com/difference-between-prokaryotic-cells-and-eukaryotic-cells
- https://wiki.bugwood.org/Xanthomonas_species
- https://microbewiki.kenyon.edu/index.php/Rickettsia
- https://en.wikipedia.org/wiki/Chlamydia
- https://www.biologydiscussion.com/micro-biology/study-notes-on-chlamydial-infection
- https://nptel.ac.in/courses/102/103/102103039/
- https://virology-online.com/general/Replication
- https://microbiologyinfo.com/techniques-of-virus-cultivation/
- https://www.apsnet.org/edcenter/disandpath/viral/pdlessons/Pages/TobaccoMosaic.aspx
- https://www.biologydiscussion.com/viruses/tobacco-mosaic-virus-tmv-structure-and-replication
- https://www.agric.wa.gov.au/potatoes/potato-virus-y-potato-crops
- https://agrihunt.com/articles/crop-diseases/potato-mosaic/
- https://www.biologydiscussion.com/bacteria/cyanophages-discovery-morphology-and-replication-microbiology/54885
- https://en.wikipedia.org/wiki/Mycovirus

- https://microbiologyinfo.com/antigen-properties-types-and-determinants-of-antigenicity/
- *https://www.msdmanuals.com › Home › Infections › Fungal Infections*
- https://www.merckmanuals.com/home/infections/fungal-infections/aspergillosis
- https://en.wikipedia.org/wiki/Aspergillosis
- https://vikaspedia.in/agriculture/agri-inputs/bio-inputs/bioinputs-for-nutrient-management/biofertilizer
- https://biologydictionary.net/archaebacteria/
- https://www.biologydiscussion.com/microbiology
- http://ecoursesonline.iasri.res.in
- https://www.caister.com/highveld/microbiology/environmental
- https://biodifferences.com/difference-between-prokaryotic-cells-and-eukaryotic-cells
- www.microrao.com
- https://foodsafetyhelpline.com
- *Boundless (2016-05-26). "Industrial Production of Antibiotics". Boundless.*
- https://sciencesamhita.com/industrial-products-from-microbes
- https://www.sare.org/Learning-Center/Books/Building-Soils
- https://en.wikipedia.org
- https://www.britannica.com/

Printed in Great Britain
by Amazon